ENVIRONMENTAL CARBON FOOTPRINTS

ENVIRONMENTAL CARBON FOOTPRINTS

Industrial Case Studies

Edited by

SUBRAMANIAN SENTHILKANNAN MUTHU

ELSEVIER

Butterworth-Heinemann
An imprint of Elsevier

Butterworth-Heinemann is an imprint of Elsevier
The Boulevard, Langford Lane, Kidlington, Oxford OX5 1GB, United Kingdom
50 Hampshire Street, 5th Floor, Cambridge, MA 02139, United States

Notices

Knowledge and best practice in this field are constantly changing. As new research and experience broaden our understanding, changes in research methods, professional practices, or medical treatment may become necessary.

Practitioners and researchers must always rely on their own experience and knowledge in evaluating and using any information, methods, compounds, or experiments described herein. In using such information or methods they should be mindful of their own safety and the safety of others, including parties for whom they have a professional responsibility.

To the fullest extent of the law, neither the Publisher nor the authors, contributors, or editors, assume any liability for any injury and/or damage to persons or property as a matter of products liability, negligence or otherwise, or from any use or operation of any methods, products, instructions, or ideas contained in the material herein.

Library of Congress Cataloging-in-Publication Data
A catalog record for this book is available from the Library of Congress

British Library Cataloguing-in-Publication Data
A catalogue record for this book is available from the British Library

ISBN: 978-0-12-812849-7

For information on all Butterworth-Heinemann publications visit our website at https://www.elsevier.com/books-and-journals

Publisher: Matthew Deans
Acquisition Editor: Ken McCombs
Editorial Project Manager: Ashlie M. Jackman
Production Project Manager: Mohanapriyan Rajendran
Designer: Victoria Pearson Esser

Typeset by TNQ Books and Journals

CONTENTS

CONTRIBUTORS

Madhoolika Agrawal
Department of Botany, Institute of Science, Banaras Hindu University, Varanasi, India

Tolulope J. Akin-Adeniyi
Department of Electrical Electronic Engineering, University of Lagos, Akoka, Nigeria

Tolulope O. Akinbulire
Department of Electrical Electronic Engineering, University of Lagos, Akoka, Nigeria

Olubayo M. Babatunde
Department of Electrical Electronic Engineering, University of Lagos, Akoka, Nigeria

Emanuele Bonamente
University of Perugia, Perugia, Italy

Debrupa Chakraborty
Netaji Nagar College, Kolkata, India

Henry I. Denwigwe
Department of Electrical Electronic Engineering, University of Lagos, Akoka, Nigeria

Sara Feijoo
University of Santiago de Compostela, Santiago de Compostela, Spain

Gumersindo Feijoo
University of Santiago de Compostela, Santiago de Compostela, Spain

Luíza S. Franca
Federal University of Rio de Janeiro, Rio de Janeiro, Brazil

Meenu Gautam
Department of Botany, Institute of Science, Banaras Hindu University, Varanasi, India

Varun Goel
NIT Hamirpur, Hamirpur, India

Sara González-García
University of Santiago de Compostela, Santiago de Compostela, Spain

Sourabh Khurana
OM Institute of Technology and Management, Hisar, India

Luis A. López
University of Castilla-La Mancha, Albacete, Spain

Madelyn Marrero
University of Seville, Seville, Spain

Alejandro Martínez-Rocamora
University of Bío-Bío, Concepción, Chile

Charles Mbohwa
University of Johannesburg, Johannesburg, South Africa

Yolanda Moldes-Diz
University of Santiago de Compostela, Santiago de Compostela, Spain

María T. Moreira
University of Santiago de Compostela, Santiago de Compostela, Spain

Himanshu Nautiyal
THDC Institute of Hydropower Engineering and Technology, Tehri, India

Oludolapo Akanni Olanrewaju
University of Johannesburg, Johannesburg, South Africa

Peter O. Oluseyi
Department of Electrical Electronic Engineering, University of Lagos, Akoka, Nigeria

Bhanu Pandey
Department of Botany, Institute of Science, Banaras Hindu University, Varanasi, India

Glaydston M. Ribeiro
Federal University of Rio de Janeiro, Rio de Janeiro, Brazil

Sara Rinaldi
University of Perugia, Perugia, Italy

Marina S.R. Rocha
Federal University of Rio de Janeiro, Rio de Janeiro, Brazil

Flavio Scrucca
University of Perugia, Perugia, Italy

Subramanian Senthilkannan Muthu
Bestseller, Kowloon, Hong Kong SAR

Venu Shree
NIT Hamirpur, Hamirpur, India

Paramvir Singh
NIT Hamirpur, Hamirpur, India

Jaime Solís-Guzmán
University of Seville, Seville, Spain

Karpagam Subramanian
The Hong Kong Polytechnic University, Kowloon, Hong Kong SAR

Carlos Vázquez-Vázquez
University of Santiago de Compostela, Santiago de Compostela, Spain

Winco K.C. Yung
The Hong Kong Polytechnic University, Kowloon, Hong Kong SAR

Jorge E. Zafrilla
University of Castilla-La Mancha, Albacete, Spain

BIOGRAPHY

Dr. Subramanian Senthilkannan Muthu is currently working for Lidl Hong Kong as a Sustainability Manager, based out of Hong Kong. He has gained his PhD from The Hong Kong Polytechnic University. He was an outstanding student throughout his studies and bagged numerous awards and medals including many gold medals in his study period. He is a renowned expert in the areas of Environmental Sustainability in Textiles and Clothing Supply Chain, Product Life-Cycle Assessment (LCA), and Product Carbon Footprint Assessment (PCF) in various industrial sectors. He has 5 years of industrial experience in textile manufacturing, research and development, and textile testing and around 7 years of experience in LCA, carbon, and ecological footprints assessment of various consumer products. He has completed many LCA, carbon footprint, and environmental assessment projects in Asia and Europe from both cradle-to-gate and cradle-to-grave stages of many products including apparels, plastics, chemicals, and packaging. He has a wide experience in environmental assessment of textiles and clothing supply chain, and he has worked on recycling of plastics and textiles, green claims, and validation of different consumer products. He has delivered extensive trainings on PCF and LCA to many external clients in India, Colombo, Bangladesh, China, and Hong Kong, apart from the delivery of many invited talks. He has delivered many invited key note speeches in various international conferences across the globe.

He has published more than 75 research publications, written numerous book chapters, and authored/edited multiple books in the areas of carbon footprint, recycling, environmental assessment, environmental sustainability, etc. (over 40 books to his credit). Famous titles of his books list include Assessment of Environmental Impacts of Textiles and Clothing Supply Chain, Assessment of Carbon Footprint in Different Industrial Sectors (two volumes), Roadmap to Sustainable Textiles and Clothing (three volumes),

Handbook of Carbon Footprint, Handbook of Life Cycle Assessment (LCA) in Textiles and Clothing, Handbook of Sustainable Apparel Production, and Textiles and Clothing Sustainability (six volumes).

He is acting as an editor, editorial board member, and reviewer for many international peer-reviewed journals of textiles and environmental science disciplines. He is also the series editor of two books series of Springer namely Textile Science and Clothing Technology and Environmental Footprints and Eco-design. He is one of the directors of Textile and Bioengineering Informatics Society, which is a charitable organization created to foster, develop, and promote all aspects of science and technology in bioengineering of materials, fibers, and textiles.

CHAPTER 1

The Need for Greenhouse Gas Analyses in Industrial Sectors

Oludolapo Akanni Olanrewaju, Charles Mbohwa
University of Johannesburg, Johannesburg, South Africa

1.1 INTRODUCTION

Modern scientific research has shown that the accumulation of greenhouse gas (GHG) emissions plays vital role in global warming (Wang et al., 2016). These gases entrap heat in the atmosphere leading to the absorption of infrared radiation, which automatically emits the thermal radiation back into the atmosphere. The global warming from the effect of increased GHG can lead to many catastrophic consequences such as the rise in sea levels, unavailability of water, as well as heat waves. Since the industrial revolution, levels of CO_2 emissions from human activities have continued to increase in the burning of fossil fuels (IPCC, 2007a,b; Wang et al., 2016), and Intergovernmental Panel on Climate Change released The Third Assessment Report in 2001 claiming human influence led to estimated 66% of climate change since the 1950s (IPCC, 2001) with an estimated 90% increase on The Fourth Assessment Report (IPCC, 2007a,b).

One of the most important features of Earth that assists life and the ecosystem in surviving and prospering is climate (Cote and Liu, 2016). Climate change remains a global issue and a matter of common concern to the international community today (Wang et al., 2016). However, climate has been affected adversely in the present generation. A fast growing economy, especially in developing countries, is notable for an increase in GHG emission (Jobert et al., 2010; Liobikiene et al., 2016). Having a proper understanding and comprehension of the nexus between energy and GHGs is critical in establishing a benchmark to track the emission history and to develop the best ways to mitigate the emission (Feng et al., 2015). Recently, assessing the way GHG emissions is being mitigated has been a concern both country-wise and globally (Hamit-Haggar, 2012). Reducing the GHG emissions led to the formation of Kyoto Protocol in Japan by 84 parties in

Environmental Carbon Footprints
ISBN 978-0-12-812849-7
http://dx.doi.org/10.1016/B978-0-12-812849-7.00001-5

1997, with the number of parties increased to 192 during its ratification in 2005. The objective of the Protocol was to reduce GHG by 5% in a 5-year period at 1990 levels (Ozturk et al., 2016). In facing the problems associated with climate change as well as the commitments made to the Kyoto calls, various policies ranging from legislative, technological, etc. have been employed by the European Union to offer various measures in various packages to see that the economy culminating from various sectors is affected (Alegria et al., 2016).

Income in the consumption of energy prices gives rise to the GHG emissions especially by our industries. Our industries are in the challenge of reducing GHGs as the plant is faced with climate change reality (Cote and Liu, 2016). Emission reduction has been regarded as the best option to mitigate and attack climate change (Liu et al., 2016b). There exist some barriers preventing transition of energy. These barriers include nongovernmental assistance and guidelines, insufficient or not-up-to standard regulations, and insufficient relevant technologies (Ren et al., 2015; Liu et al., 2016b). Among the various processes associated to energy, the energy nexus to GHG plays a major role (Liu et al., 2016b). These four policies are key to reducing GHG of any country: saving energy and improving energy efficiency policy, optimizing energy structure and developing low-carbon energy policy, updating manufacturing structure policy and improving GHG emission management for residential sector policy (Feng et al., 2015). Implementing these policies will guide in the various mitigation techniques; however, these policies have not been taking seriously to be best utilized to the advantage of the global village.

Different countries have reported steady increase in the amount of GHGs emitted. For example, Canada's GHG emissions rose by 32% from 1990 to 2007, this is the same period when energy consumed by the industrial sector rose by 28.1% (Hamit-Haggar, 2012). For United Arab Emirates, increase in emissions due to industrial processes was equivalent to 174% in 2005 from 1994 (Khondaker et al., 2016). China had an increase of industrial GHG emissions by 164.5% from 1991 to 2001 (Ouyang and Lin, 2015). With the rate at which industries emit GHGs, infrastructures, policies, and ways to address the issues must be a critical part of the industries. Most countries have their emission rate because of their economic growth. Countries from both Organization for Economic Cooperation and Development (OECD) and Brazil, Russia, India, and China are responsible for much of the global emissions. As at 2010, it was recorded that these countries emitted 76.8% of the world's emission

(IEA, 2012). South Africa also recorded a 21.1% increase in GHG emission from 2000 to 2010 (Witi and Stevens, 2014). Industrial sectors have been the major targets by most countries through emission caps and other techniques to achieve their emission targets (Liu et al., 2016b). Table 1.1 shows the amount and percentage of fuel types and sectors concerned as well as the CO_2 related emissions from the world and G20 countries in 2013 fossil fuel combustion.

The Kyoto Protocol operates under three different mechanisms; these are the emission trading, joint implementation, and clean development mechanisms. For further information on these mechanisms, interested readers can check on Alegria et al. (2016). The criticality of GHGs made its tracking a very relevant aspect in the world's climate change study with various relevant data informing national, international as well as economic, and as much as environmental negotiations (Gao et al., 2009; Liu et al., 2016b). The general knowledge that the growth in economy, especially developing countries, increases the GHG emission level has been confirmed by various researchers (Jobert et al., 2010; Liobikiene et al., 2016). The reduction of GHG is regarded as the key solution to combating climate change (Liu et al., 2016b).

Globally, climate change continues to have negative impacts economically, with increasing pressure (Heidari and Pearce, 2016). The impacts from emission are far too dangerous to let them take their effects on our everyday natural and socioeconomic systems (Heidari and Pearce, 2016). These impacts include (1) abnormally high temperatures and heat waves resulting into tens of thousands deaths as a result of hyperthermia (Fouillet et al., 2006), (2) agricultural end produce destruction (D'Amato and Cecchi, 2008), (3) intense world hunger (Parry et al., 2004), power blackouts (Klinenberg, 2008), (4) increase in sea levels resulting in gradual submerge of low-lying coastal regions (Frihy, 2003), (5) shoreline erosions (Frihy, 2003), (6) continuous flooding danger (Nicholls et al., 1999), and intrusion of saltwater (Bobba, 2002), and (7) droughts (Dai, 2013), and (8) fire (Dale et al., 2001).

Cote and Liu (2016) argued that reduction and mitigation approaches tend to be efficiently managed where there is concentration of people and industries as GHGs are mostly emitted from human activities. Policymakers regard the reduction and stabilization of CO_2 emissions as a very significant key in preventing climate change (Liobikiene et al., 2016). To develop a plan that can inform climate policies, it is very important to understand the scope around GHG emissions. Many governments do carry out an inventory of

Table 1.1 CO_2 related emissions from the world and G20 countries in 2013 fossil fuel combustion

	Total	Coal	Oil	Gas	Other	Total (%)	Coal (%)	Oil (%)	Gas (%)	Other (%)
World										
Sectors concerned	21,445	14,177	2,418	4,678	171	100	66	11	22	1
Power and heat generation[a]	13,656	9,887	887	2,753	128	63.70	46	4	12.90	0
Other energy industry own use	1,674	423	548	702	1	7.80	2	2.50	3.30	0
Manufacturing industry[b]	6,115	3,867	983	1,223	42	28.50	18	4.50	5.75	0.25
G20										
Sectors concerned	18,410	13,235	1,667	3,346	162	100	72	9	18	1
Power and heat generation[a]	11,883	9,273	488	2,001	121	64.50	50.40	2.60	10.76	0.74
Other energy industry own use	1,315	406	455	453	1	7.20	2.20	2.50	2.50	0
Manufacturing industry[b]	5,212	3,556	724	892	40	28.30	19.34	3.90	4.79	0.24

[a]includes public power and heat production.
[b]excludes emissions from nonenergy use and feedstock use of fuels.
Extracted from Olivier, J.G.G.J., Janssens-Maenhout, G., Muntean, M., Peters, J.A.H.W.W., 2016. Trends in Global CO_2 Emissions. PBL Netherlands Environmental Assessment Agency and Ispra, The Hague.

their emitted GHG before designing a plan (Feng et al., 2015). The increase in energy consumed by industries from developing countries is approximate to GHG emission problems and deteriorating supply of energy. Most of the emissions from various industrial processes are carbon dioxide, methane, and nitrous oxide. Other minute gases are emitted that stay long in the atmosphere as long as emissions do occur—these are hydrofluorocarbons, perfluorocarbons, and sulfur hexafluoride. The unwelcome situation has given rise to various mitigation measures to solving these problems (Wu et al., 2016). The existence of the benefits, which come from simultaneously abating GHGs have come to the attention of GHG emitters. From the studies of Cifuentes et al. (2001) as well as Wang and Smith (1999) to name a few, have benefits of reducing GHGs. In reducing GHGs, lots of supporting models are being and still developed and employed to that effect (Liu et al., 2016a). Various methods exist in the analyses of industrial GHGs, methods include LEAP method, MARKA, decomposition analyses methods to name a few.

Finding solution to industrial GHGs requires proper analyses of its emission. GHG analyses give insight into how emission could be mitigated, possible targets that could be achieved. When industries are considered, right policies are implemented as much as engaging in the right practice to enhance production while minimizing GHG emission. In fulfilling the mentioned, it is important to understand historical data on the emitted gases, predict possible future emission, and optimize the possible target that could be mitigated. The first technique to examine the impacts product mix changes have on energy consumed by industries was index decomposition analysis (IDA) in the late 1970s (Ang and Zhang, 2000). Ever since, it continued to be used in tracking both economic and sector based trends in energy efficiency (Liu et al., 2016b). It was introduced to energy-related emission studies in 1991 (Xiaoyan, 2013). The next section of this chapter explains the general knowledge of IDA and the most used method, which is the Logarithmic Mean Divisia Index (LMDI) in the energy-related emission studies, followed by sample applications in the industrial sector. Conclusion summarizes the basis and findings of this chapter.

1.2 DECOMPOSITION METHOD

The general practice to evaluating GHG emitted is by developing key indicators as well as tracing any change among the indicators with time (Xiaoyan, 2013). Every emission from a sector, subsector, or end-use levels are meted out as emissions from different kinds of fuel (Xiaoyan, 2013). The use of IDA provides emission information/background of every indicator/factor at various stages of emission hierarchy to be judiciously used for appropriate decisions (Xiaoyan, 2013).

The global village has faced increase in GHG emission in the past decades as a result of wealth increase, continuous energy demand plus increased fossil fuel dependency (Xiaoyan, 2013). Policy making is thus required to combat climate change (Feng et al., 2015) for this reason. Designing new policies and assessing measures needed for implementation calls for information of the parameters influencing the change of environmental input variants. Identifying those driving forces requires a decomposition analysis (Jungnitz, 2008). Energy studies require both IDA and structural decomposition analysis. This study, however, has considered various studies carried out on industrial GHG using IDA. In this chapter, LMDI has been the chosen one for literature as compared with other IDAs. The reason for this choice can be seen by Ang (2004) and Mendiluce et al. (2010) to name a few.

The formulation of the LMDI−IDA as applied to GHG studies is given below. The equations given below are attributed to the general drivers—the activity, structure intensity, and sectoral energy mix, which are the major drivers.

C—Total CO_2 emission
C_{ij}—CO_2 emissions arising from fuel j in industrial sector i
E_{ij}—Consumption of fuel j in industrial sector i, where $E_i = \sum_j E_{ij}$
$M_{ij} = E_{ij}/E_j$—The fuel-mix variable
Q_i—Value of production in sector i
Q—Total value of production $\left(Q = \sum_i Q_i\right)$
S_i—Production share of sector $i\left(S_i = \frac{Q_i}{Q}\right)$
I_i—Intensity of energy consumption in sector $\left(I_i = \frac{E_i}{Q_i}\right)$

$$C = \sum_{ij} Q\frac{Q_i}{Q}\frac{E_i}{Q_i}\frac{E_{ij}}{E_i}\frac{C_{ij}}{E_{ij}} = \sum_{ij} QS_iI_iM_{ij} \tag{1.1}$$

$$\frac{C^T}{C^0} = D_{tot} = D_{act}D_{str}D_{int}D_{mix} \tag{1.2}$$

where D_{tot} is the total CO_2 emission, D_{act} is the activity, D_{str} is the structure, D_{int} is the intensity, and D_{mix} is the sectoral energy mix.

$$D_{act} = \exp\left(\sum_{ij} \frac{\left(c_{ij}^T - c_{ij}^0\right) / \left(\ln c_{ij}^T - \ln c_{ij}^0\right)}{(c^T - c^0)/(\ln c^T - \ln c^0)} \ln\left(\frac{Q^T}{Q^0}\right)\right) \tag{1.3}$$

$$D_{str} = \exp\left(\sum_{ij} \frac{\left(c_{ij}^T - c_{ij}^0\right) / \left(\ln c_{ij}^T - \ln c_{ij}^0\right)}{(c^T - c^0)/(\ln c^T - \ln c^0)} \ln\left(\frac{S_i^T}{S_i^0}\right)\right) \tag{1.4}$$

$$D_{int} = \exp\left(\sum_{ij} \frac{\left(c_{ij}^T - c_{ij}^0\right) / \left(\ln c_{ij}^T - \ln c_{ij}^0\right)}{(c^T - c^0)/(\ln c^T - \ln c^0)} \ln\left(\frac{I^T}{I^0}\right)\right) \tag{1.5}$$

$$D_{mix} = \exp\left(\sum_{ij} \frac{\left(c_{ij}^T - c_{ij}^0\right) / \left(\ln c_{ij}^T - \ln c_{ij}^0\right)}{(c^T - c^0)/(\ln c^T - \ln c^0)} \ln\left(\frac{M_{ij}^T}{M_{ij}^0}\right)\right) \tag{1.6}$$

$$D_{tot} = D_{act} D_{str} D_{int} D_{mix} \tag{1.7}$$

The application could be either multiplicative or additive even though the choice of methodology is not important, and the key concern is the easy presentation and interpretation of results as specified by Fengling (2004). The equations presented above represent the multiplicative approach. The multiplicative approach is most preferred should the decomposition be performed on a yearly basis with a time-series energy and industrial production data. This is because the results can be easily and conveniently plotted over time. On the other hand, performing decomposition based on data of two benchmark years would lean on choosing additive decomposition as the preferred approach. This is because it would be easy to understand the results, which can either be in the form of the original aggregate unit measured or in percentage change terms.

Disentangling and identifying the key factors to explain the total change in both energy consumption and GHG emission in various industrial sectors has always been the objective of IDA from initial period 0 to final period T. IDA remains a powerful decomposition analysis research tool commonly used (Lofgren and Muller, 2010).

1.3 APPLICATIONS OF LOGARITHMIC MEAN DIVISIA INDEX

To indicate how IDA has developed in the study of GHG analysis, different case studies would be examined taking the period of application into consideration. These periods are the early 1970s to the 1980s, during the oil crisis; the 1990s to the early 2000s, when knowledge of solving the crisis intensified; and mid-2000s to present, when finding lasting solution to the crisis is of utmost importance. Change in various factors can lead to change in the intensity of carbon dioxide/GHGs. These include share of different fuels, structure, energy intensity, international structure, heat and electricity carbon emission coefficients, industrial structure shift, industrial activity, as well as carbon intensity depending on the choice of interest.

1.3.1 Early Oil Crisis

Few studies were applicable to CO_2 factors in the early 70s. Torvanger (1991) was a notable one as well as the study of Shrestha and Timilsina (1996). As Torvanger (1991) was applicable to industrial sectors, and Shrestha and Timilsina (1996) was applicable to the electricity sector. Looking at various studies during the oil crisis where IDA had been implemented to analyze GHG, this (Torvanger, 1991) analyzed the carbon dioxide emissions from the manufacturing sectors of some selected OECD countries. The study considered seven manufacturing sectors—paper and pulp; industrial chemicals; stone, clay and glass; ferrous metals; nonferrous metal; and nonraw materials. These industrial subsectors were considered from nine OECD countries between 1973 and 1987. The OECD countries are Japan, United States of America, Germany, France, Italy, United Kingdom, Denmark, Sweden, and Norway.

Using the Divisia method, factors considered in the study are emission coefficients, share of different fuels, structure (share of output from different sectors), energy intensity, and international structure (share of output from different countries). This study divided the 1973–79 into three different periods: 1973–79, 1979–87, and 1973–87. The results are depicted in Table 1.2. Divisia method was used, which has residuals after examining the changes in the factors responsible for the carbon dioxide emissions.

The result of the study showed a 21.8% reduction of CO_2 in the 1973–87 period. The 1980s recorded the most reduction. When each country within the investigated nine OECD were compared, France had the most decreased emission, whereas Denmark recorded the least in its emission reduction.

Reducing CO_2 in the nine OECD countries pointed at the reduction of energy intensity as the most important to be considered. This is believed to be achievable through increase in energy prices as well as economic

Table 1.2 Results from the analysis of Torvanger (1991)

	1973–79	1979–87	1973–87
CO_2 Intensity change	0.841	0.695	0.584
Emission coefficient	0.983	0.948	0.932
Fuel mix	1.021	1.029	1.051
Manufacturing sector	0.973	0.912	0.887
Energy intensity	0.901	0.836	0.753
International structure	0.998	1.009	1.008
Residual	−0.04	−0.06	−0.07

From Torvanger, A., 1991. Manufacturing sector carbon dioxide emissions in nine OECD countries, 1973–87. Energy Economics 13, 168–186.

growth/development, which results in technological advancement, more efficient capital equipment, and new improved industrial activities. Next to efficiently reducing CO_2 is the structural effect. Stimulating technological advancements, more efficient capital equipment, and new improved industrial activities were the implied policies envisaged from Torvanger (1991) study to limiting global warming. In descending order, the subsectors with highest intensity were ferrous metals; paper and pulp; stone; clay and glass; industrial chemicals and nonferrous metals; and nonraw materials.

Application of the Divisia method was extended to 12 Asian countries —China, Hong Kong, Taiwan, India, Pakistan, Indonesia, Malaysia, Philippines, Singapore, Thailand, and Japan between 1980 and 1990 (Shrestha and Timilsina, 1996). Their study focused on CO_2 intensity and its decomposition effect–generation mix and fuel intensity effect. From the result of the Divisia application; Japan had the lowest CO_2 intensity as opposed to India with the highest intensity. The fuel intensity was primarily responsible for the CO_2 intensity in these countries power generation and not the shifts in generation structure. For China, South Korea, India, Indonesia, Japan, Pakistan, Singapore and Taiwan—CO_2 intensity was the predominant force. For Hong Kong, both factors took their turn influencing the CO_2 intensity. Fuel intensity dominated for Malaysia between 1980 and 1986, thereafter, both factors took influencing turns. The opposite of what happened to Malaysia was applicable to Philippines as both factors were very much responsible for the CO_2 intensity from 1980 to 1986; however, fuel intensity took over predominance from 1985 to 1990. As for Thailand, both factors contributed to the CO_2 intensity from 1980 to 1986, whereas generation mix took responsibility thereafter.

1.3.2 Intensified Knowledge to Solving Greenhouse Crisis

Among the studies during the time knowledge was intensified in solving the crisis is the study of Liu et al. (2007) and Lin et al. (2006). China's industrial CO_2 was successfully analyzed using LMDI during the 1998–2005 period (Liu et al., 2007). In their study, CO_2 emission was decomposed into heat and electricity carbon emission coefficients, energy intensity, industrial structural shift, industrial activity as well as final fuel shift. 19 heavy and 17 light industrial subsectors were considered for the study.

During the whole period of investigation, industrial activity and energy intensity impacted more to the emission. Results showed three heavy industries in raw chemical materials and chemical products, nonmetal mineral products and smelting, and pressing of ferrous metals contributed 59.31% of CO_2. These three heavy industries should be considered as major focus on how to increase energy efficiency, which will impact on the energy intensity improving China's economy. From the industries investigated, coal, coke, petroleum products, natural gas, electricity, and heat in their various types were the total energies considered in their study. Table 1.3 shows the results of how each of the factors impacted on the total emission.

Industrial activity's increase of 484.07 million tons of carbon emission impacted more on China's industrial sectors in the 1998–2004 period. The 167.05 million tons' emission decrease in energy intensity was observed in the 1998–2002 period, it however increased to 79.44 million tons of carbon emission in the 2002–05 period as a result of some heavy industrial sectors energy intensity increase. With the 35.14% decrease in CO_2 emissions in the investigated period because of structural shift; smelting and pressing of ferrous metals and smelting and pressing of nonferrous metals among others increased the CO_2 emissions by 56.21 million tons' carbon of their structural shift. Such sectors are top priorities to see to the mitigation of China's CO_2 emissions.

Factors responsible for CO_2 emission were investigated by Lin et al. (2006). Their study focused on 22 Taiwan's industrial CO_2 emission from 1981 to 2001. The CO_2 was decomposed into emission factor, energy intensity, structure share, economic growth, emission changes, as well as the residuals. The decomposition result highlighted economic growth and high-energy intensity as the two main factors responsible for Taiwan's industrial emission, whereas structure share was responsible for the decrease.

Table 1.3 Impact of various drivers on China's industrial CO_2 emission (Liu et al., 2007)

	Carbon emissions	Activity effect	Energy intensity effect	Fuel mix effect	Emissions coefficient effect	Structural shift effect
1998–99	−7.5	38.7	−61.24	8.16	3.92	2.96
1999–2000	5.94	40.77	−32.11	10.22	−4.93	−8.01
2000–01	11.14	54.76	−48.2	8.74	−5.1	0.95
2001–02	41.95	63.22	−25.5	8.41	2.45	−6.64
2002–03	129.58	88.19	24.59	−1.87	12.62	6.05
2003–04	143.16	135.29	44.21	0.65	−8.14	−28.85
2004–05	95.89	63.14	10.64	8.73	−6.59	19.96
1998–2005	420.15	484.07	−87.61	43.05	−5.78	−13.58

CO_2 emission unit is million metric tons of carbon.
From Liu, L.-C., Fan, Y., Wu, G., Wei, Y.-M., 2007. Using LMDI method to analyze the change of China's industrial CO_2 emissions from final fuel use: An empirical analysis. Energy Policy 35, 5892–5900.

Their study for Taiwan was compared with the results of the United States, South Korea, Japan, Germany, and Netherlands. The comparison also confirmed economic growth as the focal factor responsible for emission within the same period. Energy intensity was also confirmed responsible for all the compared countries excluding Netherlands. Tables 1.4 and 1.5 show the decomposition results.

1.3.3 Finding Lasting Solution to the Greenhouse Situation

Among the studies that considered the three periods—early periods of oil crisis, time of intensified knowledge of solving the crisis, and time of finding lasting solution are the study of Lin and Ouyang (2014) and Shahiduzzaman et al. (2015). The study of Lin and Ouyang (2014) successfully analyzed the energy-related CO_2 emitted and the reduction potential in the Chinese nonmetallic mineral product industry using LMDI between 1986 and 2010

Table 1.4 Taiwan's decomposition results (Lin et al., 2006)

	Emission factor	Energy intensity	Structure share	Economic growth	Emission changes	Residuals (%)
1991–94	1,287	4,487	−6,450	12,854	12,147	−0.25
1994–96	2,212	−3,567	−2,207	10,586	7,020	−0.06
1996–99	7,552	−2,983	903	13,940	19,342	−0.36
1999–2001	−9,928	35,796	−11,563	4,538	18,799	−0.24
1999–2001	**3,586**	**28,454**	**−19,764**	**46,962**	**57,308**	**−3.37**

From Lin, S.J., Lu, I. J., Lewis, C., 2006. Identifying key factors and strategies for reducing industrial CO_2 emissions from a non-kyoto protocol member's (Taiwan) perspective. Energy Policy 34, 1499–1507.

Table 1.5 Decomposition results for various countries (Lin et al., 2006)

	Emission factor	Energy intensity	Structure share	Economic growth	Emission changes	Residuals (%)
United States	−112,782	10,295	−190,369	312,157	19,301	0
Japan	−35,578	85,079	−76,139	38,179	11,540	−0.01
Germany	−34,565	7,498	−35,701	22,798	−39,743	−0.57
Netherlands	−7,254	−4,051	−5,797	10,874	−6,216	−0.21
South Korea	−37,815	51,826	−30,902	68,223	50,640	−1.37

Emission unit: 10^3 metric tons.

From Lin, S.J., Lu, I. J., Lewis, C., 2006. Identifying key factors and strategies for reducing industrial CO_2 emissions from a non-kyoto protocol member's (Taiwan) perspective. Energy Policy 34, 1499–1507.

period. The emission of CO_2 from the Chinese metallic industry was decomposed into carbon intensity effect, substitution effect, energy intensity effect, industrial activity effect, and the industrial structural effect. These effects were computed according to the LMDI methodology.

The study subdivided the period into five different periods as always noted in China's 5-year plan; 1986–90, 1991–95, 1996–2000, 2001–05, and 2006–10. Table 1.6 shows the results of their analysis.

Among the factors analyzed, industrial activity contributed most to the CO_2 emission increase as can be observed from Table 1.4. Its highest contribution was during the 1986–2000 period, even though after this period it had a downward trajectory up until 2006–10. Energy intensity was hugely responsible for the decrease in CO_2 emission. The energy intensity continued to decrease from the beginning of the first period to the end. The industrial structure, however, contributed to an increase in CO_2 emission, becoming the second factor next to industrial activity.

From the study of Shahiduzzaman et al. (2015), Australia's emission was decomposed. This study was for 8 sectors and 14 subsectors from 1978 to 2010. The Australian sector was decomposed into CO_2 intensity, energy efficiency, structural change for both subsector and sector, wealth (gross domestic product per capita), and scale effect (population). The period of study was subdivided into six different periods; 1979–86, 1987–92, 1993–96, 1997–2000, 2001–05, and 2006–10. The result implies that both wealth and scale could increase the emission more than 50% that occurred over the period of study. Table 1.7 shows the result of the decomposition.

From 1979 to 1992, it can be observed that the CO_2 intensity increased as well for the energy efficiency. However, from 1993 to 2005, the intensity reduced while the energy efficiency fluctuated. From 2006 to 2010, the

Table 1.6 Decomposition analysis results from the study of Lin and Ouyang (2014)

Factors	1986–90	1991–95	1996–2000	2001–05	2006–10
Carbon intensity	−6.56	11.03	8.85	−19.55	−24.49
Substitution	−1.13	−3.2	−3.56	−2.19	−10.17
Energy intensity	−1.32	−58.38	−17.54	−76.69	−179.27
Industrial activity	30.13	95.38	224.34	217.44	155.35
Industrial structural	−2.27	22.79	−185.29	22.82	117.68
CO_2	18.86	67.62	26.79	138.84	59.1

From Lin, B., Ouyang, X., 2014. Analysis of energy-related CO_2 (Carbon dioxide) emissions and reduction potential in the Chinese nonmetallic mineral products industry. Energy 68, 688–697.

Table 1.7 Decomposition analysis results from the study of Shahiduzzaman et al. (2015)

	CO_2 intensity of energy use	**Subsectoral energy intensity (energy efficiency)**	**Structural change: subsector**	**Structural change: sector**	**Wealth**	**Scale**	**Total CO_2 emissions**
1979–86	−10.46	−26.69	6.6	1.68	31.27	24.17	26.58
1987–92	−0.43	−7.35	12.15	−10.59	28.28	21.7	43.76
1993–96	−2.15	−6.11	4.09	−11.92	33.87	13.36	31.14
1997–2000	−3.14	5.13	2.82	−24.5	39.89	14.6	34.81
2001–05	−3.74	−1.69	6.07	−21.96	35.08	22.8	36.56
2006–10	2.59	−19.65	11.15	−18.42	20.01	37.2	32.87
1979–2010	**−17.33**	**−56.35**	**42.88**	**−85.71**	**188.4**	**133.83**	**205.72**

From Shahiduzzaman, M., Layton, A., ALAM, K., 2015. Decomposition of energy-related CO_2 emissions in Australia: Challenges and Policy implications. Economic Analysis and Policy 45, 100–111.

CO_2 intensity increased, and the energy efficiency reduced. Their study confirmed wealth and scale effect as the most determining factors increasing the country's emission. The structural changes as well as the energy efficiency have played significant roles in reducing the emission, with energy efficiency contributing 17% reduction in the country's emission.

1.4 CONCLUSION

The effect of climate change has reached global attention with various organizations, academia, and government establishments looking for ways to address the issue. Pragmatic approaches have looked the more feasible in solving the various episodes due to climatic change. Analyzing GHGs emitted from the industrial sectors goes a long way to keeping the global village safe for living. Industrial sectors contribute a lot economically and as much as the economy grows from its output, it also consumes a lot of energy, which also leads to emission of harmful gases. Mitigating these gases can only mean understanding the factors or drivers responsible for these emissions, which can result in policies to address such situations.

Understanding such factors has been well interpreted by IDA. This chapter focused on the application of LMDI (mostly used approach of IDA) and on certain countries' industrial and electricity sectors. The application considered three periods; the periods of (1) early oil crisis, (2) intensified knowledge to solving greenhouse crisis, and (3) finding lasting solution to the greenhouse situation. The three applications gave similar results as to what factor to focus on the most to mitigating GHGs. The applications identified energy intensity and industrial activity mostly to be key to the reduction. These factors can assist in formulating policies needed to achieving the objective of GHG reduction.

The industry's GHG challenges are rooted in two phases: it emits a lot of harmful gases, and it does not stop to emit more. Without the right policies in place, industries will continue to emit these gases at an alarming rate. The question thus arises: What policies and actions will effectively assist industries in mitigating GHGs? Policies introduced to affect both energy intensity and industrial activity will definitely go a long way to reducing the gases emitted as it has been noticed that these two factors do affect the emission hugely. All industries across the world should have committees that see to the development of accurate policies that can have positive impacts. Right policies will always attract the right technicality for policy implementation.

REFERENCES

Alegria, I.M.D., Basanez, A., Basurto, P.D.D., Fernandez-Sainz, A., 2016. Spain's fulfillment of its kyoto commitments and its fundamental greenhouse gas (GHG) emission reduction drivers. Renewable and Sustainable Energy Reviews 59, 858–867.

Ang, B.W., 2004. Decomposition analysis for policymaking in energy: which is the preferred method? Energy Policy 32, 1131–1139.

Ang, B.W., Zhang, F.Q., 2000. A survey of index decomposition analysis in energy and environmental studies. Energy 25, 1149–1176.

Bobba, A.G., 2002. Numerical modelling of salt-water intrusion due to human activities and sea-level change in the Godavari Delta, India. Hydrological Sciences Journal 47.

Cifuentes, L., Borja-Aburto, V.H., Gouveia, N., Thurston, G., Davis, D.L., 2001. Hidden health benefits of greenhouse gas mitigation. Science 293, 1257–1259.

Cote, R.P., Liu, C., 2016. Strategies for reducing greenhouse gas emissions at an industrial park level: a case study of Debert Air Industrial Park, Nova Scotia. Journal of Cleaner Production 114, 352–361.

D'Amato, G., Cecchi, L., 2008. Effects of climate change on environmental factors in respiratory allergic diseases. Clinical and Experimental Allergy: Journal of the British Society for Allergy and Clinical Immunology 38, 1264–1274.

Dai, A., 2013. The increased risk of drought under global warming. Earthweek (Online).

Dale, V.H., Joyce, L.A., Mcnulty, S., Neilson, R.P., Ayres, M.P., Flanniganal, M.D., Lugo, A.E., 2001. Climate change and forest disturbances. BioScience 51, 723.

Feng, C., Gao, X., Wu, J., Tang, Y., He, J., Qi, Y., Zhang, Y., 2015. Greenhouse gas emissions investigation for towns in China: a case study of Xiaolan. Journal of Cleaner Production 103, 130–139.

Fengling, L., 2004. Decomposition Analysis Applied to Energy: Some Methodological Issues (A Thesis Submitted for the Degree of Doctor of Philosophy). National University of Singapore.

Fouillet, A., Rey, G., Laurent, F., Pavillon, G., Bellec, S., Guihenneuc-Jouyaux, C., Jougla, E., 2006. Excess mortality related to August 2003 heat wave in France. International Archives of Occupational and Environmental Health 80, 16–24.

Frihy, O.E., 2003. The Nile delta-Alexandria coast: vulnerability to sea-level rise, consequences and adaptation. Mitigation and Adaptation Strategies for Global Change 8, 115–138.

Gao, F., Nie, Z.R., Wang, Z.H., Li, H.M., Gong, X.Z., Zuo, T.Y., 2009. Greenhouse gas emissions and reduction potential of primary aluminium production in China. Science in China, Series E: Technological Sciences 52, 2161–2166.

Hamit-Haggar, M., 2012. Greenhouse gas emissions, energy consumption and economic growth: a panel cointegration analysis from Canadian industrial sector perspective. Energy Economics 34, 358–364.

Heidari, N., Pearce, J.M., 2016. A review of greenhouse gas emission liabilities as the value of renewable energy for mitigating lawsuits for climate change related damages. Renewable and Sustainable Energy Reviews 55, 899–908.

IEA, 2012. CO_2 Emissions from Fuel Combustion (Detailed Estimate). IEA, Paris.

IPCC, 2001. Third Assessment Report: Climate Change 2001. Intergovernmental Panel on Climate Change.

(IPCC), I.P.O.C.C, 2007a. Climate Change 2007: The Fourth Assessment Report of the Intergovernmental Panel on Climate Change. Cambridge University Press, England.

IPCC, 2007b. Fourth Assessment Report: Climate Change 2007. Intergovernmental Panel on Climate Change.

Jobert, T., Karanfil, F., Tykhonenko, A., 2010. Convergence per capita carbon dioxide emissions in the EU: legend or reality? Energy Economics 32, 1364–1373.

Jungnitz, A., 2008. Resource Productivity, Environmental Tax Reform and Sustainable Growth in Europe: Decomposition Analysis of Greenhouse Gas Emissions and Energy and Material Inputs in Germany.

Khondaker, A.N., Hasan, M.A., Rahman, S.M., Malik, K., Shafiullah, M., Muhyedeen, M.A., 2016. Greenhouse gas emissions from energy sector in the United Arab Emirates – an overview. Renewable and Sustainable Energy Reviews 59, 1317–1325.

Klinenberg, E., 2008. Are You Ready for the Next Disaster? New York Times Magazine, New York, NY.

Lin, B., Ouyang, X., 2014. Analysis of energy-related CO_2 (carbon dioxide) emissions and reduction potential in the Chinese non-metallic mineral products industry. Energy 68, 688–697.

Lin, S.J., Lu, I.J., Lewis, C., 2006. Identifying key factors and strategies for reducing industrial CO_2 emissions from a non-kyoto protocol member's (Taiwan) perspective. Energy Policy 34, 1499–1507.

Liobikiene, G., Butkus, M., Bernatoniene, J., 2016. Drivers of greenhouse gas emissions in the Baltic states: a decomposition analysis related to the implementation of Europe 2020 strategy. Renewable and Sustainable Energy Reviews 54, 309–317.

Liu, L.-C., Fan, Y., Wu, G., Wei, Y.-M., 2007. Using LMDI method to analyze the change of China's industrial CO_2 emissions from final fuel use: an empirical analysis. Energy Policy 35, 5892–5900.

Liu, M., Huang, Y., Jin, Z., Liu, X., Bi, J., Jantunen, M.J., 2016a. Estimating health co-benefits of greenhouse gas reduction strategies with a simplified energy balance based model: the Suzhou city case. Journal of Cleaner Production 1–11.

Liu, Z., Geng, Y., Adams, M., Dong, L., Sun, L., Zhao, J., Dong, H., Wu, J., Tian, X., 2016b. Uncovering driving forces in greenhouse gas emissions in China aluminium industry from the perspective of life cycle analysis. Applied Energy 166, 253–263.

Lofgren, A., Muller, A., 2010. Swedish CO_2 emissions 1993–2006: an application of decomposition analysis and some methodological insights. Environmental and Resource Economics 47, 221–239.

Mendiluce, M., Perez-Arriaga, I., Ocana, C., 2010. Comparison of the evolution of energy intensity in Spain and in the EU 15. Why Spain different? Energy Policy 38, 639–645.

Nicholls, R.J., Hoozemans, F.M., Marchand, M., 1999. Increasing flood risk and method losses due to global sea-level rise; regional and global analysis. Global Environmental Change 9, 69–87.

Olivier, J.G.J., Janssens-Maenhout, G., Muntean, M., Peters, J.A.H.W., 2016. Trends in Global CO_2 Emissions. PBL Netherlands Environmental Assessment Agency and Ispra, The Hague.

Ouyang, X., Lin, B., 2015. An analysis of the driving forces of energy-related carbon dioxide emissions in China's industrial sector. Renewable and Sustainable Energy Reviews 45, 838–849.

Ozturk, F., Keles, M., Evrendilek, F., 2016. Quantifying rates and drivers of change in long-term sector-and country-specific trends of carbon dioxide- equivalent greenhouse gas emissions. Renewable and Sustainable Energy Reviews 65, 823–831.

Parry, M.L., Rosenzweig, C., Iglesias, A., Livermore, M., Fischer, G., 2004. Effects of climate change on global food production under SRES emissions and socio-economic scenarios. Global Environmental Change 14, 53–67.

Ren, J., Tan, S., Goodsite, M.E., Sovacool, B.K., Dong, L., 2015. Sustainability, shale gas, and energy transition in China: assessing barriers and prioritizing strategic measures. Energy 84, 551–562.

Shahiduzzaman, M., Layton, A., Alam, K., 2015. Decomposition of energy-related CO_2 emissions in Australia: challenges and policy implications. Economic Analysis and Policy 45, 100–111.

Shrestha, R.M., Timilsina, G.R., 1996. Factors affecting CO_2 intensities of power sector in Asia: A Divisia decomposition analysis. Energy Economics 18, 283–293.
Torvanger, A., 1991. Manufacturing sector carbon dioxide emissions in nine OECD countries, 1973–87. Energy Economics 13, 168–186.
Wang, S., Zhou, C., Li, G., Feng, K., 2016. CO_2, economic growth, and energy consumption in China's provinces: investigating the spatiotemporal and econometric characteristics of China's CO_2 emissions. Ecological Indicators 69, 184–195.
Wang, X., Smith, K.R., 1999. Secondary benefits of greenhouse gas control: health impacts in China. Environmental Science and Technology 33, 3056–3061.
Witi, J., Stevens, L., 2014. GHG National Inventory Report South Africa 2000–2010.
Wu, J., Zhu, Q., Liang, L., 2016. CO_2 emissions and energy intensity reduction allocation over provincial industrial sectors in China. Applied Energy 166, 282–291.
Xiaoyan, X., 2013. Index Decomposition Analysis of Energy Consumption and Carbon Emissions: Some Methodological Issues. Doctor of Philosophy, National University of Singapore.

CHAPTER 2

Booming and Stagnation of Spanish Construction Sector Through the Extended Carbon Footprint Concept

Jorge E. Zafrilla, Luis A. López
University of Castilla-La Mancha, Albacete, Spain

2.1 INTRODUCTION

The Spanish construction sector was one of the growth locomotives of the Spanish economy from 2000 to 2008, during which time real estate boomed and transport and leisure infrastructure (including highways, airports, sport installations, and theme parks) was built. As a result, the construction sector reached a distorted dimension. In 2004, it represented the 17.5% of total employment, direct and indirect, ahead of the average of 11 OECD countries of 12.4% (Bielsa and Duarte, 2011). At the same time, construction is the main activity sector responsible for the reduction in the preservation of energy and the environment in the Spanish economy. In 2008, the construction sector was responsible, in total terms, for 20.28% of domestic CO_2 emissions (measured under Kyoto Protocol criteria) due to linkage effects over other inputs supplier industries, directly and indirectly, that are intensive in energy use (Zafrilla et al., 2012), including extraction, cements, ceramics, land transport, and electricity. However, the 2008 financial crisis and the following Spanish recession, which was still a fact in 2013, involved the elimination of more than 1.5 million jobs in the sector between those years, together with a reduction of the carbon footprint.

Despite these figures, the main actions and environmental policies implemented in this sector in Spain and in the European Union (EU) were focused on the reduction of the energy impact related to the energy demand required by buildings in their lifecycle. Some examples, on a

Environmental Carbon Footprints
ISBN 978-0-12-812849-7
http://dx.doi.org/10.1016/B978-0-12-812849-7.00002-7

European level, are the action plan for energy efficiency, which identifies energy efficiency in the construction sector as a priority, as a consequence of its large importance in the economy (Commission, 2009). For the Spanish case, the Spanish strategy for energy efficiency (MINETUR, 2005) identifies the consumption related to heating systems, and light installations of the whole building park in the Spanish economy, which represented 17% of total energy consumption in 2005 (MINECO, 2007).

In both examples, we find that energy and environmental policies address only a part of the whole life cycle assessment (LCA) of the sector, the typical lifecycle, i.e., once the building is built, *from gate to grave* (Sharrard et al., 2008). In the EU case, we find examples concerning the same stage of the LCA. Kuckshinrichs et al. (2010) find environmental improvements derived from a building refurbishment program that seeks to improve energy efficiency. Spain, following the European Parliament recommendations, established an energy efficiency mechanism via the royal decree in 2007 (Royal Decree 47/2007), which was too late considering that the construction sector boom began some years before. However, there is no defined strategy at the European or Spanish level to mitigate the environmental impact linked to production chain emissions of the building of houses, infrastructure, industry, etc., the stage that positions the construction sector as one of the most polluting sectors.

Our proposal is to develop an input—output life-cycle assessment model (IO-LCA) to quantify an extended carbon footprint (ECR) criterion of the construction sector for the period of 2000—12. We understand this criterion as the virtual carbon (direct and indirect emissions of CO_2) embodied in domestic and imported inputs required by the construction industry (Hendrickson et al., 1998; Joshi, 2000; Sharrard et al., 2008; Bilec et al., 2010) or, according to Zhang and Wang (2016a,b), for the building industry in China. Therefore, the carbon footprint of a sector coincides with the consumer responsibility criterion (Munksgaard and Pedersen, 2001; Andrew et al., 2009; Cadarso et al., 2012). In addition, we will add the impact of the consumption of the income generated in the production process using the Social Accounting Matrix (SAM) developed by Dejuán et al. (2013).

The construction responsibility criterion proposed considers the domestic and imported virtual carbon linked to the production phase of the sector, *from cradle to gate*. The main advantage of the IO-LCA methodology, in comparison to process-based LCA techniques, is that the former can include the whole global production chain, as Nassen et al. (2007) noted when they compared the energy required for the construction sector

(GJ/m^2) in an IO-LCA model with results obtained from 18 other papers using process-based LCA. The impact using IO-LCA was two times higher. Zhang and Wang (2016b) conclude that both methods, the process level and the input—output, have advantages and limitations.

On the other hand, the Kyoto Protocol and the 2015 Paris Agreement on Climate Changes 2015 use the Producer Responsibility criterion as the basis of emission accounting for a country and establish that a country is responsible for emissions generated in the territory of the analyzed region (total output, exports included). However, the calculus of the carbon footprint or consumer responsibility implies the implicit consideration of emissions embodied in imports. These imports are very important for the construction sector because it requires many high-polluting foreign inputs (steel imports from China, DGA, (various years)), so in environmental terms, its environmental responsibility will grow.

The use of a SAM allows us to identify and assign to any sector emissions related to consumption from those incomes generated in the sector and linked to inputs used to produce those goods attending to final demand. In this sense, we consider, as a key variable in emissions analysis, the weight of the construction sector gross added value over the whole economy, the distribution between wages and profits, and different consumption patterns of employees and entrepreneurs of the construction sector and other supplier industries. To identify these emissions, we will apply the SAM additive decomposition analysis of Stone (1985), which was applied by Llop and Pié (2011) to analyze domestic emission multipliers for Catalonia, Spain. We will differentiate between direct, indirect, and induced emissions through the dragging effect of sectorial income expenditure.

This chapter continues as follows: Section 2.2 describes the methodological foundation and database; Section 2.3 examines the main results; and Section 2.4 provides a summary of our conclusions.

2.2 METHODOLOGY AND DATABASE

2.2.1 Extended Carbon Footprint

The literature about carbon footprint estimation using input—output models (Arto and Dietzenbacher, 2014; Hubacek et al., 2014; Peters et al., 2016) and using LCA (Alvarez and Rubio, 2015; Hertwich et al., 2015; Wiedmann et al., 2011) is vast. In this chapter, we propose an extension of the carbon footprint concept combining both approaches and adding emissions generated because of the expenditure of income generated by industry. The concept of an *ECR* for an industry includes direct, indirect, and induced CO_2 emissions to supply goods to the autonomous final demand. Therefore, we include emissions related to the domestic and imported input consumption of each sector (carbon footprint), which is similar, in terms of LCA, to *cradle to gate* emissions (*CF*), plus all emissions, domestic and imported, linked to the consumption of income (added value) generated in each sector (*EY*).

$$ECR = CF + EY \tag{2.1}$$

The introduction of emissions linked to gross added value (income) consumption in a sector is the main contribution of this chapter, in which we assign responsibilities at a sectorial level. With this methodology, the importance of the environmental impact of the construction sector depends on the weight of its added value over the whole economy added value, on the profit and wage distribution, and on the different consumption patterns of workers and entrepreneurs from the sector and from other supplier industries.

2.2.2 Total Emissions and Social Accounting Matrix Multipliers

The input–output methodology allows the estimation of total emissions, direct and indirect, related to the domestic or imported consumption of each industry. Matrix multipliers of the *k* energy goods considered are quantified, in domestic terms (M^d), by columns direct, indirect, and induced needs to produce goods to meet final demand:

$$M^d = \left(1 - A^{*d}\right)^{-1} \tag{2.2}$$

where A^{*d} is the extended matrix (working with a SAM structure) of domestic technical coefficients, with households endogenized. The main characteristics of the SAM, where we endogenize households' final consumption without considering nonresident consumption, can be consulted in Dejuán et al. (2013).

One step further allows us to define the domestic emission multiplier matrix, multiplying the emission factor of each energy good (ε_k) diagonalized, tCO_2 per M€ combusted, by expression (2.2):

$$EM^d = \varepsilon_k\left(1 - A^{*d}\right)^{-1} \tag{2.3}$$

The emission factors in expression (2.3) are compiled in Zafrilla et al. (2012).

We propose the decomposition of the extended multiplier matrix to isolate the environmental impacts of the household consumption endogenization in a similar way as Pyatt and Round (1979, 2003). The notation used is similar to that used in Miller and Blair (2009).

Applying the Stone's additive decomposition Stone (1985),

$$EM^d = \varepsilon_k\left[I + \left(M_1^d - I\right) + \left(M_2^d - I\right)M_1^d + \left(M_3^d - I\right)M_2^d M_1^d\right] \quad (2.4)$$

where

$$M_1^d = (I - Z)^{-1} \quad (2.5)$$

$$M_2^d = (I + T) \quad (2.6)$$

$$M_3^d = \left(I - T^2\right)^{-1} \quad (2.7)$$

where we define Z as $\begin{pmatrix} A^d & 0 \\ 0 & 0 \end{pmatrix}$ and T as $(I - Z)^{-1}R$, where $R = \begin{pmatrix} 0 & C^H \\ V^i & 0 \end{pmatrix}$, with C^H being the household final consumption vector and V^i being the proportion of added value endogenized in the SAM. The decomposed multipliers are as follows:

$$M_1^d = (I - Z)^{-1} = \begin{pmatrix} \left(I - A^d\right) & 0 \\ 0 & I \end{pmatrix}^{-1} = \begin{pmatrix} \left(I - A^d\right)^{-1} & 0 \\ 0 & I \end{pmatrix} \quad (2.8)$$

M_1^d matrix is a Leontief's multiplier matrix, the so-called own effects matrix, without including the multiplier effects derived from the added value and derived household consumption endogenization. In our case, these own effects are related to the impulse of the autonomous demand, considering public administration consumption, tourist consumption, investment, and exports.

$$M_2^d = I + T = I + (I + Z)^{-1}R = \begin{pmatrix} I & \left(I - A^d\right)^{-1}C^H \\ V^i & I \end{pmatrix} \quad (2.9)$$

Matrix M_2^d is known as the indirect multipliers matrix or open loop effect because it shows the effects of the exogenous inputs over income (wages plus the part of the profits consumed) that households receive, but it does not show the feedback effects produced when those incomes are destined for consumption. In our case, the impact of M_2^d is null because of the characteristics of our emission factor methodology. We obtain results about emissions only when energy goods are combusted/consumed. There is no endogenized consumption in M_2^d, only the generation of incomes. The consumption of the income and therefore the emissions are shown in M_3^d.

$$M_3^d = (I + T^2)^{-1} = \left(I - \left[(I - Z)^{-1}R\right]^2\right)^{-1} \tag{2.10}$$

$$M_3^d = \begin{pmatrix} \left[I - (I - A^d)^{-1}C^H V^i\right]^{-1} & 0 \\ 0 & \left[I - V^i(I - A^d)^{-1}C^H\right]^{-1} \end{pmatrix} \tag{2.11}$$

M_3^d is the circular effects matrix capturing the feedback effects derived from the circular flow of income because of exogenous impacts. For example, an increase of exports (exogenous) generates an increase of the production to attend to the final demand, followed by an increase in the household's income, which will generate an increase in final demand, etc. To discuss data and results, we will assume that induced effects M_2^d and M_3^d come from multipliers, and M_1^d will be Leontief's output multiplier. Concretely, if we are referring to a determined sector, such as construction, the importance of that amount of the feedback effect will be high because of the linkage effects of this sector with those derived from the consumption of a high range of intermediate goods and those derived from the consumption of incomes generated in the sector.

The results that we obtain, M_3^d, for this multiplier will be conditioned by the structure of the distribution of income in the construction sector. In this sense, the application of measures to increase taxes places pressure on profits and more exhaustive control over the black economy and bank credits, which will reconfigure the consumption patterns and environmental impacts, would had been recommended.

The responsibility criterion used considers emissions linked to the import of goods. Because of this, we will use the domestic technical assumption (DTA) to account for emissions abroad. We assume that imported goods will be produced with Spanish technology. This assumption is restrictive but is commonly used in the literature (Munksgaard and Pedersen, 2001; Peters and Hertwich, 2006; Cadarso et al., 2010; Serrano and Dietzenbacher, 2010; Zhang and Wang, 2016a). However, other papers use multiregional input–output models (MRIO) to estimate emissions linked to trade (Lenzen et al., 2004; Peters and Hertwich, 2008; Monsalve et al., 2016). The DTA assumption is applicable for the Spanish economy. Andrew et al. (2009) show that the differences in the accounting of the carbon footprint between a DTA model and an MRIO model for the Spanish economy in 2001 was only 1%. The estimation of the direct and indirect emissions abroad was done in a similar way as the expressions shown before but with a large difference; in this case, we are not working with a SAM, only IOT. We consider only emissions related to production processes under the DTA assumption; we do not consider induced emissions abroad. Initially, we estimate the total multiplier matrix (M^t):

$$M_k^t = (1 - A^t)^{-1} \tag{2.12}$$

where A^t is the matrix of total technical coefficients in this case, without taking into account the endogenization of households but considering all international rounds of production. The total emission multiplier matrix is obtained multiplying the emissions factors of expression (2.3) by expression (2.11):

$$EM_k^t = \varepsilon_k (1 - A^t)^{-1} \tag{2.13}$$

2.2.3 The Estimation of the Extended Carbon Footprint

The ECR criterion of each sector includes the direct and indirect CO_2 emissions necessary to the supply final demand and those relative to the consumption of energy goods (in and outside the country) plus domestic-induced emissions derived from income consumptions. The expression used to quantify those emissions by sector and for the total amount of k energy goods considered is as follows:

$$RM^E = \underbrace{\sum_{j=1}^{k} M_k^d \widehat{y}^d}_{2.14.1} + \underbrace{\sum_{j=1}^{k} M_k^t \left(A^m (I - A^d)^{-1} \widehat{y}^d + \widehat{y}^m \right)}_{2.14.2} - \underbrace{\sum_{j=1}^{k} \varepsilon_k i_k' \widehat{y}_k^s}_{2.14.3} \tag{2.14}$$

where $\widehat{y}^d$ is the diagonalized vector of autonomous final demand (public administrator expenditure, tourist consumption, investments, and exports) and $\widehat{y}_k^s$ is a diagonalized vector of k energy goods stocks variations. The result in expression (2.14) is a matrix with *k* rows and *n* columns, where columns quantify total emissions related to the production of each sector to attend to autonomous final demand. The sum by columns will be the ECR of a sector.

It is interesting to note that the emissions related to the ERR of a sector are sensible to the methodology used in the construction of the compact SAM because the configuration of intermediate and final demands will be altered. The composition of endogenous and exogenous components of the model, induced and autonomous, will condition the results. In our case, we are endogenizing all types of consumption (except the nonresident). In this case, for example, the autonomous consumption related to pensions would be exogenous; emissions related to this consumption would not go to the sectors where the income was generated.

The proposed measure of ECR is especially useful for quantifying the carbon footprint of the construction sector. The reason is that the majority of its production is distributed as final demand (54.9% in 2005), and 35.8% are sales to other industries from the same construction sector (intraindustrial commerce), so direct and indirect emissions are still assigned to the same industry. Only 9.3% are sales to other sectors, which would be out of the ECR of the construction sector. This measure would not be adequate if we were measuring the environmental responsibility of the electricity sector because this sector transfers, via intermediate consumption, more than 80% of emissions to other sectors. The emission spreading complicates accurate estimation of the sector's extended responsibility.

The extent of the environmental impact depends on the direct and indirect energy mix used. The first addendum (2.14.1) shows those domestic emissions linked to the combustion of energy goods (oil products, gas, and coal) needed to satisfy the domestic final demand for all the sectors and households. The second addendum (2.14.2) quantifies sectorial and households' direct and indirect emissions linked to input imports. Finally, in the third addendum (2.14.3), direct emissions linked to the use of stock variations are subtracted.

The Kyoto Protocol in the past, as well as the Paris Agreement now, establishes producer responsibility as the key emissions allocation criterion. A basic definition of this criterion establishes that a country is responsible of all emissions generated within the borders of the country, regardless of where the consumption of the produced goods is held. The same result is found by a vast number of papers (Munksgaard and Pedersen, 2001; Lenzen et al., 2004; Sánchez-Choliz and Duarte, 2005; Serrano and Dietzenbacher, 2010). The producer responsibility expression is as follows:

$$RM^{\text{Prod}} = \underbrace{\sum_{j=1}^{k} M_k^d \widehat{\gamma}^d}_{2.15.1} + \underbrace{\sum_{j=1}^{k} \varepsilon_k i_k' (Z^{*m} + \widehat{\gamma}^m)}_{2.15.2} - \underbrace{\sum_{j=1}^{k} \varepsilon_k i_k' \widehat{\gamma}_k^x}_{2.15.3} - \underbrace{\sum_{j=1}^{k} \varepsilon_k i_k' \widehat{\gamma}_k^s}_{2.15.4} \tag{2.15}$$

where $\widehat{\gamma}_k^x$ is a diagonalized matrix that includes autonomous final demand exports of energy input k. Elements 2.15.1 and 2.15.4 are part of the extended responsibility. On the other hand, addendum 2.15.2 shows direct emissions linked to energy input imports because the combustion of those inputs will occur within the borders. Addendum 2.15.3 subtracts direct emissions of exported energy inputs because the combustion will occur outside our borders.

Under the consumer responsibility criterion, a country will account for direct and indirect emissions linked to domestic production sold within the borders, plus total emissions linked to the production of all intermediate and final imported goods to attend to domestic demand without considering emissions embodied in exports (Munksgaard and Pedersen, 2001, Gallego and Lenzen, 2005; Cadarso et al., 2012). The expression is as follows:

$$\begin{aligned} RM^{\text{Cons}} = {} & \sum_{j=1}^{k} M_k^d \widehat{\gamma}^d + \sum_{j=1}^{k} M_k^t \left(A^m (I - A^d)^{-1} \widehat{\gamma}^d + \widehat{\gamma}^m \right) - \sum_{j=1}^{k} M_k^d \widehat{\gamma}^x \\ & - \sum_{j=1}^{k} \varepsilon_k i_k' \widehat{\gamma}_k^s - \sum_{j=1}^{k} M_k^t \left(A^m (I - A^d)^{-1} \widehat{\gamma}^x \right) \end{aligned} \tag{2.16}$$

The main difference between expressions (2.16) and (2.15) is that in consumer responsibility, the second element of the expression considers total virtual carbon, direct and indirect, linked to all the goods and services imports of the country, whereas producer responsibility considers only emissions linked to energy input imports. Moreover, expression (2.16) includes a fifth addendum where the total virtual carbon imported to be exported is subtracted. Emissions derived from vertical specialization processes (Hummels et al., 2001) are not included. However, the construction sector has a peculiarity; there are no exports, so expressions (2.13) and (2.15) are the same for this sector.

2.2.4 Databases

This chapter is focused in the analysis of the booming and stagnation processes of the Spanish economy, together with a period of energy efficiency and environmentally friendly legislation improvements. The period considered is 2000–12, where both sides of the economic cycle were well drawn (growth and crisis), and all the environmental policy implementations were concentrated. The databases used are the input–output framework from the National Accounts provided by the Statistical National Institute of Spain (INE, Various years) for the years 2000 and 2008. Data from 2008 to 2012 are taken from Zafrilla et al. (2012) and Dejuán et al. (2013), where an extended environmental model is proposed to measure the emission evolution of the Spanish economy.

2.3 MAIN RESULTS

2.3.1 Carbon Footprint of the Spanish Economy

Economic and population growth shown by the Spanish economy in the period of 2000–08 drove the huge increase of domestic CO_2 emissions. However, since 2005, a slowdown has been shown in the evolution of emissions because of the consolidation of environmentally friendly legislation and, since 2008, the financial crisis. In particular, a 58.9% emission reduction from 2008 to 2012 came as a result of policy implementation and 41.1% came because of the economic crisis (Zafrilla et al., 2012).

The concept of an ECR allows us to identify the final autonomous demand supplier industries or sectors that are responsible for an amount of virtual carbon linked to the production of domestic or imported inputs and to the consumption of the income generated in those industries. In this sense, sectors such as capital goods, nonmarket services, construction, and market services (Fig. 2.1) stand out, similar to Tarancón et al. (2011) or Serrano and Dietzenbacher (2010), until the year 2000, although those authors considered only Spanish domestic emissions and the latter did not include induced emissions. Throughout the carbon footprint by industries, emissions are transferred to those sectors that demand emissions-intensive inputs or to sectors with dragging effects over other industries. The sectorial reallocation of emissions will be more intensive to those sectors with higher dragging effects over the rest of the industries. The construction sector shows relevant domestic linkages as is present in the 19 domestic important coefficients and 3 imported in 2005 in a total of 73 sectors

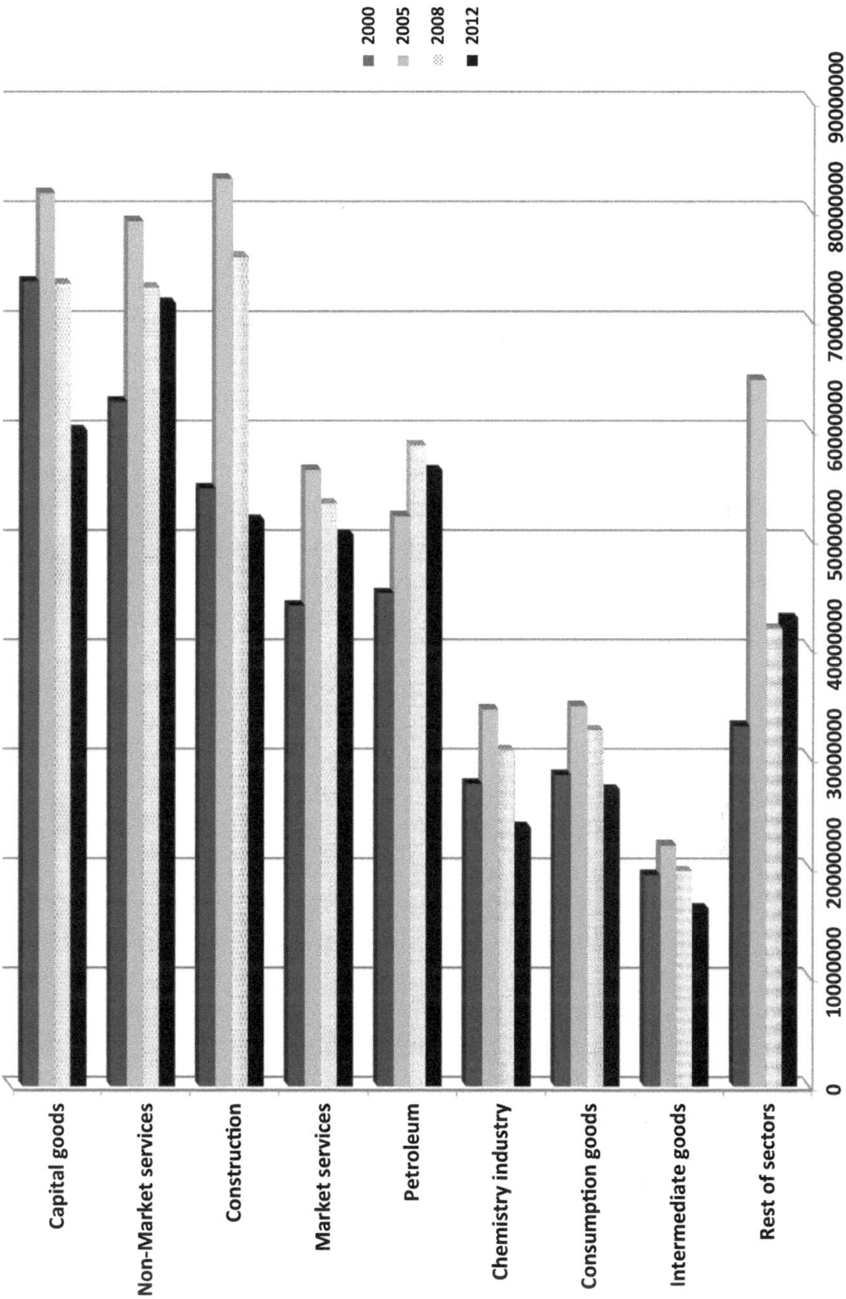

Figure 2.1 Extended carbon footprint by sector.

considered (López et al., 2011). For all the industries highlighted, the effects of environmental policies and the economic crisis are evident because the virtual carbon of those industries has fallen since 2005. The exception is the oil sector; Spain has a high dependency on road transport, which explains this result (Mendiluce, 2010). Emissions continued growing with the strong Spanish economic growth until 2008. Only during the economic crisis was a small reduction observed in the ECR.

Focusing on the construction sector, in the year 2000 it emerged as the third main polluter sector in the Spanish economy in terms of ECR. The same results were found by Sánchez-Choliz and Duarte (2005) and Tarancón and del Río (2007) for the year 1995 but only in terms of domestic emissions. The results in 2005 are quite different. After the boom of the construction sector bubble in the Spanish economy, the construction sector rose to first place in terms of emissions, showing the highest growth rate in the Spanish economy since the year 2000. Mainly since 2005, environmental policy improvements have reduced construction sector emissions; however, in the year 2008, it was still the most polluting sector in terms of environmentally extended responsibility. In 2012, after the hardest years of the financial crisis and subsequent construction sector crash, emissions fell considerably, from 92 million of tCO_2 in 2008 to 60 million in 2012. This is the most pronounced reduction compared to the rest of the sectors; however, it is still placed on the podium of the main polluters sectors in the third position. In summary, the amount of construction sector emission responsibility represents 24.2% of the total emissions in the year 2000, 30.0% in 2005, 29.25% in 2008, and 22.8% in 2012 (Fig. 2.1).

This high emissions responsibility of the construction sector relies on the following characteristics. First, while directly it is not a high pollutant sector, indirectly it is an intensive user of domestic and imported energy inputs coming from sectors such as metallurgy (iron, steel, etc.), nonmetallic minerals (cement, concrete, lime, gypsum, glass, ceramic, bricks, etc.), and transport (Gutiérrez et al., 2011). Second, this industry distributes its production between final demand (54.9% in 2005) and within the same industry (35.8%); sales to other industries are quite low (9.3%). This means that it does not transfer to other industries; for the energy or emission embodied expenditures, almost all is integrated in the production to attend to final demand. Third, the construction sector boom implied a huge autonomous demand in all those years (construction investment reach 59.5% of total investments in 2005) linked to second homes in coastal areas and to public investments in infrastructure funded by regional governments or by the EU. Fourth, Spanish economic growth was followed by the arrival of more than four million immigrants between 2000 and 2008, which was linked to the creation of more than 3.5 million jobs. The induced demand of first homes by all those immigrants grew because of the expenditure of all the generated incomes.

2.3.2 Decomposition of Extended Carbon Footprint of Construction Sector

Direct emissions of the construction sector (1) represent a minimum part of the total emissions of the sector (Table 2.1). In the year 2000, those emissions reached the highest values but remained under 5% of the total. In 2012, it represented only 1.15%. Acquaye and Duffy (2010) figure is slightly different: 17% of direct emissions in the Irish construction sector; however, they consider direct emissions those generated in the consumption of electricity. Those emissions are indirect in our case.

Regarding virtual carbon linked to the consumption of inputs in the construction industry, domestic total emissions (2) represent 54.3% of the total responsibility. This figure is stable in the period analyzed. The reasons are related to the high dependency of highly polluting inputs, such as metallurgy or nonmetallic minerals, which use energy products intensively (Gutiérrez et al., 2011). Specifically, in 2012, the construction industry consumed 33.7% of intermediate inputs supplied by the national extraction sector and 24.7% of the intermediate goods sector. Other authors, such as Zhang and Wang (2016a,b), find that materials manufacturing is the main contributor to building embodied emissions, accounting for 80%–90% of the total domestic and imported emissions in the case of China. The potential reduction of the construction sector carbon footprint would be indirect and would be benefited by those policies that incentivize environmental efficiency in both industries. Although both industries were included in the European Union Emissions Trading Scheme (EU-ETS), the effects were not as good as expected because of the free allocation system of emission allowances held at the beginning (Clò, 2010). Directive 2009/29/EC (Commission, 2009) has tried to solve this problem since 2013 by establishing a general criterion of emissions allowance auctions for most of the industries and only free allocation for those industries under risk of carbon leakage. Even so, the economic crisis and the stagnation of the construction sector led to a price reduction of emission allowances

Table 2.1 Extended carbon footprint (ECR) of the construction sector (tCO_2)

	2000	2005	2008	2012
(1) Direct domestic CO_2	3,810,665	3,056,561	2,066,019	881,060
% over ECR (5)	7.00%	3.70%	2.70%	1.70%
(2) Total domestic CO_2	34,523,976	51,920,082	47,559,473	32,646,467
% over ECR (5)	63.20%	62.60%	62.70%	63.10%
(3) Total imported CO_2	20,108,610	31,012,838	28,244,477	19,099,917
% over ECR (5)	36.80%	37.40%	37.30%	36.90%
(4) Induced domestic CO_2	21,476,419	33,329,792	30,838,828	21,026,259
% over ECR (5)	39.30%	40.20%	40.70%	40.60%
(5) Total construction ECR [(2)+(3)]	54,632,587	82,932,920	75,803,950	51,746,384
(5) over total CO_2 ECR—Spain	14.10%	16.30%	16.50%	12.90%
Total CO_2 ECR—Spain	388,070,121	507,828,040	458,774,057	399,800,419
(5) over total CO_2 PR—Spain	19.60%	23.90%	24.10%	19.70%
Total CO_2 PR—Spain	278,628,471	347,300,449	315,025,969	262,847,673
Construction producer responsibility	47,304,305	73,170,281	63,899,767	40,973,176
Construction consumer responsibility	54,628,643	82,929,717	75,801,016	51,744,370

while production decreased, followed by the necessity to buy allowances in the market.

The virtual carbon embodied in input imports has a higher weight than domestic carbon without considering induced emissions (~62% of emissions are imported in the whole period). Those imported emissions come from intermediate goods, 31% of the total imports of the construction sector in 2012, and 43.3% are from capital goods imports. The distribution of domestic and imported virtual carbon is similar to the findings of Acquaye and Duffy (2010). They found that for Ireland almost 50% of emissions are linked to imports, although they did not consider induced emissions. The amount of emissions would be higher without the DTA assumption. More emissions could be masked by the use of an average emissions factor for the rest of the world lower than using the Chinese emissions factor to account for the production of a large amount of steel and iron imported from that country.

The addition of direct plus domestic and imported indirect emissions represents the consumer responsibility or the carbon footprint of the construction sector in Spain (60.071.079 ktCO_2 in 2012), as this sector does not have exports. Consumer responsibility is higher than the construction sector producer responsibility (27.120.044 ktCO_2). These results are similar to those found by Sánchez-Choliz and Duarte (2005) for 1995 and Serrano and Dietzenbacher (2010) between 1995 and 2000. As we can see, the construction sector is responsible for a large amount of emission leakage, which implies a transfer of environmental costs to other countries. This carbon leakage helped Spain come closer to fulfilling the Kyoto Protocol commitments in 2012 (Zafrilla et al., 2012). However, it was not enough. Spain had to go to international markets to buy emissions allowances.

The best way to estimate emissions and to design mitigation policies would be to consider a wider criterion, the consumer criterion, taking into account the amount of imports of the sector to account adequately for the carbon footprint. Directive 2009/29/EC (Commission, 2009) allows the free allocation of emission allowances to those industries at risk of carbon leakage; some of them supply inputs to the construction sector. However, this allocation does not incentivize domestic emission reductions because it does not encourage technological improvements or promote energy source substitution (gas instead of coal). Another possibility would be to include those importers in EU-ETS. In this case, both domestic producers and importers would have incentives to reduce their emission impact. In addition, the implementation of a border tax accounting for the virtual carbon embodied in intermediate and capital goods by the construction sector would be an alternative to reducing the carbon footprint (Atkinson et al., 2011).

The estimation of the induced emissions in the carbon footprint allowed us to obtain the ECR. In the Spanish economy, induced emissions reached 31.9% in the year 2000 and 35% in 2012 of the total extended emissions in the sector. The negative effect on induced emissions of the economic crisis is noteworthy. A deeper analysis of induced emission results is shown in Fig. 2.2. The relevant weight of this amount of induced emissions in the construction sector is shown compared with the rest of the sectors. As we can see, in the year 2000, the main sectors with high induced emissions are market services, construction, and nonmarket services, representing 64.7% of total induced emissions. Services account for a huge amount of employment in direct terms and construction in indirect terms because of the dragging effect over other sectors. In both cases, there is a high level of income generation, so there is a subsequent amount of emissions that come from the consumption of those incomes. In the years 2008 and 2012, the list of sectors is the same; however, the total share grew to 71.5% in those years with a higher presence of service sectors because of the stagnation of the construction sector. Llop and Pié (2011) and Lima et al. (2004) show comparable results to ours. In both papers studying regional figures in Spain, they showed weights of induced emissions of the services and construction sectors reaching 62% of total emissions.

On the one hand, it is worth noting that the emissions increased in pointed sectors compared to the industrial and hotel and restaurant sectors. Both have increased induced emissions but not as fast as services and construction. On the other hand, it is also worth noting the effect of the economic crisis over those induced emissions. Only the construction sector showed a notable reduction of induced emissions, which was motivated by the direct and indirect job destruction after the stagnation of the sector. The economic crisis effect was shown to be less pronounced in the rest of the sectors with a reduction in emissions and a reduction in the production of those sectors.

2.3.3 Construction Sector Emission Multipliers

The study of the emission multipliers allows us to analyze the evolution of environmental efficiency of production processes in the construction sector. Those multipliers are used to estimate the direct, indirect, and induced

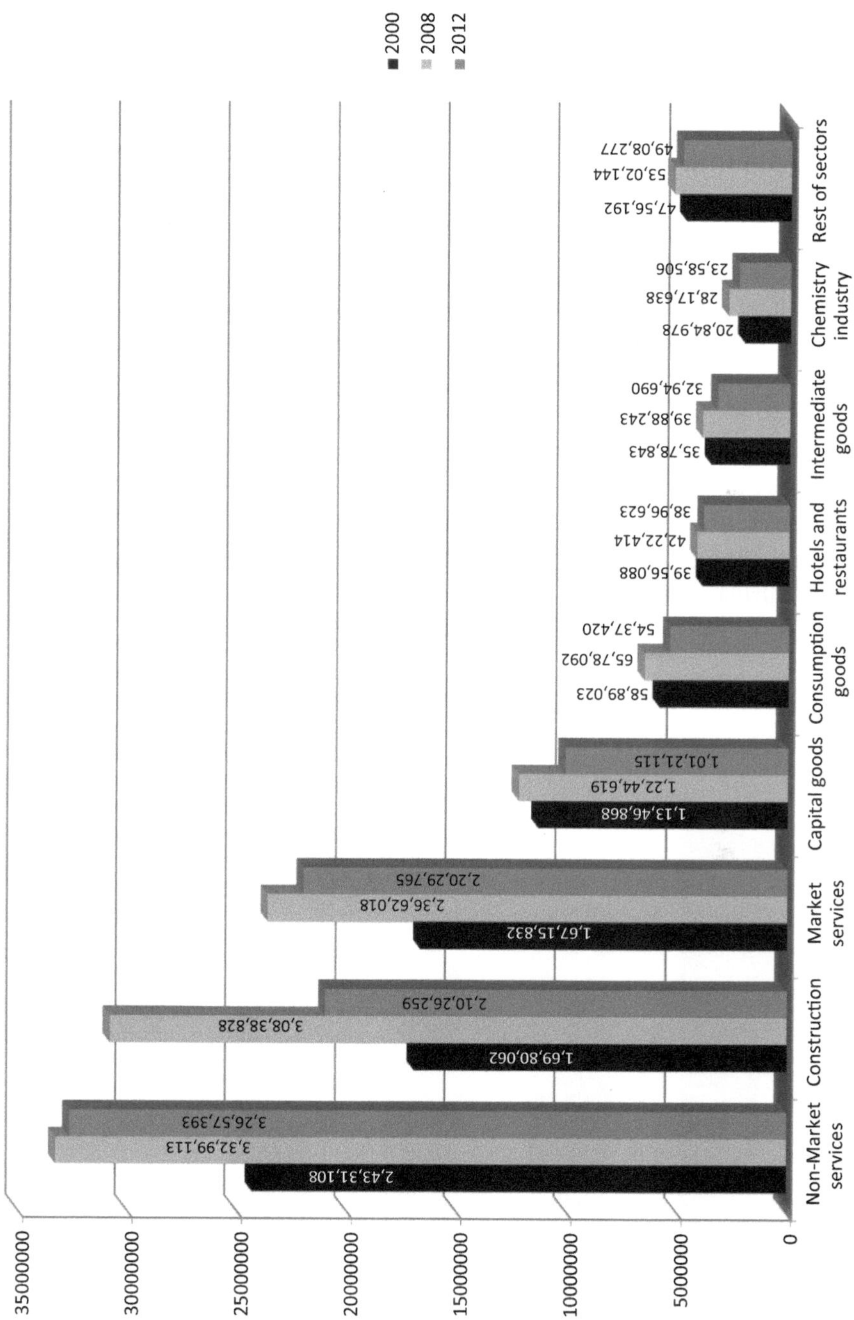

Figure 2.2 Induced emissions by sector: 2000, 2008, and 2012 (tCO_2).

requirements to production in each sector. In domestic terms, there is a relevant improvement of the environmental efficiency (Fig. 2.3) in the construction sector because the emission multiplier decreased from 438.1 tCO_2/M€ in 2000 to 287.1 tCO_2/M€ in 2012, a reduction of 34.5%. The evolution of the national energy mix, specifically the electricity mix, will determine environmental results in this sector (Gustavsson and Joelsson, 2010). The construction sector uses a huge amount of electricity indirectly; the improvement in the emissions coefficient in the electricity sector (30%) has yielded, indirectly, improvements in the construction sector coefficients. The positive electricity results in terms of environmental efficiency since 2002 are highlighted also by Cansino et al. (2012).

The methodology proposed here allows us to analyze the improvement by energy goods. The requirements of oil products are decreasing. The multiplier changed from 256.1 tCO_2/M€ in 2000 to 124.4 tCO_2/M€ in 2012, a reduction of 51.1%. Coal is another energy good with a drastic reduction in the emission multiplier (78.4%). Technological improvements, but mainly the gradual disappearance from the electricity mix, guide this result (Zafrilla, 2014). Gas presents a different evolution. It is the only energy good with a growing multiplier in the period from 78.6 tCO_2/M€ in 2000 to 140.5 tCO_2/M€ in 2012, an increase of 78.8%. Zafrilla et al. (2012) show that since 2005, Spain has experienced a gasification of the energy mix, mainly the electricity mix as a measure to substitute less

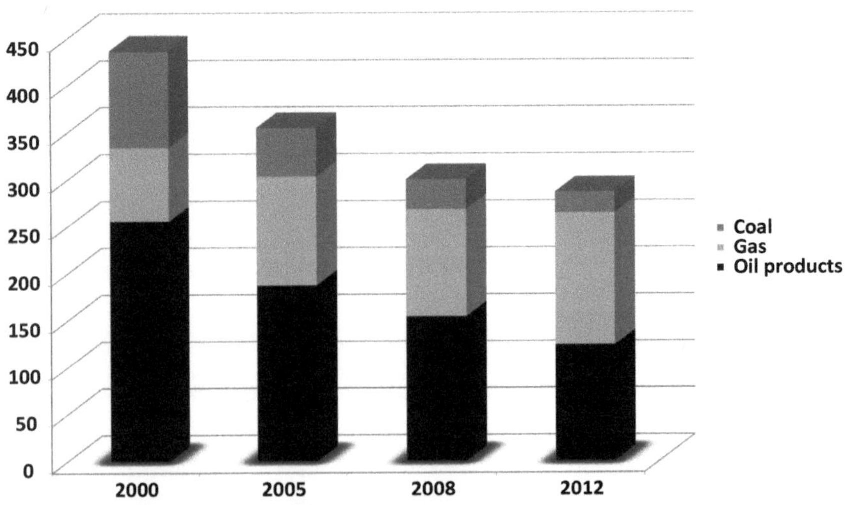

Figure 2.3 Domestic emission multipliers of the construction sector by type of energy good (tCO_2/M€).

efficient technologies (such as coal power thermal plants) but also a boom in the construction sector because most of the heating systems of new houses use gas as fuel. The boom of renewable energies in Spain has been the main development since 2005, with legislation making it possible to increase the presence of this new technology in the electricity mix (Del Río, 2008). This fact determines the results recently described about fossil energy source multipliers. The spread of renewables helped to reduce the fossil fuel dependency in Spain. The case of gas is special because the environmental efficiency of those energy goods, mainly electricity generation, is high, which is why gas is part of the energy cleanliness strategy (Tarancón Morán et al., 2008).

In Fig. 2.4, a multiplier decomposition analysis is shown. This decomposition includes the domestic multiplier or output multiplier (IOT EM), domestic-induced multipliers (Induced EM), and the imported multiplier (Import EM). The inclusion of the imported multiplier increases the total multiplier by 80% for all the years, showing the relevance of these imported effects in the construction sector. In the year 2000, the total multiplier reached 825.1 tCO_2/M€, and in 2012, it reached 502.5 tCO_2/M€, showing a reduction of 39.1%. This is the most suitable multiplier for quantifying the environmental efficiency of the construction sector because it considers all the inputs that are part of the production process. The IOT

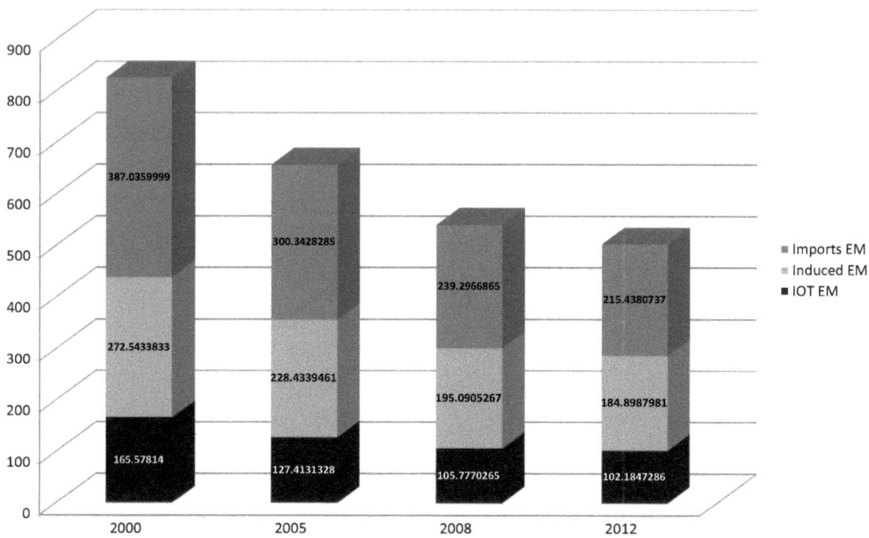

Figure 2.4 Domestic and induced emission multipliers decomposition.

EM multipliers represent 20% of the total multiplier in the period, whereas imported multipliers (Import EM) represent 46.9% and 42.9% of the total multiplier. Although the imported emission multiplier is estimated under the DTA assumption, the relevance compared to the domestic one is explained by considering (1) all the international production layers and (2) that the construction sector imports a huge amount of energy and pollution-intensive inputs.

Regarding Induced EM, they represent more than 60% of the total domestic efficiency, as shown in Table 2.1. Compared with all multipliers, induced effects represent more than 33% of the total effect. Income distribution is a key factor for evaluating the environmental impacts of a sector. The cyclical behavior of the income of construction sector workers, the higher profits for those from other sectors, and the relevance of the underground economy in the sector contribute greatly to uncontrolled expenditures and a subsequent huge carbon footprint. These results are similar to those found by Lima et al. (2004) for a Spanish region. Those high-induced effects show the existence of an intensive sectorial interaction within the Spanish economy, mainly in the construction sector, as we have concluded by analyzing previous results.

2.3.4 Advantages and Limitations of This Study

The ECR concept that we present in this chapter is based on the use of the IO-LCA models. Input—output analysis has some advantages. The information about interindustry connections with the economy is huge. In the sense, when the carbon footprint is estimated using IO analysis, the method allows us to account for emissions from the whole supply chain. This perspective is the most suitable measure to trace environmental responsibilities.

About the limitations some of them are well defined in the literature (Peters et al., 2016). (1) There is an assumption of fixed proportionality between inputs and output, but linearity is a problem resolved in this chapter as we use a dynamic model to solve this proportionality for energy goods. (2) The sectoral level of aggregation is another limitation in contrast with LCA that considers thousands of products. However, the LCA approach does not cover the whole supply chain. (3) The implicit prices in each sector are another limitation. The IO tables are measured in monetary flows which assume only one price for all the products per sector. Moreover, because each industry has only one price, we have to assume that each industry produces only one kind of product.

With our extension of the conventional IO-LCA model, the ECR concept comes with the construction of a SAM derived from the national accounts and the proper IO tables data. The environmentally extended SAM that we built allows us to account for induced emissions coming from the income-generated expenditure. The endogenization of the household consumption allows us to compute both industrial emissions and household emissions. The importance of the amount of induced emissions will depend on the weight of industry added value over the whole economy added value, the income distribution pattern, and the different consumption patterns of salaried employers. The SAM extension is useful for the purpose of this chapter; however, its main limitation is the necessity for a big amount of concise data about the income distribution and redistribution in the economy that will be analyzed to build up an accurate matrix.

2.4 CONCLUSIONS

Since the late 1990s, Spain began a booming economic process based significantly on the private and public expansion of the construction sector, promoted by friendly legislation that determined the economic and environmental transformation in the next decade. Following an inflation scenario in the new and second-hand housing market from 1997 to 2007 of ~200% (Febrero and Dejuán, 2009), lubricated by easy credit access, in 2008, with the financial crisis, the pillars of the Spanish economy collapsed, sinking the country into one of the hardest economic crises in the history of Spain.

This chapter proposes an ECR criterion for the construction sector to adequately account for and allocate the environmental responsibility of the built-up phase and attend to growing final demand from 2000 to 2012. On the one hand, we will account for the relevant induced emissions that come from the expenditure of the income generated in the analyzed sector. The results show how those emissions represent more than 60% of total domestic emissions in the period. On the other hand, we consider imported emissions showing the growing internationalization process of some of the most polluting industries in the last few years. In the period analyzed, imported emissions generated between 46.9% and 42.9% of total emissions under the extended responsibility criterion. This account will allow us to allocate responsibilities to those industries that may be using mechanisms of the environmental legislation stated in the framework of the Kyoto Protocol to avoid the emission reduction commitments throughout input imports.

The construction sector serves as the engine of Spanish economic growth; however, in the midterm, negative economic, social, and environmental consequences have arisen. In social terms, Spanish unemployment rates were higher than 20% from 2010 to 2016, reaching a peak of 26%. After the construction sector stagnated, the subsequent financial crisis reshaped the financial markets. One of the worst consequences in social terms was the extinction of a huge number of regional and rural savings banks, which limited the access to credit to thousands of people and firms. In environmental terms, the economic growth based on the excessive production of the construction sector meant a huge increase of emissions because this sector has one of the highest carbon footprints in Spain. The disproportionate boom of the construction sector was confirmed as a resource overspend. Some examples include the following: more than 500,000 houses have been empty since 2008; all the toll motorways built during those years had to be nationalized because of bankruptcy, e.g., private radial highways in Madrid; and some new public airports have not operated since 2008.

In terms of energy and environmental policy, the proposed analysis allows us to identify which industries and energy goods explain the environmental responsibility of the construction sector in Spain and to design adequate sectorial policies to reduce the impact (Guggemos and Horvarth, 2003). In the period analyzed, the construction sector indirectly benefited from improvements in the global Spanish pollution structure, especially the electricity mix environmental efficiency. The substitution of fossil energy sources such as coal with renewable energy or other more efficient fossil fuels such as gas is an explaining factor. The result is an environmental efficiency improvement in the period, which led to a reduction in emission multipliers in the construction sector of 34.5%. Despite this, Spain still exceeded the emissions commitment of the Kyoto Protocol in 2012 by 8.3% (Zafrilla et al., 2012). More environmental efforts for energy-intensive industries that supply inputs to the construction sector would be an adequate action. Actions on nonmetallic and metallic mineral industries are required, although the inclusion of these industries in the EU-ETS complicates the emission controls in the national and international transport sector, which is used extensively to transport the huge amounts of physical inputs needed by the construction industry. The inclusion of the air transport industry in the EU-ETS and the commitment of the aviation international agency to reduce emissions in 2015 are a positive step to mitigate the environmental impact. This path should be followed by maritime transport because it is the main international means of transport.

ACKNOWLEDGMENT

We thank the Regional Government of Castilla-La Mancha for funding the research project "PPII-2014-006-P", which led to this chapter.

REFERENCES

Acquaye, A.A., Duffy, A.P., 2010. Input – output analysis of Irish construction sector greenhouse gas emissions. Building and Environment 45, 784–791.
Alvarez, S., Rubio, A., 2015. Compound method based on financial accounts versus process-based analysis in product carbon footprint: a comparison using wood pallets. Ecological Indicators 49, 88–94.
Andrew, R., Peters, G.P., Lennox, J., 2009. Approximation and regional aggregation in multi-regional input-output analysis for national carbon footprint accounting. Economic Systems Research 21, 311–335.
Arto, I., Dietzenbacher, E., 2014. Drivers of the growth in global greenhouse gas emissions. Environmental Science & Technology 48 (10), 5388–5394.
Atkinson, G., Hamilton, K., Ruta, G., Van Der Mensbrugghe, D., 2011. Trade in 'virtual carbon': empirical results and implications for policy. Global Environmental Change 21, 563–574.
Bielsa, J., Duarte, R., 2011. Size and linkages of the Spanish construction industry: key sector or deformation of the economy? Cambridge Journal of Economics 317–334.
Bilec, M.M., Ries, R.J., Matthews, H.S., 2010. Life-cycle assessment modeling of construction processes for buildings. Journal of Infrastructures Systems 16.
Cadarso, M.-Á., López, L.-A., Gómez, N., Tobarra, M.-Á., 2010. CO_2 emissions of international freight transport and offshoring: measurement and allocation. Ecological Economics 69, 1682–1694.
Cadarso, M.-Á., López, L.-A., Gómez, N., Tobarra, M.-Á., 2012. International trade and shared environmental responsibility by sector. An application to the Spanish economy. Ecological Economics 83, 221–235.
Cansino, J.M., Cardenete, M.A., Ordóñez, M., Román, R., 2012. Economic analysis of greenhouse gas emissions in the Spanish economy. Renewable and Sustainable Energy Reviews 16, 6032–6039.
Clò, S., 2010. Grandfathering, auctioning and carbon leakage: assessing the inconsistencies of the new ETS directive. Energy Policy 38, 2420–2430.
Commission, E., 2009. In: Parliament, E. (Ed.), Directive 2009/29/EC of the European Parliament and of the Council of 23 April 2009.
Dejuán, Ó., López, L.A., Tobarra, M.Á., Zafrilla, J., 2013. A post-keynesian age model to forecast energy demand in Spain. Economic Systems Research 25, 321–340.
Del Río, P., 2008. Ten years of renewable electricity policies in Spain: an analysis of successive feed-in tariff reforms. Energy Policy 36, 2917–2929.
DGA various years. Base de datos de comercio exterior. Dirección General de Aduanas de España.

Febrero, E., Dejuán, O., 2009. Endeudamiento familiar y crecimiento económico: un patrón de crecimiento insostenible. Ekonomiaz 72, 80–97.

Gallego, B., Lenzen, M., 2005. A consistent input-output formulation of shared producer and consumer responsibility. Economic Systems Research 17, 365–391.

Guggemos, A.A., Horvarth, A., 2003. Strategies of extended preducer responsibility for buildings. Journal of Infrastructure Systems 9.

Gustavsson, L., Joelsson, A., 2010. Life cycle primary energy analysis of residential buildings. Energy and Buildings 42, 210–220.

Gutiérrez, F., Marrero, G.M., Puch, L.A., 2011. Los efectos medioambientales del boom y el parón inmobiliario en España. Economía Industrial 379.

Hendrickson, C., Horvath, A., Joshi, S., Lave, L., 1998. Economic input-output models for environmental life-cycle assessment. Environmental Science and Technology Analysis 32, 184–191.

Hertwich, E.G., Gibon, T., Bouman, E.A., Arvesen, A., Suh, S., Heath, G.A., Bergesen, J.D., Ramirez, A., Vega, M.I., Shi, L., 2015. Integrated life-cycle assessment of electricity-supply scenarios confirms global environmental benefit of low-carbon technologies. Proceedings of the National Academy of Sciences 112 (20), 6277–6282.

Hubacek, K., Feng, K., Minx, J.C., Pfister, S., Zhou, N., 2014. Teleconnecting consumption to environmental impacts at multiple spatial scales. Journal of Industrial Ecology 18, 7–9.

Hummels, D., Ishii, J., Yi, K.-M., 2001. The nature and growth of vertical specialization in world trade. Journal of International Economics 54, 75–96.

INE Various years. Marco Input-Output. Instituto Nacional de Estadística.

Joshi, S., 2000. Product environmental life-cycle assessment using input-output techniques. Journal of Industrial Ecology 3, 95–120.

Kuckshinrichs, W., Kronenberg, T., Hansen, P., 2010. The social return on investment in the energy efficiency of buildings in Germany. Energy Policy 38, 4317–4329.

Lenzen, M., Pade, L.-L., Munksgaard, J., 2004. CO_2 multipliers in multi-region input-output models. Economic Systems Research 16, 391–412.

Lima, M.A., Cardenete, M.A., Hewings, G.D., Vallés, F., 2004. A structural analysis of a regional economy using social accounting matrices: 1990–1999. Investigaciones Regionales 5, 113–138.

Llop, M., Pié, L., 2011. Decomposition of emission multipliers in a national accounting matrix including environmental accounts: the case of Catalonia. Journal of Industrial Ecology 15, 206–216.

López, L.A., López, J., Zafrilla, J., 2011. Fragmentación, offshoring y vertebración de la economía española. Revista de Economía, ICE 859, 141–161.

Mendiluce, M., 2010. In: Energy, E.f. (Ed.), Análisis de la evolución de la intensidad energética en España (Vigo).

Miller, R.E., Blair, P.D., 2009. Input-Output Analysis: Foundations and Extensions.

MINECO, 2007. Spanish Energy Efficiency Strategy, Action Plan 2008–2012. http://www.idae.es (Madrid).

MINETUR, 2005. Estrategia española de eficiencia energética. Ministerio de Industria, E.y.T.

Monsalve, F., Zafrilla, J.E., Cadarso, M.-Á., 2016. Where have all the funds gone? Multiregional input-output analysis of the European Agricultural Fund for Rural Development. Ecological Economics 129, 62–71.

Munksgaard, J., Pedersen, K.A., 2001. CO_2 accounts for open economies: producer or consumer responsibility? Energy Policy 29, 327–334.

Nassen, J., Holmberg, J., Wadeskog, A., Nyman, M., 2007. Direct and indirect energy use and carbon emissions in the production phase of buildings: an input-output analysis. Energy 32, 1593–1602.

Peters, G., Andrew, R.M., Karstensen, J., 2016. Global Environmental Footprints. A Guide to Estimating, Interpreting and Using Consumption-Based Accounts of Resource Use and Environmental Impacts. Copenhagen, Denmark.

Peters, G., Hertwich, E., 2008. Post-Kyoto greenhouse gas inventories: production versus consumption. Climatic Change 86, 51–66.

Peters, G.P., Hertwich, E.G., 2006. Pollution embodied in trade: the Norwegian case. Global Environmental Change 16, 379–387.

Pyatt, G., Round, J.I., 1979. Accounting and fixed price multipliers in a social accounting matrix framework. The Economic Journal 89, 850–873.

Round, J., 2003. Social Accounting Matrices and SAM-Based Multiplier Analysis, The Impact of Economic Policies on Poverty and Income Distribution: Evaluation Techniques and Tools. World Bank Publications.

Sánchez-Choliz, J., Duarte, R., 2005. CO_2 emissions embodied in international trade: evidence for Spain. Energy Policy 32, 1999–2005.

Serrano, M., Dietzenbacher, E., 2010. Responsibility and trade emissions balances: an evaluation of approaches. Ecological Economics 69, 2224–2232.

Sharrard, A., Matthews, S., Ries, R.J., 2008. Estimating construction project environmental effects using an input-output-based hybrid life-cycle assessment model. Journal of Infraestructures Systems 14, 327–336.

Stone, R., 1985. The disaggregation of the household sector in the national accounts. In: Pyatt, G., Round, J.I. (Eds.), Social Accounting Matrices: A Basis for Planning. The World Bank, Washington DC, pp. 145–185.

Tarancón, M.A., del Río, P., 2007. CO_2 emissions and intersectoral linkages. The case of Spain. Energy Policy 35, 1100–1116.

Tarancón, M.A., del Río, P., Callejas, F., 2011. Determining the responsibility of manufacturing sectors regarding electricity consumption. The Spanish case. Energy 36, 46–52.

Tarancón Morán, M.Á., del Río, P., Albiñana, F.C., 2008. Tracking the genealogy of CO_2 emissions in the electricity sector: an intersectoral approach applied to the Spanish case. Energy Policy 36, 1915–1926.

Wiedmann, T.O., Suh, S., Feng, K., Lenzen, M., Acquaye, A.A., Scott, K., Barrett, J.R., 2011. Application of hybrid life cycle approaches to emerging energy technologies – the case of wind power in the UK. Environmental Science & Technology 45, 5900–5907.

Zafrilla, J.E., 2014. The mining industry under the thumb of politicians: the environmental consequences of the Spanish coal decree. Journal of Cleaner Production 84, 715–722.

Zafrilla, J.E., López, L.A., Cadarso, M.Á., Dejuán, Ó., 2012. Fulfilling the Kyoto protocol in Spain: a matter of economic crisis or environmental policies? Energy Policy 51, 708–719.

Zhang, X., Wang, F., 2016a. Assessment of embodied carbon emissions for building construction in China: comparative case studies using alternative methods. Energy and Buildings 130, 330–340.

Zhang, X., Wang, F., 2016b. Hybrid input-output analysis for life-cycle energy consumption and carbon emissions of China's building sector. Building and Environment 104, 188–197.

CHAPTER 3

The Environmental Impact of Magnetic Nanoparticles Under the Perspective of Carbon Footprint

Sara Feijoo, Sara González-García, Yolanda Moldes-Diz, Carlos Vázquez-Vázquez, Gumersindo Feijoo, María T. Moreira
University of Santiago de Compostela, Santiago de Compostela, Spain

3.1 INTRODUCTION

Nanotechnology is highlighted as one of the most important research and development frontiers in modern science (Salamanca-Buentebello et al., 2005), with potential application in a wide range of fields, such as development of drug-delivery nanosystems in the medical sector; manufacture of electronic devices: nanotransistors, nanodiodes, etc.; and production of nanotubes, aerogels, nanoparticles (NPs), fuel cells, and solar cells, etc. (Barreto et al., 2011; Bruce et al., 2008; Li et al., 2008; Qu et al., 2013).

Nanomaterials are defined as materials with at least one characteristic dimension in the range of 1−100 nm (Colvin, 2003). Two main factors cause the properties of nanomaterials to differ significantly from other materials: increased relative surface area and quantum effects. These factors can change or enhance properties such as reactivity, strength, optical, magnetic, and electrical characteristics in comparison with bulk or dissolved forms (Michalet et al., 2005).

NPs can be classified according to the shape: spherical, tubular, or irregular; degree of aggregation: fused, aggregated, or agglomerated; and origin: nonengineered and engineered (Nowack et al., 2011). Engineered nanoparticles are intentionally produced by using different materials, and most of them belong to two general classes, depending on whether they are carbon-based (e.g., carbon nanotubes (CNTs) and fullerenes) or metal-containing (e.g., Ag, TiO_2, CeO_2, Fe) (Grillo et al., 2015). The methods of

Environmental Carbon Footprints
ISBN 978-0-12-812849-7
http://dx.doi.org/10.1016/B978-0-12-812849-7.00003-9

production are numerous and include photocatalytic reduction (Chang et al., 2006), chemical reduction (Li et al., 2010), photochemical or radiation-chemical reduction, metallic wire explosion, sonochemical, polyols (Nersisyan et al., 2003a,b,c), matrix chemistry (Ayyad et al., 2010), photoreduction (Courrol et al., 2007), reverse micelle-based methods (Xie et al., 2006), and biologically synthesis (Kalishwaralal et al., 2010; Sathishkumar et al., 2010; Zeiri et al., 2002).

Among the different engineered nanomaterials (ENMs), magnetic nanoparticles (mNPs) have received a great deal of attention, especially for various biomedical applications (Babic et al., 2008; Das et al., 2013). The straightforward separation of mNPs by means of a magnetic field offers a clear advantage over other NPs. Chemical coprecipitation is probably the most common method to prepare mNPs from a mixture of ferrous and ferric salts suspended in alkaline medium (Laurent et al., 2008). The parameters of this process have been carefully studied to demonstrate the influence of the alkaline solution, pH value, cations and the ratio of Fe^{2+}/Fe^{3+} on the precipitation yield, and the diameter and polydispersity of the NPs obtained (from 4 to 20 nm). Microemulsion is another major alternative based on the formation of nano-sized water droplets dispersed in an oil phase and stabilized by surfactant molecules at the water/oil interface (López-Quintela et al., 2004).

Beyond the alternatives that render into stabilized mNPs, the coating of magnetite NPs surface with inorganic or organic protective coatings according to core/shell structures considers the subsequent addition of silica, metal oxides, polyelectrolyte, and micelles, among others (Gupta and Gupta, 2005). The magnetic core ensures a strong magnetic response, and the shell provides favorable functional groups. Among the different types of coating materials, silica is preferred because of its unique properties such as chemical stability, biocompatibility, strong affinity towards magnetite, and reactivity with various coupling agents (Kulkarni et al., 2014).

Surface functionalization of coated NPs opens the possibility of biotechnological applications because the desired molecules such as proteins could be immobilized on conjugated particles, enabling targeted biological applications such as site-specific markers, bioconjugates, and sensors. Moreover, with the increasing attention to the enzymatic cascade reaction, it is possible to develop the coimmobilization of multienzymes on this type of NPs. Enzyme immobilization relies on two main approaches: ionic interaction between the surface charges between the NP and the enzyme and covalent bonding that requires that chemical groups on the support surface are activated and react with nucleophilic groups on the protein (Arroyo, 1998). Most enzymes are covalently attached using their lysine amino groups because of their frequent presence on the protein surface and high reactivity (Brady and Jordaan, 2009).

Given the early stages of development of nanoscience, new materials are being synthesized on small scales (hundreds of milligrams or less) for testing specific physical properties. As a result, the examination of any unintended properties of the material or concerns about efficiencies of the production process is often deferred. In view of that possibility, we need to address the impacts of nanotechnology from a prospective perspective, not only from the technological complexity of a production system but also from the potential health and environmental risks associated. The assessment of the environmental impacts of nanoscale materials before they are accepted as mature technologies presents an opportunity to minimize presumed negative consequences from the very beginning of the concept design stage and ultimately leads to the implementation of green chemistry processes.

The methodology that can be applied for the holistic assessment of nanotechnologies can be based on a life cycle perspective (Grieger et al., 2012). The application of life cycle assessment (LCA) to nanotechnology would include all aspects of activities during the life of a product, such as extraction of raw materials and resources, production processes, use of products, recovery, recycling of some fractions, and final disposal at the end-of-life stage. However, there are inherent difficulties when it comes to perform a comprehensive analysis of nanomaterials. The major obstacles include the uncertainty arising from the immature nature of the technology, the availability of extensive information on processes subjected to patent protection, and the lack of characterization factors for LCA impact assessment (Tsuzuki, 2013). Although these drawbacks can be a major obstacle to overcome, the development of new nanomaterials can be carried out with sustainability in mind.

The available LCA studies on ENMs are focused on nanomaterials such as CdTe, carbon nanofibres (CNFs) and nanotubes, nanoclay, nanoscale Pt-group metals, nanocrystaline-Si, Ag, titanium, and titanium oxide. Hischier and Walser (2012) and Gavankar and Suh (2012) reviewed all studies applying LCA to ENMs, whereas Upadhyayula et al. (2012) specifically focused on CNTs and CNFs. The main outcomes from these studies report that energy and chemicals use during ENM synthesis contributes a dominant share of total life cycle environmental impacts

(Walser et al., 2011; Eckelman et al., 2012). However, the life cycle inventories (LCIs) cannot be classified as comprehensive because they often lack ENM-specific data related to the outputs of the processes (Hischier and Walser, 2012).

In this chapter, we aim to examine how the application of LCA principles to nanoscience can be helpful in the decision-making process to guide technological progress within this emerging field. Different approaches of mNPs synthesis were evaluated: from the production of oleic acid stabilized magnetite to production schemes where the coating of a shell on the preformed NPs is developed: polyacrylic acid (PAA), polyethylenimine (PEI), and silica-coated mNPs. Moreover, additional criteria for selection would also identify the performance of the different mNPs for the immobilization of the target enzyme.

3.2 MATERIALS AND METHODS

LCA is a methodology that aims to analyze products, processes, and/or services from an environmental point of view and should be part of the decision-making process toward sustainability (Baumann and Tillman, 2004). This approach has been followed in this study for the comprehensive assessment of the impacts of different strategies of mNPs production. Thus, the guidelines established by ISO standards (ISO 14040, 2006) have been considered to perform the LCA study.

3.2.1 Goal and Scope Definition

3.2.1.1 Objectives

The aim of this study is to assess the environmental profiles of different mNPs production routes. To do so, the comparative analysis is based on primary data from the production scenarios of mNPs developed by the NanoMag group at the University of Santiago de Compostela. Fig. 3.1 displays the production sequences under environmental evaluation. The detailed description of the four scenarios is reported below.

3.2.1.2 Functional Unit

The function of this study is to produce mNPs under six different production protocols, which can be viable supports for the immobilization of proteins, although all of them are with different sizes and functionalization groups on the surface. The size distribution of the different types of NPs is summarized in Table 3.1.

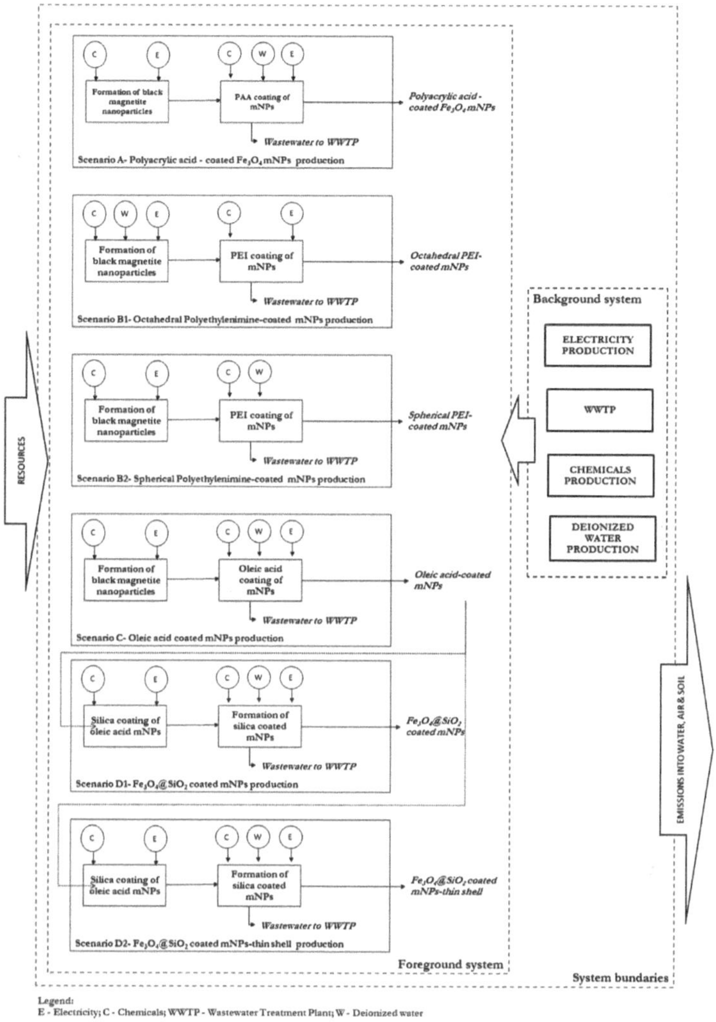

Figure 3.1 Block flow diagrams depicting the different production pathways of magnetic nanoparticles (mNPs). *PAA*, polyacrylic acid; *PEI*, polyethylenimine.

Table 3.1 Size distribution of the different magnetic nanoparticles

Magnetic nanoparticles	Size (nm)
Fe_3O_4 at PAA	10.1
Octahedral Fe_3O_4 at PEI	50−100
Spherical Fe_3O_4 at PEI	10
Fe_3O_4 at OA	8−10
Fe_3O_4 at SiO_2	21.5
Fe_3O_4 at SiO_2 (thin layer)	17

PAA, polyacrylic acid; *PEI*, polyethylenimine.

Therefore, the LCA functional unit must be selected carefully to allow comparisons between the production systems. Having in mind the use of mNPs, all the environmental results will be reported and discussed per gram of mNP.

3.2.1.3 *Description of the mNPs Production Scenarios*

Six different mNPs production scenarios have been defined and proposed for assessment, considering a cradle-to-gate approach, i.e., considering the extraction of raw materials to produce the required inputs and the synthesis of mNPs but not the phases of their use and/or final disposal. This perspective has been assumed since the production protocols have been developed at semipilot scale, and these mNPs are not available in the markets yet. Finally, the environmental results will be compared between them to identify the best production route from an environmental perspective. Moreover, their consideration within the system boundaries should increase the uncertainty of results. A detailed description of the different production routes is reported below.

3.2.1.3.1 Preparation of Polyacrylic Acid-Coated Fe_3O_4 mNPs (Scenario A)

PAA-coated Fe_3O_4 NPs have been prepared by coprecipitation of Fe^{2+} and Fe^{3+} salts, following the Massart method (1981). Initially, in a typical synthesis, 24.3 g of $FeCl_3 \cdot 6H_2O$ and 16.7 g of $FeSO_4 \cdot 7H_2O$ (molar ratio Fe^{3+}/Fe^{2+} of 1.5) are placed in a 500 mL round-bottom flask and dissolved in 200 mL of 0.01 M HCl solution with mechanical stirring. Temperature is increased to 60°C, and 60 mL of ammonia solution (28%) is added to the mixture, observing the immediate formation of black magnetite NPs. Secondly, 3.9 g of PAA is added 30 s after the addition of the ammonia solution and the reaction is allowed to continue for 1 h. After cooling down to room temperature, pH is acidified to 4 by the addition of HCl (9% wt). Finally, the mNPs are magnetically separated and washed four times with deionized water and finally redispersed in water.

3.2.1.3.2 Preparation Polyethylenimine-Coated mNPs (Scenario B): Octahedral PEI-Coated mNPs (Scenario B1) and Spherical PEI-Coated mNPs (Scenario B2)

PEI is a branched polymer with a high-density amine group, (CH_2CH_2NH) n. Because of the different pKa values of the primary, secondary, and tertiary amino groups, PEI has the ability to capture protons at different pH conditions, which is known as the "proton sponge" mechanism. PEI molecules bounded to particles by electrostatic interaction and the negative charges on the surface of the particles were converted to positive charges. Because of its unique properties, PEI appears to be one of the most appropriate molecules for the surface modification of mNPs for biological applications.

The production process of the PEI-coated mNPs by oxidative precipitation is carried out following the method described by Goon et al. (2010). Initially, nitrogen gas is bubbled in deionized water to remove dissolved oxygen. Solutions of $FeSO_4 \cdot 7H_2O$, NaOH, and KNO_3 prepared in deionized and deoxygenated water are mixed and after 5 min, temperature is increased up to 90°C to complete the reaction after 2 h. Next, the mNPs are magnetically separated and washed three times with deionized water and immediately redispersed in water. Finally, 10 mL of PEI solution (20 g/L) is added to the magnetite NPs obtained previously. The mNPs are magnetically shaken (500 rpm, 1 h) and dried for 12 h at 75°C to improve the polymer absorption on the NPs.

The production process of spherical PEI-coated mNPs started with the formation of the black magnetite NPs obtained as aforementioned in Scenario A (Massart, 1981). PEI (2 g) was added to the magnetite NPs. After the reaction and cooling down to room temperature, pH was acidified to 4 with HCl. Then, the mNPs were magnetically separated and washed four times with deionized water and finally redispersed in water.

3.2.1.3.3 Preparation of Oleic Acid mNPs (Scenario C)

Oleic acid is a surfactant used to stabilize the mNPs synthesized by the traditional coprecipitation method. Oleic acid forms a waterproofing shell around the mNPs. The procedure to produce oleic acid-coated Fe_3O_4 mNPs starts with the incubation of black magnetite NPs (as indicated in Scenario A) with oleic acid. The reaction is allowed to continue for 1 h before being heated on a hot plate until flocculation occurred. The mNPs are washed twice with deionized water to remove residual concentration of the surfactant molecules and finally redispersed in cyclohexane.

3.2.1.3.4 Preparation of Fe_3O_4 at SiO_2-Coated mNPs (Scenario D)

The Fe_3O_4 at SiO_2-coated mNPs was prepared through a water-in-cyclohexane reverse microemulsion starting from the oleic acid mNPs previously obtained in Scenario C (Fang et al., 2011; Han et al., 2008). The reverse microemulsion method considered to coat mNPs with silica requires neither the use of silane coupling agents as surface primers nor a polymer stabilizer.

3.2.1.3.5 Fe_3O_4 at SiO_2-Coated mNPs (Scenario D1)

In a typical synthesis, polyoxyethylene(5)nonylphenyl ether (Igepal CO-520) and cyclohexane are mechanically stirred for 15 min before the addition of oleic acid magnetite NPs (2.5% wt in cyclohexane). Finally, ammonium hydroxide solution and tetraethyl orthosilicate (TEOS) are added consecutively under stirring to form a transparent red solution of reverse microemulsion until complete reaction after 18 h. The core−shell NPs are precipitated with isopropanol (IPA) to disrupt the reverse microemulsion and washed four times with IPA and deionized water. Finally, they are subjected to several cycles of centrifugation (9000 rpm, 120 min) and washed with deionized water until no foam is observed. Then, the core−shell NPs were redispersed in deionized water to a concentration ca. 0.5% wt (determined by thermogravimetric analysis).

3.2.1.3.6 Fe_3O_4 at SiO_2-Coated mNPs (Silica Thin Shell) (Scenario D2)

The production of mNPs with a thin silica coating was produced following the procedure previously mentioned in Scenario D1. In this case, the concentration of cyclohexane added was 0.5% wt and the core−shell NPs were redispersed in deionized water to a concentration c.0.75% wt (determined by thermogravimetric analysis).

3.2.2 Life Cycle Inventory Analysis

A reliable environmental assessment requires the collection of high-value LCI data. In this study, inventory data for the foreground systems (direct inputs and outputs for each scenario) such as electricity requirements (estimated with power and operational data from the different units: reactors, dryers, heaters) as well as use of chemicals and water were average data from semipilot scale experiments. These inventory data consisted of representative data obtained by on-site measurements of production processes developed for a period of 3 months. A summary of the inventory data corresponding to the foreground systems is displayed in Tables 3.2−3.7 for the different scenarios.

Concerning the background system, the inventory data corresponding to the different inputs (production of electricity, chemicals, and water) as well as the wastewater treatment process were taken from Ecoinvent database v3 (Dones et al., 2007; Althaus et al., 2007; Doka, 2003). It is important to remark that regarding electricity production, the process defined in the database has been updated managing data for the average electricity generation and import/export data from Spain in 2014 (Red Eléctrica de España, 2014).

Table 3.2 Life cycle inventory of polyacrylic acid (PAA)-coated Fe_3O_4 magnetic nanoparticles (mNPs) (Scenario A)

Inventory	Amount	Unit
Inputs from technosphere		
Formation of mNPs		
$FeCl_3 \cdot 6H_2O$ (99% wt)	24.3	G
$FeSO_4 \cdot 7H_2O$ (99% wt)	16.7	g
NH_4OH (28% wt)	60	mL
HCl (37% wt)	98.5	mg
Electricity for agitation	441	kJ
Electricity for heating (60°C)	1134	kJ
PAA coating of mNPs		
PAA	3.9	g
Electricity for heating (200°C)	648	kJ
Electricity for ultrasonic bath	192	kJ
Deionized water	0.50	L
Outputs to technosphere		
PAA-coated Fe_3O_4 mNPs	5	g
Emissions to treatment facility		
Wastewater	0.4	L

Table 3.3 Life cycle inventory of octahedral polyethylenimine (PEI)-coated magnetic nanoparticles (mNPs) (Scenario B1)

Inventory	Amount	Unit
Inputs from technosphere		
Formation of magnetite nanoparticles		
$FeSO_4 \cdot 7H_2O$ (99% wt)	14	g
NaOH	8	g
KNO_3	40	g
Electricity from agitation	588	kJ
Electricity from heating (90°C)	1296	kJ
Electricity from ultrasonic bath	96	kJ
Deionized water	0.5	L
PEI coating mNPs		
PEI	4	g
Electricity	4125	kJ
Outputs to technosphere		
PEI-coated nanoparticles	6.4	g
Emissions to treatment facility		
Wastewater	0.35	L

Table 3.4 Life cycle inventory of spherical polyethylenimine (PEI)-coated magnetic nanoparticles (mNPs) (Scenario B2)

Inventory	Amount	Unit
Inputs from technosphere		
Formation of mNPs		
$FeCl_3 \cdot 6H_2O$ (99% wt)	12.2	g
$FeSO_4 \cdot 7H_2O$ (99% wt)	8.4	g
HCl (37% wt)	98.5	mg
NH_4OH (28% wt)	30.0	mL
Electricity for agitation	441	kJ
Electricity for heating (60°C)	1134	kJ
PEI coating mNPs		
PEI	2	g
HCl (9% wt)	35	mL
Deionized water	0.5	L
Outputs to technosphere		
PEI-coated nanoparticles	7–7.5	g
Emissions to treatment facility		
Wastewater	0.5	L

Table 3.5 Life cycle inventory of oleic acid-coated magnetic nanoparticles (mNPs) (Scenario C)

Inventory	Amount	Unit
Inputs from technosphere		
Formation of mNPs		
$FeCl_3 \cdot 6H_2O$ (99% wt)	24.3	g
$FeSO_4 \cdot 7H_2O$ (99% wt)	16.3	g
HCl (37% wt)	197	mg
NH_4OH (28% wt)	60	mL
Electricity for agitation	441	kJ
Electricity for heating (60°C)	1134	kJ
Oleic acid coating mNPs		
Oleic acid	4.0	g
Electricity for heating (200°C)	648	kJ
Deionized water	0.5	L
Cyclohexane	120	mL
Outputs to technosphere		
Oleic acid-coated nanoparticles	22.5	g
Emissions to treatment facility		
Wastewater	0.4	L

Table 3.6 Life cycle inventory of Fe_3O_4 at SiO_2-coated magnetic nanoparticles (mNPs) (Scenario D1)

Inventory	Amount	Unit
Inputs from technosphere		
Silica coating of oleic acid mNPs		
Oleic acid-coated nanoparticles	1	g
Igepal CO-520	150	g
Cyclohexane	1.8	kg
NH_4OH (28% wt)	21	mL
Tetraethyl orthosilicate	24	mL
Electricity for agitation	189	kJ
Electricity for mechanical agitation	4032	kJ
Formation of silica-coated mNPs		
Isopropanol	7.1	kg
Deionized water	0.5	L
Electricity for agitation	210	kJ
Electricity for ultrasonic bath	192	kJ
Outputs to technosphere		
Fe_3O_4 at SiO_2-coated nanoparticles	7.5	g
Emissions to treatment facility		
Wastewater	0.7	L

Table 3.7 Life cycle inventory of Fe_3O_4 at SiO_2-coated magnetic nanoparticles thin shell (Scenario D2)

Inventory	Amount	Unit
Inputs from technosphere		
Silica coating of oleic acid mNPs		
Oleic acid-coated nanoparticles	1	g
Igepal CO-520	150	g
Cyclohexane	360	g
NH_4OH (28% wt)	21	mL
Tetraethyl orthosilicate	24	mL
Electricity for agitation	189	kJ
Electricity for mechanical agitation	4032	kJ
Formation of silica-coated mNPs		
Isopropanol	7.1	kg
Deionized water	0.5	L

Continued

Table 3.7 Life cycle inventory of Fe_3O_4 at SiO_2-coated magnetic nanoparticles thin shell (Scenario D2)—cont'd

Inventory	Amount	Unit
Electricity for agitation	210	kJ
Electricity for ultrasonic bath	192	kJ
Outputs to technosphere		
Fe_3O_4 at SiO_2-coated nanoparticles	7.5	g
Emissions to treatment facility		
Wastewater	0.7	L

3.2.3 Impact Assessment Methodology

Among the steps defined within the life cycle impact assessment of the LCA methodology (ISO 14040, 2006), only classification and characterization steps were undertaken representing the most objective approach as it includes no interpretation of values when judging the importance of different impact categories. The mNPs production scenarios under assessment involve emissions to the environment and consumption of resources throughout the global systems. The environmental impacts of these emissions and resources consumption can be computed in terms of different impact categories. In this study, the environmental assessment was conducted using characterization factors from ReCiPe midpoint methodology (Goedkoop et al., 2008) and the following impact categories have been considered in the analysis: climate change (CC), ozone depletion (OD), terrestrial acidification (TA), freshwater eutrophication, marine eutrophication (ME), human toxicity (HT), photochemical oxidant formation (POF), terrestrial ecotoxicity (TET), freshwater ecotoxicity (FET), marine ecotoxicity (MET), and fossil depletion (FD). Moreover, special attention will be paid on the carbon footprint profiles corresponding to the different production routes considering the results obtained in CC. To do so, SimaPro 7.3.3 has been the software used for the computational implementation of the LCI data and the computation of the environmental profiles (PRé Consultants, 2017).

3.3 ENVIRONMENTAL RESULTS

The aim of this study is to analyze from an environmental perspective the environmental profiles associated with four different mNPs production pathways to identify the one with the best environmental perspective. The performance of the different types of mNPs as support for immobilization of laccase was also investigated as additional criteria and a balance between the environmental impacts associated to mNP production and the efficiency of enzyme immobilization must determine which one is the most suitable support for the intended use of laccase immobilization.

The different production routes rendered into viable mNPs, whose superparamagnetic characteristics were evidenced under a magnetic field. Characterization of the NPs was also performed following the standard procedures of transmission electron microscopy to determine shape, size, and uniformity of the particles under optimum conditions. The magnetite NPs were predominantly spherical with variable size as depicted in Table 3.1. Once when the different routes provided different types of mNPs, the processes were environmentally assessed according to the stages and processes depicted in Fig. 3.1.

Considering that detailed inventory data directly from industrial production are unavailable, the evaluation of the environmental impacts associated to NPs production was based on data from laboratory and semipilot scale.

The life cycle impact assessment (LCIA) supplies indicators as well as the basis for analyzing the likely contributions of the resource extraction, wastes, and emissions from an inventory to a selection of impacts, considering the goal and scope of the study (González-García et al., 2014). Thus, the result of the LCIA is an evaluation of a product life cycle, on a functional unit basis, in terms of several impact categories (Baumann and Tillman, 2004). Firstly, the LCI data are assigned and aggregated into common impact groups. This first step is known as "classification." Next, the assessment of the magnitude of potential impacts of each inventory flow into its corresponding environmental impact is estimated by means of the "characterization" step. Therefore, characterization provides a way to directly compare the inventory results within each category. To do so, characterization factors are commonly referred to as equivalency factors. As previously reported, characterization factors from ReCiPe midpoint methodology (Goedkoop et al., 2008) have been managed. The characterization results for the different production pathways as shown in Table 3.8.

The worst profile obtained corresponded with Scenario D1, performing the production of Fe_3O_4 at SiO_2-coated mNPs. Although all Scenarios include one production stage in common, i.e., the reaction leading to the formation of magnetite NPs, remarkable differences were identified on the basis of the requirements of chemicals for the formulation of the final NPs.

Table 3.8 Environmental impacts in the production of nanoparticles per functional unit (1 g of magnetic nanoparticle)

Impact category	Scenario						
	Unit	A	B1	B2	C	D1	D2
CC	kg CO_2 eq	7.94×10^{-2}	2.37×10^{-1}	3.74×10^{-2}	2.65×10^{-2}	1.61	1.15
OD	kg CFC-11 eq	7.45×10^{-9}	1.25×10^{-8}	4.49×10^{-9}	2.19×10^{-9}	8.13×10^{-8}	5.33×10^{-8}
TA	kg SO_2 eq	6.32×10^{-4}	1.45×10^{-3}	2.94×10^{-4}	1.70×10^{-4}	6.29×10^{-3}	4.47×10^{-3}
FE	kg P eq	3.83×10^{-5}	6.76×10^{-5}	1.99×10^{-5}	1.13×10^{-5}	3.78×10^{-4}	2.25×10^{-4}
ME	kg N eq	3.52×10^{-5}	6.23×10^{-5}	2.30×10^{-5}	9.00×10^{-6}	2.15×10^{-4}	1.47×10^{-4}
HT	kg 1,4-DB eq	3.85×10^{-2}	6.11×10^{-2}	2.34×10^{-2}	1.09×10^{-2}	3.73×10^{-1}	2.50×10^{-1}
POF	kg NMVOC	3.09×10^{-4}	7.31×10^{-4}	1.45×10^{-4}	1.16×10^{-4}	7.53×10^{-3}	5.14×10^{-3}
TET	kg 1,4-DB eq	3.56×10^{-6}	6.94×10^{-6}	3.26×10^{-6}	1.09×10^{-6}	5.11×10^{-5}	3.61×10^{-5}
FET	kg 1,4-DB eq	9.36×10^{-4}	1.32×10^{-3}	5.11×10^{-4}	2.90×10^{-4}	1.34×10^{-2}	9.20×10^{-3}
MET	kg 1,4-DB eq	9.08×10^{-4}	1.35×10^{-3}	5.03×10^{-4}	2.80×10^{-4}	1.26×10^{-2}	8.56×10^{-3}
FD	kg oil eq	2.25×10^{-2}	5.03×10^{-2}	1.08×10^{-2}	1.15×10^{-2}	1.12	8.06×10^{-1}

CC, climate change; *FD*, fossil depletion; *FE*, freshwater eutrophication; *FET*, freshwater ecotoxicity; *HT*, human toxicity; *ME*, marine eutrophication; *MET*, marine ecotoxicity; *OD*, ozone depletion; *POF*, photochemical oxidant formation; *TA*, terrestrial acidification; *TET*, terrestrial ecotoxicity.

Differences are also evidenced in the second stage of production, focused on the coating of the previously formed NPs. This stage also requires magnetic separation, washing, and redispersion, as can be seen in A, B1, B2, and C. However, in Scenario D1 and D2, the production of Fe_3O_4 at SiO_2-coated mNPs and Fe_3O_4 at SiO_2-coated mNPs thin shell is based on the use of oleic acid-coated mNPs (from Scenario C) as main raw material, and they require IPA for formation of the silica-coated NPs. Thus, activities such as precipitation, centrifugation, washing, and redispersion are also included in this stage.

The effect of these production differences is reflected not only on the environmental results but also on the mNPs production yields. According to the inventory data displayed in Tables 3.2–3.7, a large range of production values is obtained from 5 g of mNPs in scenario A to 22.5 g of mNPs in Scenario C. Global results show that, across synthesis routes, impacts associated with the consumption of energy are dominant but also chemicals used in the formulations and for redispersion are relevant.

In line with Fig. 3.2 and Table 3.8, the production of 1 g of oleic acid-coated mNPs (Scenario C) should report the best profile regardless the

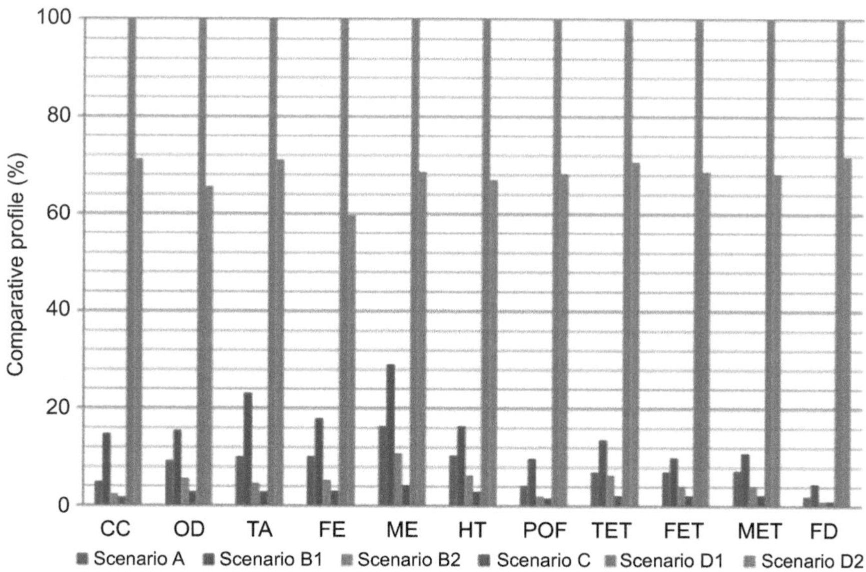

Figure 3.2 Comparative environmental profiles of the different magnetic nanoparticles production routes. *CC*, climate change; *FD*, fossil depletion; *FE*, freshwater eutrophication; *FET*, freshwater ecotoxicity; *HT*, human toxicity; *ME*, marine eutrophication; *MET*, marine ecotoxicity; *OD*, ozone depletion; *POF*, photochemical oxidant formation; *TA*, terrestrial acidification; *TET*, terrestrial ecotoxicity.

impact category with reduction of 97%–99% regarding the worst option (Scenario D1 in all categories). According to these global comparative results, it is unquestionable that increasing the complexity of the mNPs production route, as it is performed in Scenarios D1 and D2, would have to be discussed from an environmental perspective due to the highest environmental burdens obtained per g of mNPs. Moreover, and although it will be analyzed in more detail below, Scenario D2 does not achieve the highest level of laccase immobilization, despite the worse environmental profile. To identify the main responsible input of these negative results, the different production routes will be separately assessed.

3.3.1 Scenario A

The production of PAA-coated Fe_3O_4 mNPs involves two stages carried out in the pilot plant (formation of magnetite NPs and PAA coating) as well as the treatment of the wastewater. Fig. 3.3 displays the contributions to the global environmental profile from the different stages involved.

According to Fig. 3.3, the formation of magnetite NPs is the most important environmental stage, and it reports contributions ranging from 38% to 79% depending on the category. The large contribution from this

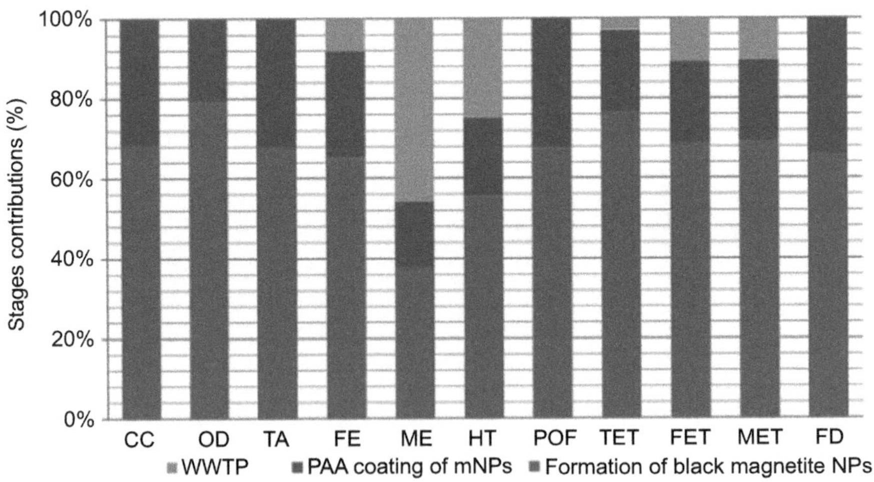

Figure 3.3 Contributions to the environmental impacts for the stages in Scenario A. *CC*, climate change; *FD*, fossil depletion; *FE*, freshwater eutrophication; *FET*, freshwater ecotoxicity; *HT*, human toxicity; *mNPs*, magnetic nanoparticles; *ME*, marine eutrophication; *MET*, marine ecotoxicity; *NPs*, nanoparticles; *OD*, ozone depletion; *PAA*, polyacrylic acid; *POF*, photochemical oxidant formation; *TA*, terrestrial acidification; *TET*, terrestrial ecotoxicity; *WWTP*, wastewater treatment plant.

subsystem can be expected because in this step, 65% of total electricity requirements take place as well as the largest consumption of chemicals. Contributions from the coating stage are specifically remarkable in categories such as CC, TA, POF, and FD. All these categories are inherently related and highly dependent on electricity requirements, which represent contributing ratios ranging from 80% to 97% depending on the category. The treatment of the wastewater produced in the synthesis is only outstanding in two categories: ME and HT (46% and 25% of total contributions). If the environmental burdens derived from the formation of magnetite NPs are assessed in more detail, Fig. 3.4 displays the ratios from the responsible processes of these results.

According to Fig. 3.4, electricity requirements in this stage can be considered as the environmental *hotspot*. Production of electricity contributes to more than 63% of total life cycle impacts (except OD, TET, FET, and MET), and heating at 60°C is the most important contributor. Within the contributions from the chemical requirements, special attention should be paid to the ammonia solution because the use of this chemical is responsible of ratios ranging from 28% to 66% of impacts derived from chemicals.

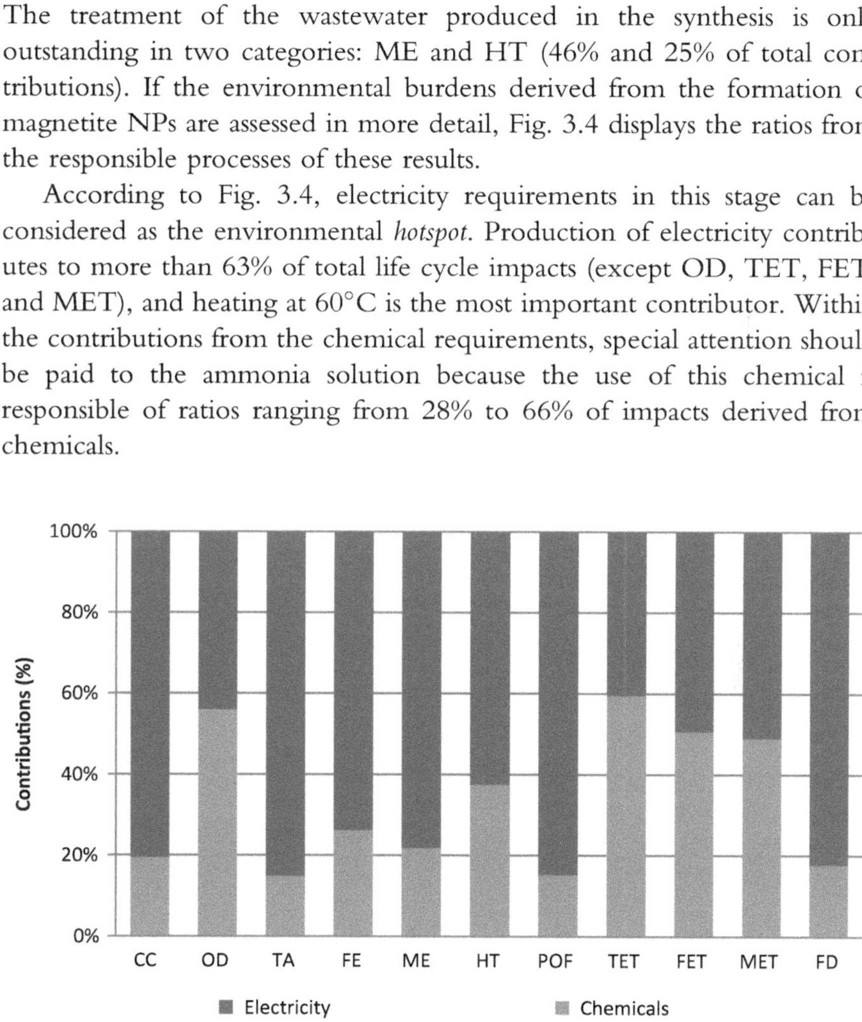

Figure 3.4 Environmental impacts associated to the formation of magnetite nanoparticles. *CC*, climate change; *FD*, fossil depletion; *FE*, freshwater eutrophication; *FET*, freshwater ecotoxicity; *HT*, human toxicity; *ME*, marine eutrophication; *MET*, marine ecotoxicity; *OD*, ozone depletion; *POF*, photochemical oxidant formation; *TA*, terrestrial acidification; *TET*, terrestrial ecotoxicity.

3.3.2 Scenario B1

This scenario is focused on the production of PEI-coated mNPs under an oxidative precipitation scheme. Once again, two production stages are carried out in the pilot plant: formation of magnetite NPs and PEI coating. In addition, the wastewater stream has to be treated in a further stage. The ratios to the global environmental profile from steps involved in the production systems are displayed in Fig. 3.5.

In this scenario, the two steps performed within the pilot plant are responsible of 82%—100% of the environmental impacts. Thus, contributions from the wastewater treatment plant (WWTP) are minor. However, and in contrast to Scenario A, the largest electricity requirement takes place in the coating stage: 68% of total electricity consumption due to the drying step at 75°C. Contribution from electricity consumption to PEI coating of mNPs step represents ~98% (on average) of the total impacts. Regarding the formation of magnetite NPs, the environmental profile derived from this subsystem is assigned to chemicals and electricity requirements. Thus, the contribution from deionized water requirements is negligible as it is displayed in Fig. 3.6, regardless the impact category.

As displayed in Fig. 3.6, the required chemicals are also an environmental *hotspot*, and special attention should be paid to the consumption of

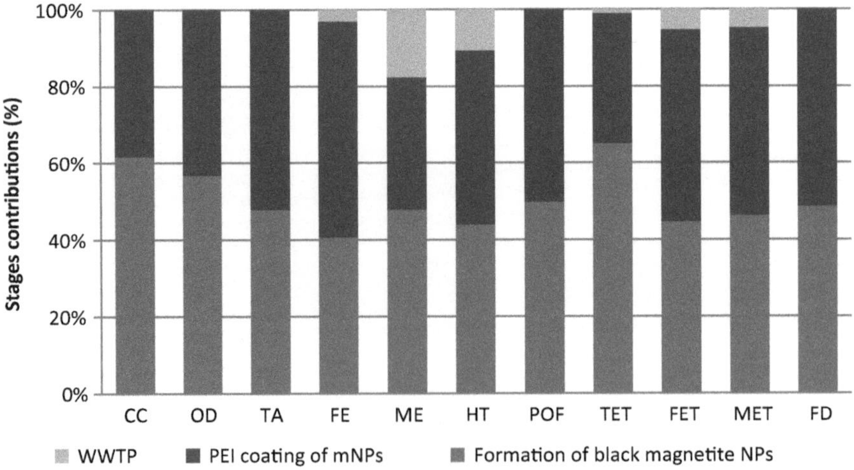

Figure 3.5 Contributions to the environmental impacts for the stages in Scenario B1. *CC*, climate change; *FD*, fossil depletion; *FE*, freshwater eutrophication; *FET*, freshwater ecotoxicity; *HT*, human toxicity; *mNPs*, magnetic nanoparticles; *ME*, marine eutrophication; *MET*, marine ecotoxicity; *NPs*, nanoparticles; *OD*, ozone depletion; *PEI*, Polyethylenimine; *POF*, photochemical oxidant formation; *TA*, terrestrial acidification; *TET*, terrestrial ecotoxicity; *WWTP*, wastewater treatment plant.

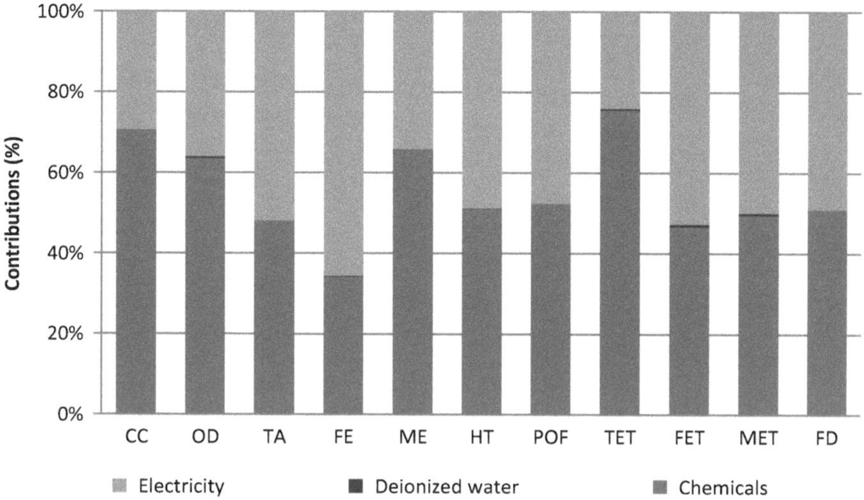

Figure 3.6 Environmental impact distribution for the formation of magnetite nanoparticles stage. *CC*, climate change; *FD*, fossil depletion; *FE*, freshwater eutrophication; *FET*, freshwater ecotoxicity; *HT*, human toxicity; *ME*, marine eutrophication; *MET*, marine ecotoxicity; *OD*, ozone depletion; *POF*, photochemical oxidant formation; *TA*, terrestrial acidification; *TET*, terrestrial ecotoxicity.

potassium nitrate because the impacts derived from this chemical add up to 76%–97% of the burdens. Thus, further improvement activities should be focused on searching alternative chemicals with better environmental profile or even on reducing its use. Electricity is also remarkable in this stage, being outstanding the heating process at 90°C.

In global, the environmental *hotspots* in this scenario are the large electricity requirements (specifically in the drying process) as well as the consumption of potassium nitrate.

3.3.3 Scenario B2

This scenario is focused on the production of spherical PEI-coated mNPs. Some differences were identified in the production system regarding octahedral PEI-coated mNPs (Scenario B1). The main differences are linked to the chemicals used in the formation of magnetite NPs step because ammonium hydroxide is used instead of potassium nitrate as well as in the PEI coating mNPs step, where no heating is required in contrast to Scenario B1. Fig. 3.7 displays the contributions to the global environmental profile from the involving stages.

According to Fig. 3.7, undoubtedly the production of magnetite NPs is the environmental critical stage because in all the categories assessed, except

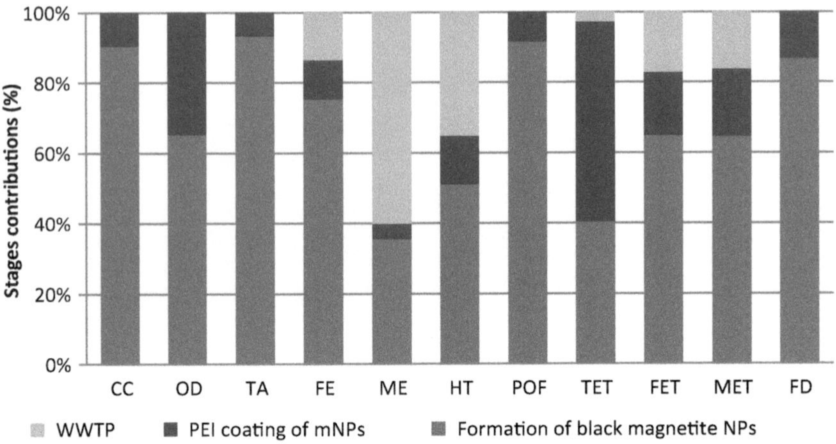

Figure 3.7 Contributions to the environmental impacts for the stages in Scenario B2. *CC*, climate change; *FD*, fossil depletion; *FE*, freshwater eutrophication; *FET*, freshwater ecotoxicity; *HT*, human toxicity; *mNPs*, magnetic nanoparticles; *ME*, marine eutrophication; *MET*, marine ecotoxicity; *NPs*, nanoparticles; *OD*, ozone depletion; *PEI*, Polyethylenimine; *POF*, photochemical oxidant formation; *TA*, terrestrial acidification; *TET*, terrestrial ecotoxicity; *WWTP*, wastewater treatment plant.

in terms of ME and TET, contributing ratios higher than 50% are achieved. Beyond the analysis of this critical step, Fig. 3.8 displays the ratios from the responsible processes of these environmental results.

According to Fig. 3.8, the electricity requirements can be once again considered as environmental *hotspot*. This factor contributes to 61%–92% of total impacts derived from this stage (depending on the category), being the heating process (at 60°C) the most important contributor. Regarding spherical PEI coating of mNPs stage, 100% of impacts are linked to the chemicals required because deionized water consumption reports negligible contributions. Within these chemicals, 82%–97% of the environmental burdens derive from HCl consumption.

3.3.4 Scenario C

In this route, oleic acid-coated mNPs are produced considering oleic acid as a surfactant used to stabilize the mNPs synthesized by the traditional coprecipitation method as previously reported. Two production stages are carried out in the pilot plant, all of them common to the other scenarios, although under specific conditions: formation of magnetite NPs and oleic acid coating of mNPs. In addition, the wastewater stream is treated in a

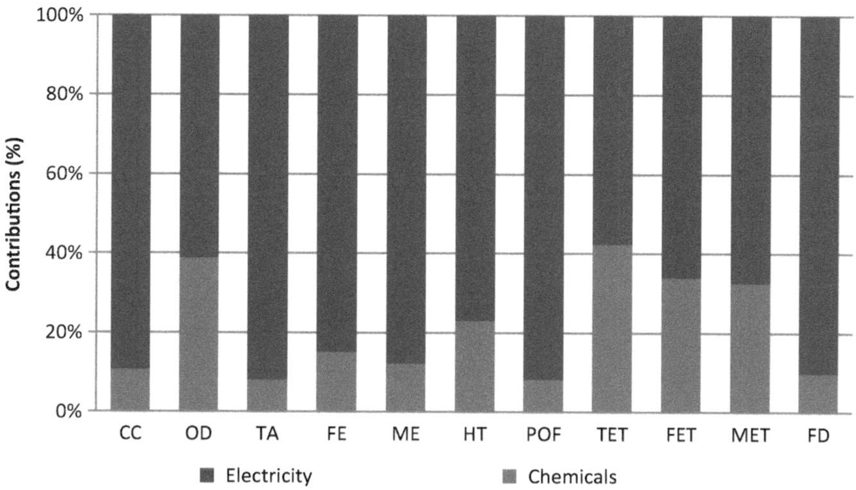

Figure 3.8 Environmental impact distribution regarding processes involved in the formation of magnetite nanoparticles stage. *CC*, climate change; *FD*, fossil depletion; *FE*, freshwater eutrophication; *FET*, freshwater ecotoxicity; *HT*, human toxicity; *ME*, marine eutrophication; *MET*, marine ecotoxicity; *OD*, ozone depletion; *POF*, photochemical oxidant formation; *TA*, terrestrial acidification; *TET*, terrestrial ecotoxicity.

further stage. Fig. 3.9 displays the contributions to the global environmental profile from the involving stages.

According to Fig. 3.9, similar contributing ratios can be identified from the stages performed at the pilot plant, being contribution from the WWTP only remarkable in terms of ME and HT. Both stages require chemicals and electricity, although electricity requirements are higher in the first stage than the second one (71% vs. 29%).

Beyond the analysis of the stage of reaction for the formation of magnetite NPs, Fig. 3.10 displays the ratios from the responsible processes of these environmental results.

According to Fig. 3.10, the production of electricity requirements is once again considered as environmental *hotspot* with remarkable contributions in almost all the categories (41%–85%). Once again, the heating activity (in this case at 60°C) is the main responsible factor of these results. Within the contributions from the chemical requirements, special attention should be paid to the ammonium hydroxide because reports impart ratios of environmental impacts in this stage.

Regarding the second step, oleic acid coating of mNPs, Fig. 3.11 displays the contributions from processes involved, and according to it, the

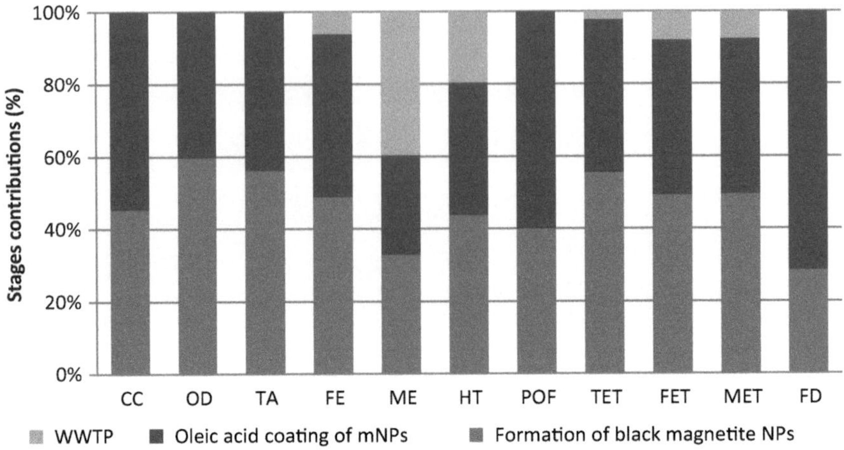

Figure 3.9 Contributions to the environmental impacts from the stages in Scenario C. *CC*, climate change; *FD*, fossil depletion; *FE*, freshwater eutrophication; *FET*, freshwater ecotoxicity; *HT*, human toxicity; *mNPs*, magnetic nanoparticles; *ME*, marine eutrophication; *MET*, marine ecotoxicity; *NPs*, nanoparticles; *OD*, ozone depletion; *POF*, photochemical oxidant formation; *TA*, terrestrial acidification; *TET*, terrestrial ecotoxicity; *WWTP*, wastewater treatment plant.

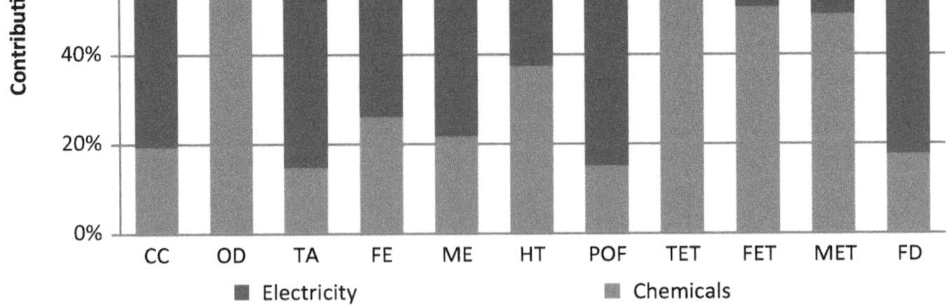

Figure 3.10 Environmental impact distribution regarding processes involved in the formation of magnetite nanoparticles step. *CC*, climate change; *FD*, fossil depletion; *FE*, freshwater eutrophication; *FET*, freshwater ecotoxicity; *HT*, human toxicity; *ME*, marine eutrophication; *MET*, marine ecotoxicity; *OD*, ozone depletion; *POF*, photochemical oxidant formation; *TA*, terrestrial acidification; *TET*, terrestrial ecotoxicity.

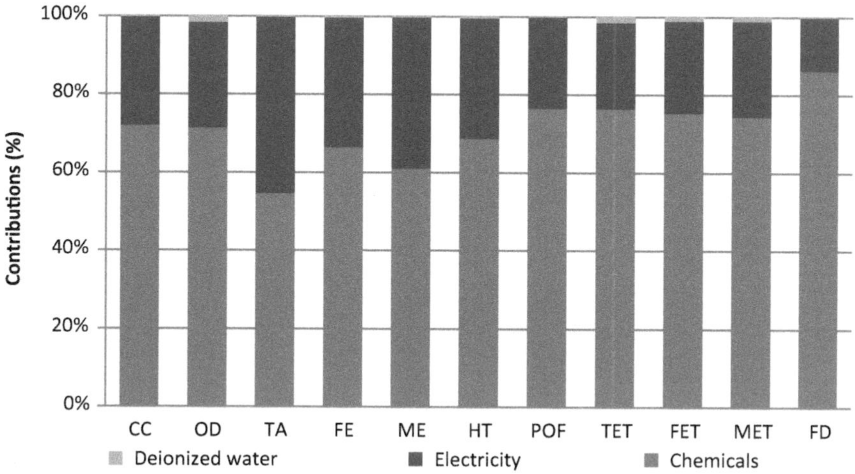

Figure 3.11 Environmental impact distribution of the processes involved in the formation of magnetite nanoparticles step. *CC*, climate change; *FD*, fossil depletion; *FE*, freshwater eutrophication; *FET*, freshwater ecotoxicity; *HT*, human toxicity; *ME*, marine eutrophication; *MET*, marine ecotoxicity; *OD*, ozone depletion; *POF*, photochemical oxidant formation; *TA*, terrestrial acidification; *TET*, terrestrial ecotoxicity.

consumption of chemicals is the environmental *hotspot*, being the cyclohexane required in the redispersion is responsible for more than 97% of impacts from chemicals.

3.3.5 Scenario D1

This scenario aims to produce Fe_3O_4 at SiO_2-coated mNPs considering oleic acid mNPs as raw material. The production scheme is completed with four addition steps: silica coating of oleic acid mNPs and formation of silica-coated mNPs. As difference to previous routes, oleic acid-coated mNPs are the core—shell NP, previously redispersed with cyclohexane. Next, the core—shell NPs are precipitated with IPA and redispersed in deionized water. Fig. 3.12 displays the environmental profile per impact categories in terms of contributions from steps or activities involved in the production system.

According to the results displayed in Fig. 3.12, the silica coating stage of oleic acid mNPs is unquestionably an environmental *hotspot* because this stage is responsible for more than 50% of contributions to almost all the different impact categories (except in terms of CC, FET, MET, and FD). In this step, the oleic acid mNPs are mixed in a reactor with chemicals (Igepal

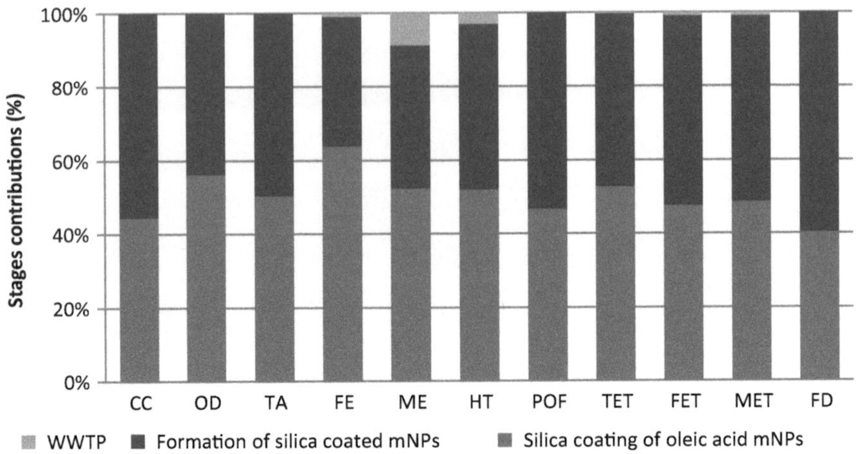

Figure 3.12 Contributions to the environmental impacts from the stages of Scenario D1. *CC*, climate change; *FD*, fossil depletion; *FE*, freshwater eutrophication; *FET*, freshwater ecotoxicity; *HT*, human toxicity; *mNPs*, magnetic nanoparticles; *ME*, marine eutrophication; *MET*, marine ecotoxicity; *OD*, ozone depletion; *POF*, photochemical oxidant formation; *TA*, terrestrial acidification; *TET*, terrestrial ecotoxicity; *WWTP*, wastewater treatment plant.

CO-520, cyclohexane, TEOS, and NH_4OH) to produce the core–shell mNPs.

Fig. 3.13 displays the distribution of environmental burdens from the reactor step between the main factors involved: chemicals, oleic acid-coated mNPs, and electricity consumptions. Contributions from chemicals are higher than 75%. The cyclohexane required in the formulation is the environmental critical chemical because it is responsible for more than 86% of burdens derived from chemicals contributions regardless the impact category. However, in the majority of impact categories assessed, contributions from electricity are higher than 50%.

Regarding electricity, it is consumed in this step in the agitation activities, being remarkable the mechanical agitation required to obtain the transparent red solution of reverse microemulsion until complete reaction (97% of total electrical requirements). Regarding chemicals, the cyclohexane required in the formulation is the environmental critical chemical because it is responsible for more than 85% of burdens derived from chemicals contributions regardless the impact category.

Regarding the stage related to the formation of the silica-coated mNPs, contributions range from 35% to 60% depending on the category. In this

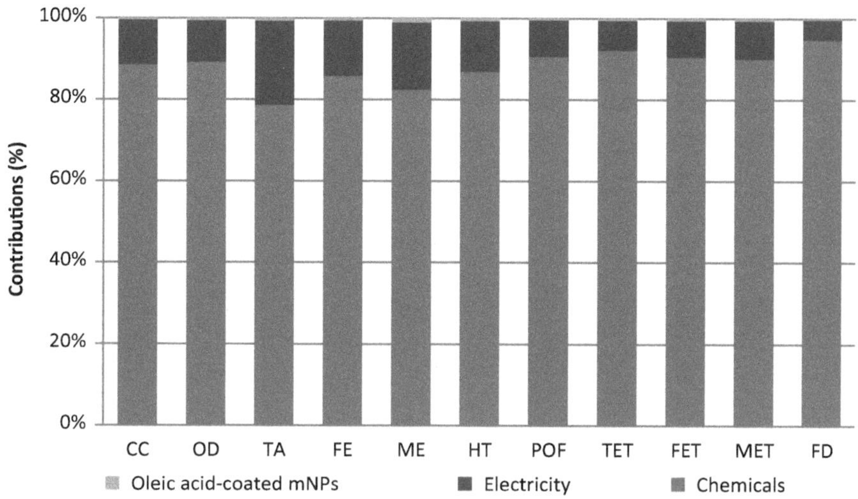

Figure 3.13 Environmental impacts distribution of the silica coating process of oleic acid magnetic nanoparticles (mNPs). *CC*, climate change; *FD*, fossil depletion; *FE*, freshwater eutrophication; *FET*, freshwater ecotoxicity; *HT*, human toxicity; *ME*, marine eutrophication; *MET*, marine ecotoxicity; *OD*, ozone depletion; *POF*, photochemical oxidant formation; *TA*, terrestrial acidification; *TET*, terrestrial ecotoxicity.

step, the core–shell mNPs are precipitated with IPA to break the reverse microemulsion prepared in the previous stage, and finally, washed and redispersed with deionized water. Taking a look into the derived environmental profile, the contributing actors are presented in Fig. 3.14.

According to the profile reported in Fig. 3.14, the consumption of IPA can be considered as an environmental *hotspot*. Thus, special attention should be paid on the requirements of IPA and cyclohexane in this type of mNP production system to improve its environmental profile.

3.3.6 Scenario D2

This scenario aims to produce Fe_3O_4 at SiO_2-coated mNPs, also considering oleic acid mNPs as raw material but with a lower consumption of cyclohexane to produce mNPs with a thin silica coating. The main remarkable difference regarding Scenario D1 is the lowest requirement of cyclohexane that is five times lower than that in Scenario D1. Because cyclohexane is an environmental *hotspot* as previously indicated, environmental improvements of around 28%–40% are expected for this type of mNPs because the same amount of mNPs is produced per batch (see Table 3.8 and Fig. 3.2). Fig. 3.15 displays the environmental profile per

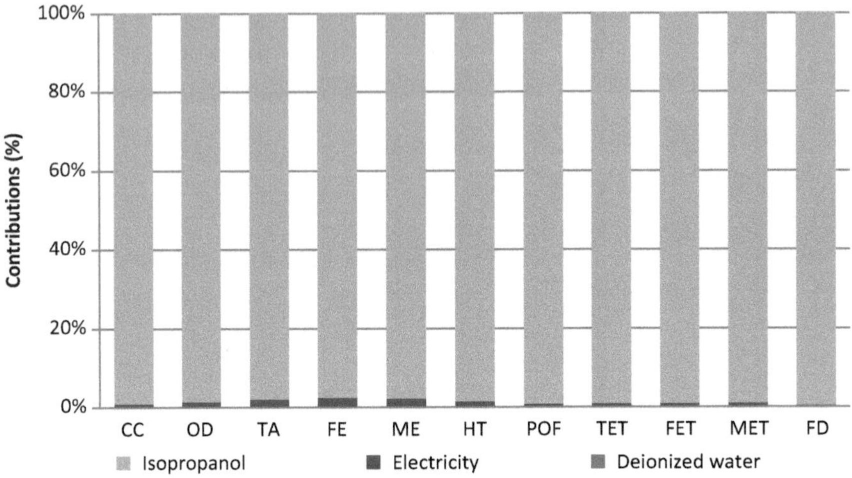

Figure 3.14 Environmental impacts distribution of the formation of the silica-coated magnetic nanoparticles. *CC*, climate change; *FD*, fossil depletion; *FE*, freshwater eutrophication; *FET*, freshwater ecotoxicity; *HT*, human toxicity; *ME*, marine eutrophication; *MET*, marine ecotoxicity; *OD*, ozone depletion; *POF*, photochemical oxidant formation; *TA*, terrestrial acidification; *TET*, terrestrial ecotoxicity.

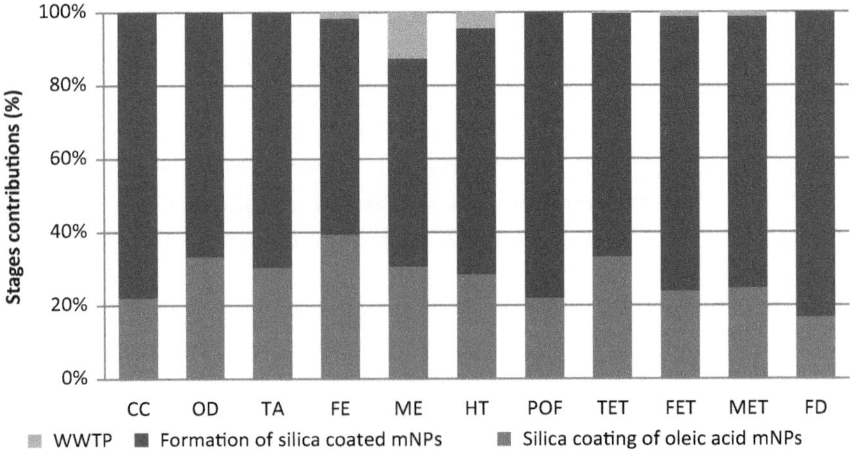

Figure 3.15 Contributions to the environmental impacts from the stages of Scenario D2. *CC*, climate change; *FD*, fossil depletion; *FE*, freshwater eutrophication; *FET*, freshwater ecotoxicity; *HT*, human toxicity; *mNPs*, magnetic nanoparticles; *ME*, marine eutrophication; *MET*, marine ecotoxicity; *OD*, ozone depletion; *POF*, photochemical oxidant formation; *TA*, terrestrial acidification; *TET*, terrestrial ecotoxicity; *WWTP*, wastewater treatment plant.

impact categories in terms of contributions from steps or activities involved in the production system.

In this scenario and as different to Scenario D1, both stages carried out with the pilot plant considerably change their behavior in the environmental profile. Thus, the effect from the silica coating of oleic acid mNPs step is not as remarkable as in Scenario D1, being the formation of silica-coated mNPs the environmental critical stage. The reduction on cyclohexane requirements is the main responsible factor of the environmental improvements. Assessing in more detail the contributions from this step (silica coating of oleic acid mNPs), Fig. 3.16 displays the contributing ratios and as different to Scenario D1, the environmental *hotspot* is related with chemicals requirements, and cyclohexane consumption is also the most important chemical from an environmental perspective with contributing ratios of 54%–75% (depending on the category) of impacts derived from chemicals. However, the electricity consumption is only responsible for less than of 40% of impacts derived from this stage.

Because no differences have been identified in the second step (formation of the silica-coated mNPs) between Scenario D1 and Scenario D2, the same behavior and profile (Fig. 3.14) is expected.

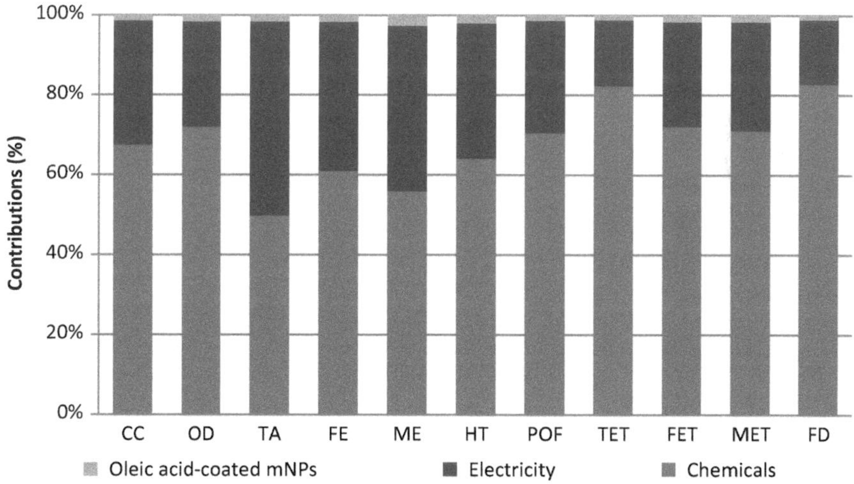

Figure 3.16 Environmental impacts distribution of the process of silica-coated magnetic nanoparticles (mNPs). *CC*, climate change; *FD*, fossil depletion; *FE*, freshwater eutrophication; *FET*, freshwater ecotoxicity; *HT*, human toxicity; *ME*, marine eutrophication; *MET*, marine ecotoxicity; *OD*, ozone depletion; *POF*, photochemical oxidant formation; *TA*, terrestrial acidification; *TET*, terrestrial ecotoxicity.

3.3.7 Environmental Analysis per Unit Immobilization Yield

As previously indicated, the environmental profiles for the different mNPs production routes are displayed considering as functional unit 1 g of mNPs. Nevertheless, and although these mNPs can be used for enzyme immobilization (the main purpose), they do not present the same immobilization yields and in line with experimental assays, they are totally different (Table 3.9).

According to it, Scenario D1 (Fe_3O_4 at SiO_2-coated mNPs) should be preferred in terms of units of enzyme immobilized per batch, and Scenario A (PAA-coated Fe_3O_4 mNPs) should be the worst option. If the production routes are assessed from an environmental perspective, the consideration of the unit of laccase immobilized as functional unit entirely changes the environmental profiles as reported in Fig. 3.17. This alternative functional unit considers both the mNPs production yield and the enzyme unit immobilization yield.

Taking into account the environmental profiles displayed in Fig. 3.17, Scenario A and Scenario D2 should be the worst options regardless the impact category considered for assessment (as different to the results reported per gram of mNP, where Scenario D1 is undoubtedly the undesirable route). However, Scenario B2 (spherical PEI-coated mNPs) should be preferred because reductions of around ~97% could be obtained with respect to Scenario A and Scenario D2 (depending on the category). Thus, spherical PEI-coated mNPs could be considered as the best option not only from technical but also from environmental perspective. Paying special attention to carbon footprint, i.e., to greenhouse gases emissions, the sequence of preference of mNPs production routes to be chosen entirely changes when the results are reported per unit of laccase immobilized and gram of mNP produced.

Table 3.9 Enzyme load per mg of magnetic nanoparticle (mNP) and per batch of production

Scenario	Enzyme load (U mg^{-1}mNPs)	Enzyme batch (U batch mNPs)
A	0.107	535
B1	0.811	5,190
B2	1.54	11,165
C	0.2	4,500
D1	2.66	19,950
D2	1.14	8,550

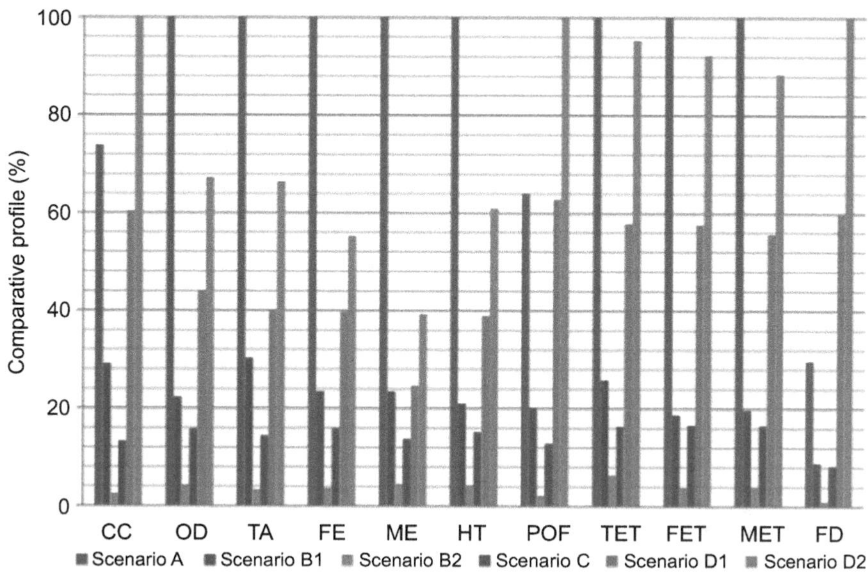

Figure 3.17 Comparative environmental profiles for the different scenarios considering 1 U of enzyme immobilized. *CC*, climate change; *FD*, fossil depletion; *FE*, freshwater eutrophication; *FET*, freshwater ecotoxicity; *HT*, human toxicity; *ME*, marine eutrophication; *MET*, marine ecotoxicity; *OD*, ozone depletion; *POF*, photochemical oxidant formation; *TA*, terrestrial acidification; *TET*, terrestrial ecotoxicity.

In terms of carbon footprint, the preferred sequence should be the following (Table 3.10):

Scenario C < Scenario B2 < Scenario A < Scenario B1 < Scenario D2 < Scenario D1

When expressed in terms of units of laccase immobilized, the sequence would be (Table 3.11):

Scenario B2 < Scenario C < Scenario B1 < Scenario D1 < Scenario A < Scenario D2

According to these results, differences on the selection of the best production sequence are clear as well as on the magnitude of the environmental burden. Regarding contributing processes, there is a clear factor of contributions to carbon footprint regardless the scenario under assessment: the large electricity requirements. However, it should be highlighted that the production systems have been assessed at pilot scale, and optimizations should be required at large scale. Moreover, special attention should be paid on the functional unit selected to report the environmental results because it could clearly affect decision-making strategies.

Table 3.10 Carbon footprint results per gram of magnetic nanoparticle (mNP) for the different production routes

Production route	Carbon footprint
Scenario C	26.5 g CO_2 eq/g mNP
Scenario B2	37.4 g CO_2 eq/g mNP
Scenario A	79.4 g CO_2 eq/g mNP
Scenario B1	237 g CO_2 eq/g mNP
Scenario D2	1.15 kg CO_2 eq/g mNP
Scenario D1	1.61 kg CO_2 eq/g mNP

Table 3.11 Carbon footprint results per unit of laccase immobilized (U) for the different production routes

Production route	Carbon footprint
Scenario B2	24.3 mg CO_2 eq/U
Scenario C	0.13 g CO_2 eq/U
Scenario B1	0.29 g CO_2 eq/U
Scenario D1	0.61 g CO_2 eq/U
Scenario A	0.74 g CO_2 eq/U
Scenario D2	1.01 g CO_2 eq/U

3.4 CONCLUSIONS

This work provides the first detailed LCI and environmental assessment for the different alternatives of mNPs production. In addition to highlighting again the importance of energy and chemicals use, the conducted assessment demonstrates the validity of the assumptions for the LCI. Thus, changes in the analyzed key parameters result in limited variations of the obtained environmental profile with respect to the baseline scenario.

Spherical PEI-coated mNPs are the most reliable option over the other supports because of the highest amount of laccase units immobilized per production batch. In addition, the environmental impacts associated to its production support its selection as the most suitable synthesis route.

ACKNOWLEDGMENTS

This work was financially supported by the Spanish Ministry of Economy and Competitiveness (CTQ2013-44762-R and CTQ2016-79461-R, programme cofunded by FEDER). The authors (SF, SG-G, YM-D, GF, MTM) belong to the Galician Competitive Research Group GRC 2013-032 as well as to CRETUS (AGRUP2015/02), and the author CV-V belongs to GRC 2013-044, programs funded by FEDER and Xunta de Galicia. Dr. S. González-García would like to express her gratitude to the Spanish Ministry of Economy and Competitivity (Grant reference RYC-2014-14984) for financial support.

REFERENCES

Althaus, H.J., Chudacoff, M., Hischier, R., Jungbluth, N., Osses, M., Primas, A., 2007. Life Cycle Inventories of Chemicals. Ecoinvent Report No. 8 v2.0 EMPA. Swiss Centre for Life Cycle Inventories, Düberdorf.

Arroyo, M., 1998. Inmovilización de enzimas. Fundamentos, métodos y aplicaciones. Ars Pharmaceutica 39 (2), 23–39.

Ayyad, O., Muñoz-Rojas, D., Oro-Sole, J., Gomez-Romero, P., 2010. From silver nanoparticles to nanostructures through matrix chemistry. Journal of Nanoparticle Research 12, 337–345.

Barreto, J.A., O'Malley, W., Kubell, M., Graham, B., Stephan, H., Spiccia, L., 2011. Nanomaterials: applications in cancer imaging and therapy. Advanced Materials 23 (12), 18–40.

Babic, A., Lindner, A.B., Vulic, M., Stewart, E.J., Radman, M., 2008. Direct visualization of horizontal gene transfer. Science 319, 1533–1536.

Baumann, H., Tillman, A.M., 2004. The Hitch Hilker's Guide to LCA. An Orientation in Life Cycle Assessment Methodology and Application. Studentlitteratur, Lund, Sweden, ISBN 9144023642.

Brady, D., Jordaan, J., 2009. Advances in enzyme immobilisation. Biotechnology Letters 31, 1639.

Bruce, P.G., Scrosati, B., Tarascon, J.-M., 2008. Nanomaterials for rechargeable lithium batteries. Agewandte Chemie International Edition 47 (16), 2930–2946.

Chang, L.H., Sasirekha, N., Chen, Y.W., Wang, W.J., 2006. Preferential oxidation of CO in H-2 stream over Au/MnO_2-CeO_2 catalysts. Industrial Engineering Chemical Research 45, 4927–4935.

Colvin, V.L., 2003. The potential environmental impact of engineered nanomaterials. Nature Biotechnology 21 (10), 1166–1170.

Courrol, L.C., Silva, F., Gomes, L., 2007. A simple method to synthesize silver nanoparticles by photo-reduction. Colloids and Surfaces A: Physicochemical and Engineering Aspects 305, 54–57.

Das, S., Mitra, S., Khurana, S.M.P., Debnath, N., 2013. Nanomaterials for biomedical applications. Frontiers in Life Science 7 (3), 90–98.

Doka, G., 2003. Life Cycle Inventories of Waste Treatment Services. Ecoinvent Report No. 13. Swiss Centre for Life Cycle Inventories, Dübendorf.

Dones, R., Bauer, C., Bolliger, R., Burger, B., Faist, E.M., Frischknecht, R., Heck, T., Jungbluth, N., Röder, A., Tuchschmid, M., 2007. Life Cycle Inventories of Energy Systems: Results for Current Systems in Switzerland and Other UCTE Countries. Ecoinvent Report No. 5. Swiss Centre for Life Cycle Inventories, Dümbendorf.

Eckelman, M., Mosher, M., Gonzalez, A., Sherman, J., 2012. Comparative life cycle assessment of disposable and reusable laryngeal mask airways. Anesthesia and Analgesia 114 (5), 1067–1072.

Fang, Y., Loc, W.S., Lu, W., Fang, J., 2011. Synthesis of In_2O_3@SiO_2 core–shell nanoparticles with enhanced deeper energy level emissions of In_2O_3. Langmuir 27 (23), 14091–14095.

Gavankar, S., Suh, S., 2012. Life cycle assessment at nanoscale: review and recommendations. The International Journal of Life Cycle Assessment 17 (3), 295–303.

Goedkoop, M.J., Heijungs, R., Huijbregts, M., Schryver, A.D., Strujis, J., Zelm, R., 2008. A Life Cycle Impact Assessment Method Which Comprises Harmonised Category Indicators at the Midpoint and the Endpoint Level. Report I: Characterisation ReCiPe, first ed. (revised).

González-García, S., Feijoo, G., Moreira, M.T., Berg, S., Agnemo, R., 2014. Environmental impact assessment of forest operations and pulp manufacture (Chapter 18). In: The Management of Industrial Forest Plantations. Theoretical Foundations and Applications. Part III. ISBN 978-94-017-8898-4, pp. 517–535.

Goon, I., Zhang, C., Lim, M., Gooding, J.J., Amal, R., 2010. Controlled fabrication of polyethylenimine-functionalized magnetic nanoparticles for the sequestration and quantification of free Cu^{2+}. Langmuir 26 (14), 12247–12252.
Grieger, K.D., Laurent, A., Miseljic, M., Christensen, F., Baun, A., Olsen, S.I., 2012. Analysis of current research addressing complementary use of life-cycle assessment and risk assessment for engineered nanomaterials: have lessons been learned from previous experience with chemicals? Journal of Nanoparticle Research 14, 958.
Grillo, R., Rosa, A.H., Fraceto, L.F., 2015. Engineered nanoparticles and organic matter: a review of the state-of-the-art. Chemosphere 119, 608–619.
Gupta, A.K., Gupta, M., 2005. Synthesis and surface engineering of iron oxide nanoparticles for biomedical applications. Biomaterials 26 (18), 3995–4021.
Han, Y., Jiang, J., Lee, S.S., Ying, J.Y., 2008. Reverse microemulsion-mediated synthesis of silica-coated gold and silver nanoparticles. Langmuir 24 (11), 5842–5848.
Hischier, R., Walser, T., 2012. Life cycle assessment of engineered nanomaterials: state of the art and strategies to overcome existing gaps. Science of Total Environment 425, 271–282.
ISO 14040, 2006. Environmental Management. Life Cycle Assessment-Principles and Framework. Geneve, Switzerland.
Kalishwaralal, K., BarathManiKanth, S., Pandian, S.R.K., Deepak, V., Gurunathan, S., 2010. Silver nano-A trove for retinal therapies. Journal of Controlled Released 145, 76–79.
Kulkarni, S.A., Sawadh, P.S., Palei, P.K., 2014. Synthesis and characterization of superparamagnetic Fe_3O_4@SiO_2 nanoparticles. Journal of the Korean Chemical Society 58, 100–104.
Laurent, S., Forge, D., Port, M., Roch, A., Robic, C., Est, L.V., Muller, R.N., 2008. Magnetic iron oxide nanoparticles: synthesis, stabilization, vectorization, physicochemical characterizations, and biological applications. Chemical Reviews 108 (6), 2064–2110.
Li, Y., Gao, W., Ci, L., Wang, C., Ajayan, P.M., 2010. Catalytic performance of Pt nanoparticles om reduced graphene oxide for methanol electro-oxidation. Carbon 48, 1124–1130.
Li, Q., Mahebdra, S., Lyon, D.Y., Brunet, L., Liga, M.V., Li, D., Alvarez, P.J.J., 2008. Antimicrobial nanomaterials for water disinfection and microbial control: potential applications and implications. Water Research 42, 4591–4602.
López-Quintela, M.A., Tojo, C., Blanco, M.C., García-Río, L., Leis, J.R., 2004. Microemulsion dynamics and reactions in microemulsions. Current Opinion in Colloid and Interface Science 9, 264–278.
Massart, R., 1981. Preparation of aqueous magnetic liquids in alkaline and acidic media. IEEE Transactions on Magnetics 17 (2), 1247–1248.
Michalet, X., Pinaud, F.F., Bentolilla, L.A., Trsay, J.M., Doose, S., Li, J.J., Sundaresan, G., Wu, A.M., Gambhir, S., Weiss, S., 2005. Quantum dots for live cells, in vivo imaging and diagnostics. Science 307 (5709), 538–544.
Nersisyian, H.H., Lee, J.H., Lee, S.I., Won, C.W., 2003a. The role of the reaction medium in the self propagating high temperature synthesis of nanosized tantalum powder. Combustion and Flame 135, 539–545.
Nersisyian, H.H., Lee, J.H., Son, H.T., Won, C.W., Maeng, D.Y., 2003b. A new and effective chemical reduction method for preparation of nanosized silver powder and colloid dispersion. Materials Research Bulleting 38, 949–956.
Nersisyian, H.H., Lee, J.H., Won, C.W., 2003c. Combustion of TiO_2-Mg and TiO_2-Mg-C systems in the presence of NaCl to synthesize nanocrystaline Ti and TiC powders. Materials Research Bulleting 38, 1135–1146.
Nowack, B., Krug, H.F., Height, M., 2011. 120 years of nanosilver history: implications for policy makers. Environmental Science & Technology 45 (4), 1177–1183.

PRé Consultants, 2017. http://www.pre.nl.
Qu, X., Alvarez, P.J.J., Li, Q., 2013. Applications of nanotechnology in water and wastewater treatment. Water Research 47, 3391–3946.
Red Eléctrica de España, 2014. Avance del informe del Sistema Eléctrico Español [Online] Available at: http://www.ree.es/sites/default/files/downloadable/avance_informe_sistema_electrico_ 2014.pdf.
Salamanca-Buentebello, F., Persad, D.L., Court, E.B., Martin, D.K., Daar, A.S., Singer, P.A., 2005. Nanotechnology and the developing world. PLoS Medicine 2 (5), e97.
Sathishkumar, M., Sneha, K., Yun, Y.S., 2010. Immobilization of silver nanoparticles synthesized using *Curcuma longa* tuber powder and extract on cotton cloth for bactericidal activity. Bioresource Technologies 101, 7958–7965.
Tsuzuki, T., 2013. Nanotechnology Commercialisation. Pan Stanford Publishing, Singapore.
Upadhyayula, V.K.K., Meyer, D.E., Curran, M.A., Gonzalez, M.A., 2012. Life cycle assessment as a tool to enhance the environmental performance of carbon nanotubes products: a review. Journal of Cleaner Production 26, 37–47.
Walser, T., Demou, E., Lang, D.J., Hellweg, S., 2011. Prospective environmental life cycle assessment of nanosilver T-shirts. Environmental Science and Technology 45 (10), 4570–4578.
Xie, Y., Ye, R., Liu, H., 2006. Synthesis of silver nanoparticles in reverse micelles stabilized by natural biosurfactants. Colloids and Surfaces A: Physicochemical and Engineering Aspects 279 (1), 175–178.
Zeiri, L., Bronk, B.V., Shabtai, Y., Czégé, J., Efrima, S., 2002. Silver metal induced surface enhanced Raman of bacteria. Colloids and Surfaces A: Physicochemical and Engineering Aspects 208 (1), 357–362.

CHAPTER 4

Carbon Footprint of Municipal Solid Waste Considering Selective Collection of Recyclable Waste

Luíza S. Franca, Marina S.R. Rocha, Glaydston M. Ribeiro
Federal University of Rio de Janeiro, Rio de Janeiro, Brazil

4.1 INTRODUCTION

According to the Global Waste Management Outlook Report of 2015, which was published by the International Solid Waste Association with the United Nations Environment Programme, the waste generation annually grows around 2 billion tons/year (UNEP and ISWA, 2015). Specifically the urban waste, which takes into account the waste produced by commercial, industrial and construction sectors, could grow from 7 to 10 billion tons/year (UNEP and ISWA, 2015). The waste from construction sector has the major participation on the total waste produced, with 36%, followed by the residential waste (24%) and the industrial waste (21%) (UNEP and ISWA, 2015).

In relation to the waste disposal, uncontrolled disposal, through open dumping and open burning, was the norm everywhere until the 1960s and, according to the World Bank (2016), it is still the norm in most developing countries. A significant progress is being made by some cities in middle-income countries, with controlled disposal rates often in the range of 70%–95%, though there is a lot of variation both within and between countries. The situation is much worse in low-income countries, where controlled disposal rates are often well below 50% in urban areas and around 0% in rural areas (UNEP, 2015). Considering this context combined with the 2014 data for world population by country income groups, it can be

Environmental Carbon Footprints
ISBN 978-0-12-812849-7
http://dx.doi.org/10.1016/B978-0-12-812849-7.00004-0

estimated that at least 3 billion people worldwide still lack access to controlled waste disposal facilities, depending on dumps (UNEP, 2015).

The waste disposal in dumps leads to the leachate percolation on the soil surface, which might pollute groundwater. That could, consequently, impact on the quality of the water supplying of other cities more distant to the dump. Besides that, this practice could attract disease vectors such as insects, rodents, and vultures. Lastly, the waste decomposition produces the methane gas, one of the main greenhouse gases (GHG), going directly to the atmosphere without treatment.

The Intergovernmental Panel on Climate Change (IPCC) estimates that the GHG produced by waste management, mainly from sanitary landfills, is around 3% of the emissions on 2010 (UNEP and ISWA, 2015).

Considering this scenario, developing countries need to adapt to the challenges of the Agenda 21, which were, among others, the mitigation of climate changes and the access to basic sanitation to the entire population.

In the case of Brazil, the total amount of municipal solid waste (MSW) generated in 2014 was more than 200,000 tons/day (Abrelpe, 2014). This type of waste is formed of a mixture of household waste (residential and commercial waste), sweepings from streets, parks and gardens, and sludge (Loureiro et al., 2013).

In terms of waste management, 99.96% of Brazilian municipalities offer public waste management, but 50.75% of them dispose the waste in dumps (MMA, 2015). Since the establishment of the Law no 12.305/2010 that institutes the National Policy for Solid Waste (NPSW), Brazil has started to invest in waste selection practices to reduce the amount disposed in landfills and to increase waste reuse rates (MMA, 2015).

One of the major problems of the sector is the lack of investments. To receive financial resources from the national government, the municipalities need to formulate and implement municipal plans for waste management. Thus, the Brazil Recycle Program, which was encouraged by the Environment Ministry (MMA, acronym in Portuguese) since 2010, intends to provide the necessary resources to do sanitary landfills to cover 73% of the urban population (118 million of inhabitants). Besides, the Program aims to promote selective collection to attend 59% of the Brazilian inhabitants (94 million of inhabitants) (Cempre, 2013). Nowadays, only 27 million of inhabitants are covered by the selective collection. However, one of the big challenges is the reduction costs of the recyclable waste logistic chain allowing the sector to become financially independent. In 2012, the selective collection was 4.5 times more expensive in relation to the conventional collection of solid waste (Cempre, 2013).

Specifically in Rio de Janeiro, an important Brazilian megacity, the amount of MSW generated in 2014 was 21,834 tons/day, with a rate of 1.307 kg/inh/day. Moreover, 68.4% of the total amount of MSW generated was destined to sanitary landfills, 21.7% to controlled landfills, and 9.9% to dumps (Abrelpe, 2014).

The benefits of recycling, such as resource conservation, are undeniable. However, it is not so evident how better waste management practices can contribute to reduce anthropogenic GHG emissions (Fitzgerald et al., 2012).

In 2012 the emissions inventory showed that Rio de Janeiro's city produced 22.6 millions of tons of CO_2eq, of which 1.6 millions of tons are related to urban waste disposal, 0.5 millions of tons came from domestic and commercial sewage, and 0.13 millions of tons were because of industrial effluent (City Hall of Rio de Janeiro, 2015).

The integrated management of solid waste may lead to opportunities to reduce the GHG emissions. The use of recyclable waste as an input to the industrial sector could reduce the GHG emissions because of the increase of life span of sanitary landfills. In the other hand, the recycling process consumes energy and generates atmospheric pollutants emissions, which might generate trade-offs in terms of net GHG emissions (IPEA, 2010).

Also, as stated by Bing et al. (2016) to improve MSW recycling management operational research should be integrated with other disciplines such as life cycle assessment (LCA), which would support decision makers in a more thorough and holistic manner. In this sense, a carbon footprint (CF) approach seems adequate to evaluate how changes in MSW practices contribute to climate change.

This methodology is based on life cycle thinking and, though it does not have a consensus on how it should be applied, it usually considers non-carbon emissions and use carbon dioxide (CO_2) equivalent indicators, similarly of what is used in LCA studies (Weidema et al., 2008). Many studies to analyze possible configurations of MSW management system have used CF as a supporting tool to drive conclusions on better MSW schemes considering climate change prerogatives (Friedrich and Trois, 2016; Cifrian et al., 2013; Eisted et al., 2009).

Therefore, this chapter intends to evaluate the CF of selective collection of recyclable waste, which is still poorly implemented in Brazil. To analyze the results in practice, a case study of a company responsible of the MSW collection from commercial establishments in Rio de Janeiro was conducted. Rio de Janeiro is the second megacity of Brazil. The city has a population of about 6.3 million inhabitants, a population density of 5265.81 hab/km^2, a GDP per capita of R$ 28,405.95 and an HDI of 0.84 (City Hall of Rio de Janeiro, 2012). Recently, Rio de Janeiro hosted the last Olympic and Paralympics games on 2016.

Currently, the MSW is collected and sent directly to transfer stations, and after to a sanitary landfill, without any segregation of the recyclable waste. So, the business-as-usual (BAU) CF is evaluated, and it is compared to scenarios where the selective collection is being implemented. Six scenarios are developed to program the selective collection by regions of the city as the segregation at source starts to exist. Thus, these scenarios are compared with the base scenario to see the differences on the CF when this logistic chain, which enables recycling and composting, is set up.

So, this chapter is divided into seven sections, followed by this Introduction. The logistic chain of recyclable waste, which is in practice in Brazil and in Rio de Janeiro, is presented on Section 4.2. Section 4.3 explores the Brazilian waste management outlook, with its characteristics of MSW generation and management. Section 4.4 presents the case study. The methodology of CF and how it was applied to the case study is presented in Section 4.5. Section 4.6 indicates the results and their discussions. Lastly, the study conclusions are presented in Section 4.7.

4.2 LOGISTIC CHAIN OF RECYCLABLE WASTE

The stages of waste management are the elements of the logistic chain. So, the logistic chain is divided into the waste demand region; the transfer stations, which are intermediary points; and final destinations, which represent the process of treatment and decomposition of the waste. Fig. 4.1 illustrates these elements.

Transfer stations are units located near by the demand regions, so that the vehicle used for the waste collection could discharge the waste on that local and could return as soon as possible to the route of collection. Usually these facilities are installed in situations where the distance between the landfill and the demand region is larger than 25 km. So, transfer stations have equipments or structures, which pick up the waste and place it in vehicles with bigger capacities. These vehicles are responsible to transfer the waste to the landfills (IBAM, 2001).

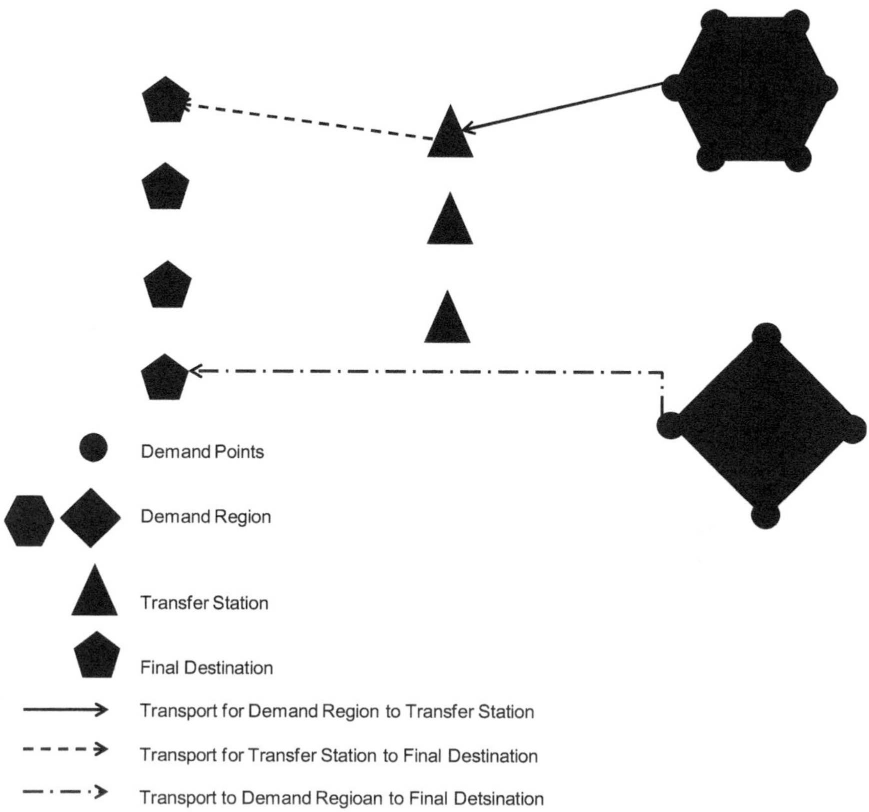

Figure 4.1 Logistic chain of solid waste.

On the other hand, in the logistic chain of recyclable waste, it is necessary to segregate the waste according to its different characteristics and types. At least the differentiation between the wet and dry waste need to be done by the waste generators (City Hall of Rio de Janeiro, 2012). In this case, the intermediary points in the logistic chain are sorting units instead of transfer stations, which perform the segregation and processing of each type of recyclable waste (glass, plastic, paper and metal, among others). These intermediary points are responsible to send the recyclable waste to recycling units, which finally transforms the material (Ferri et al., 2015). For the organic fraction, the collection needs to be done by another vehicle, and the treatment is done on composting units.

On the sorting units, each type of recyclable waste is separated from another through their disposal on treadmills. These materials could be manually or mechanically separated, which is possible to obtain an efficiency around 3%–6% in relation to the weight, depending on the manual or mechanical technology of the unit (IBAM, 2001). Then, the materials are grouped and/or pressed, and they are ready to go to the recycling unit, in which they will suffer a chemical and physical modification on their structure.

In the case of the composting unit, the organic fraction is submitted to a biological decomposition by the action of microorganisms. The composting process could be aerobic or anaerobic, depending on the presence of oxygen. In an anaerobic composting, the process is done with low temperatures, demanding more time and generating strong odors. For the aerobic composting, the temperature can go to 70°C, and the process is faster. The organic compost, resulted from these processes, is rich in nutrients and could be used in agriculture sector (IBAM, 2001). Specifically in case of Rio de Janeiro, the Municipal Law no 3273/2001 indicates that all commercial establishments, which produce a quantity of waste bigger than 120 L or 60 kg/day need to hire a private company (PC) to collect the waste. In this case, the public sector is only responsible to collect the residential waste.

4.3 BRAZILIAN WASTE MANAGEMENT OUTLOOK

4.3.1 The National Perspective

According to Abrelpe (2014), the total urban waste generated in Brazil in 2014 was about 78.6 million tons, which represents an increase of 2.9% in comparison with the urban waste generation on the year before. This rate is greater than the population growth rate in the country, which was 0.9% in the same year.

In relation to the urban waste disposal, SNIS (2016) indicates the total of 52.0% of waste collected was sent to sanitary landfills in 2014. The controlled landfills are the second type of waste final destination in Brazil, with 13.1%, followed by the significant 12.4% that were sent to the open dumps. The sorting and composting units are represented, respectively, by 2.5% and 0.4% of the total waste collected. Fig. 4.2 shows the different types of urban waste final destination adopted by each municipality in Brazil. It is possible to observe that the North, Northeast, and Midwest regions still present high incidence of dumps, whereas on South and Southeast regions the main waste final destination is the sanitary landfill.

In relation of Brazilian legislation, the NPSW determined the objectives and instruments of the integrated management of solid waste. According to

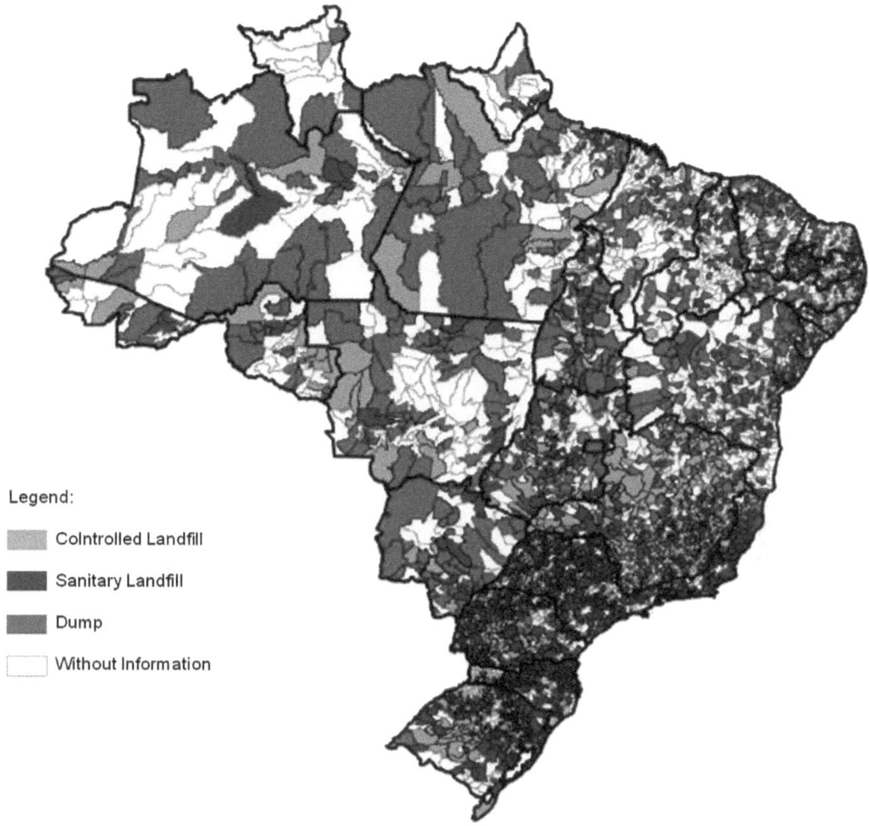

Figure 4.2 Types of waste final destination in Brazil. *(From SNIS, 2016. Diagnóstico do Manejo de Resíduos Sólidos Urbanos – 2014. Sistema Nacional de Informações sobre Saneamento (SNIS) – Ministério das Cidades, Brasil.)*

article 6°, one of its principles is the holistic approach of waste management, which takes into account the sustainable development. The waste is to be considered an economic good with social value. Besides, article 7° indicates the objectives must be, in that order: nongeneration, reduction, reutilization, and recycling of the waste and, in last case, the controlled disposal (Law no 12.305/2010).

SNIS (2016) also highlights that the selective collection of recyclable waste has been in practice on 1322 Brazilian municipalities, considering the total of 3765 municipalities, which has been consulted by the SNIS research, representing 30% of Brazil's urban population. Thus, 35.3% of these municipalities offer the door-to-door service. Also, 43.5% of the total recyclable waste is collected by the informal sector, 37.8% by companies, which are hired by the City Halls, and 18.7% are provided directly by the City Hall (SNIS, 2016).

The informal sector is an important stakeholder group on the market of recyclable waste in Brazil. However, until recently they were not recognized as a legitimate and, consequently, not considered in the municipal solid waste management. As waste management is an important social, economic, and environmental requirement, public–private partnership schemes to MSW should incorporate the informal sector. Cooperation between the informal and the formal sector can be part of the solution in solid waste management, but it requires support as the selective collection is performed both on a door-to-door schedule or adopting small volume delivery stations (PEVs, acronym in Portuguese) that can be combined for garbage collection (Ferri et al., 2015).

SNIS (2016) indicates that the Brazilian public sector, in general, does not receive a sufficient income to manage all the services, which are part of the waste management system. Because of that, the public sector allows an accreditation for private companies to do the service of waste collection and transportation to final destination, besides the management of transfer stations and landfills.

In terms of recyclable waste, the quantity of aluminum recycled in 2012 was 508 millions of tons. The Brazilian aluminum recovery rate was 30.4%, which could be compared with Italy (52.4%), the United Kingdom (51.5%), and the United States (47.5%). In the case of paper recycling, it was possible do recover 45.7% in 2012, which is a lower position in comparison with other countries like South Korea (91.6%), Germany (84.5%), and Japan (79.6%). Finally, in relation to plastic recycling, Brazil recovered 58.9% of this material in 2012 (Abrelpe, 2014). It is worth mentioning that the selective collection is still incipient in the country. IPEA (2010) states that 0.02 kg/inh/day or 1.2 millions of tons of recyclable waste is collected per year, which represents only 2.4% of the total waste collected by the municipalities in Brazil.

4.3.2 Panorama of Rio de Janeiro

The city of Rio de Janeiro has a population of about 6.3 million inhabitants, a population density of 5265.81 hab/km^2, a GDP per capita of US$ 9162.03 and an HDI 0.84 (City Hall of Rio de Janeiro, 2012). Because of its size, the city is divided into five planning areas, which are indicated on Fig. 4.3.

The waste generated on Rio de Janeiro was 10,815 tons/day taking as reference the month of December 2011. Considering this amount,

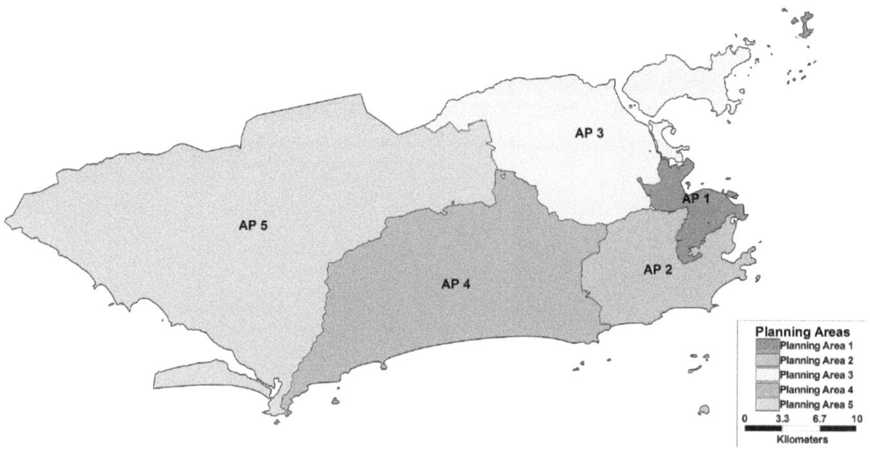

Figure 4.3 Planning areas of Rio de Janeiro. *(From Ferrari, T., 2017. Modelagem Matemática para Localização de Bases de Despacho de Veículos de Resgate: Um Estudo de Caso no Município do Rio De Janeiro. Master dissertation from COPPE. Rio de Janeiro.)*

4777 tons/day were from residential waste (49.4%), 3139 tons/day (32.5%) were collected from the streets, and 1155 tons/day (11.9%) were from commercial establishments (City Hall of Rio de Janeiro, 2012). Lastly, 6.2% were from emergency service provided by the City Hall. Also, the Preliminary Diagnostic of Solid Waste from Rio de Janeiro indicated the participation of private companies on the waste collection. They collected 133.20 tons/day in shopping malls, 195.0 /day in hotels, and 200.0 tons/day in supermarkets in 2014 (City Hall of Rio de Janeiro, 2015).

The waste collected without separation of the recyclable materials is sent to transfer stations, which can be managed by the government or by private companies. In the case of the public sector, the transfer stations are located on the districts: Santa Cruz (1094 tons/day), Bangu (1832 tons/day), Penha (1458 tons/day), Caju (3281 tons/day), besides Marechal Hermes, Taquara, and Jacarepaguá (729 tons/day each one). The municipal sanitary landfill is situated on the district of Seropedica, which receives a quantity of 9843 tons/day, in an area of 220 ha (City Hall of Rio de Janeiro, 2012).

In addition, there are transfer stations and sanitary landfills, which are responsibility of private companies. They are located around the metropolitan region of Rio de Janeiro, such as the transfer station in Jardim Gramacho and landfills in Nova Iguaçu and Alcântara. Fig. 4.4 illustrates the locations of both private and public transfer stations and landfills.

Figure 4.4 Locations of transfer stations and sanitary landfills in Rio de Janeiro. *(From Google Earth, 2017. Location of Facilities in Rio de Janeiro. https://www.google.com.br/googleearth.)*

With respect to recyclable waste, Fig. 4.5 shows the composition of the residential waste collected in the city. A significant slice of 40.99% corresponds to the recyclables, such as metals, plastics, glass, paper, and cardboard. The other fraction corresponds to the organic (52.68%) and the

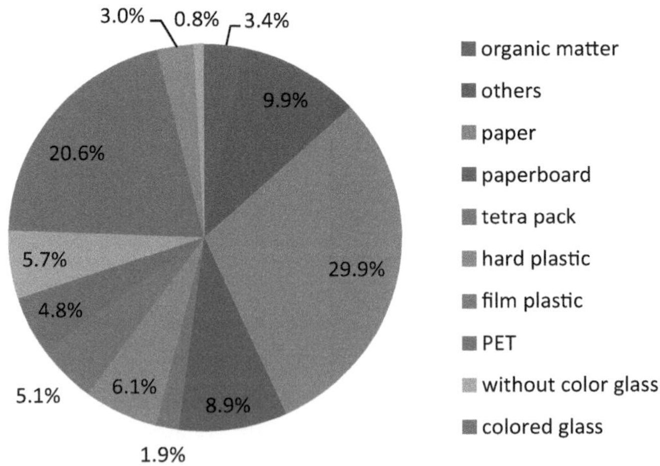

Figure 4.5 Recyclable waste that is collected by the public sector. *(From City Hall of Rio de Janeiro, 2015. Preliminary Diagnostico f Solid Waste of Rio de Janeiro. Rio de Janeiro, Brazil.)*

remnants of wood and fabric, among other wastes (6.33%). Of the total of recyclables in the domestic waste, plastic, in its several forms (PVC, PET, PE, among others) and paper and cardboard are almost 90% of these materials (City Hall of Rio de Janeiro, 2012).

The selective collection is set by the Municipal Act no. 30.624/2009, which designates the recyclable waste to municipal management, besides small companies. This act also stimulates the development of cooperatives and associations of waste pickers to support the waste management system (City Hall of Rio de Janeiro, 2012).

Big wholesalers, which purchase the recyclable waste from these small companies and cooperatives, are responsible to send the materials directly to recycling plants, normally located out of the Rio de Janeiro state. The municipal sector reported 1000 tons/day being sent to recycling in 2014 by 47 registered companies (City Hall of Rio de Janeiro, 2015).

The Program for Extension of Selective Collection in Rio de Janeiro (*Programa de Amplicação da Coleta Seletiva*, in Portuguese) aims to extend the services to all 160 neighborhoods of the city, promoting the social and productive inclusion of 1500 informal workers and ensuring the recycling of 5% of the potentially recyclables present in the domestic waste. To make the Program feasible, a contract was signed, in December 2010, between the Rio de Janeiro local government and Brazilian Development Bank involving funds of ~52 million. The residential collection of recyclables was to be performed twice a week, by a door-to-door operation, to avoid the access of the homeless to the segregated materials by the population. The program covered the construction of a total of six sorting centers, three with capacity to process 30 tons/day, which are located at Centro, Bangu, and Campo Grande neighborhoods. Other three centers are planned to be constructed at Penha, Irajá, and Jacarepaguá districts and will process 20 tons/day each one (City Hall of Rio de Janeiro, 2012).

The municipal government also invested in a composting unit, located in Caju, which generates organic compost called FERTILURB. This compost is used for reforestation projects on the entire city of Rio de Janeiro (City Hall of Rio de Janeiro, 2012).

And, lastly, the Municipal Act of Climate Changes, which is established by Municipal Act no. 5.248, sets goals of the reduction of GHG emissions on Rio de Janeiro City: reduction of 8% in 2012; reduction of 16% in 2016, and reduction of 20% in 2020 of the emissions. As shown in Fig. 4.6, the GHG emissions from waste management represent 14% of the total emissions generated in Rio de Janeiro. With incentives such as the selective

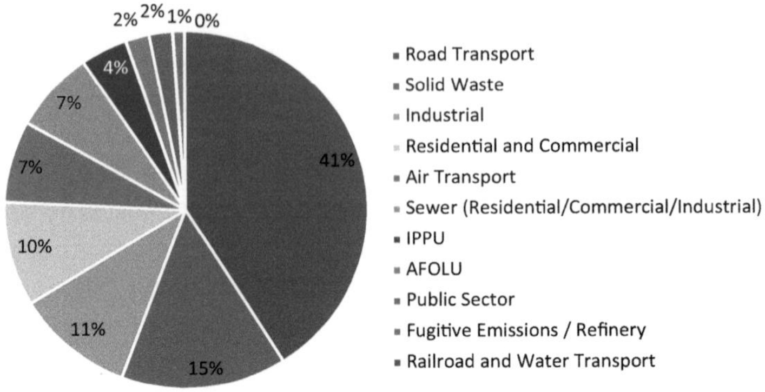

Figure 4.6 Greenhouse gases emissions by sector. *(From City Hall of Rio de Janeiro, 2012. Municipal Plan for Integrated Management of Solid Waste of Rio de Janeiro. Rio de Janeiro, Brazil.)*

collection, the reuse and the recycling, ensures the reduction of the solid waste generation (City Hall of Rio de Janeiro, 2012).

4.4 CASE STUDY

4.4.1 Characteristics of the Private Company

The PC, which is the object of this work, provides waste collection and transportation in the city of Rio de Janeiro and is located on the neighborhood of Bonsucesso. Around 2000 clients are collected by the company, which are commercial establishments such as hotels, restaurants, schools, churches, supermarkets, and among others. PC also collects waste from some industries. However, these are not industrial waste, they are office and restaurant waste from those facilities located in the industries. The quantity of waste collected per day represents ~2% of the residential waste collected by the municipal company.

The waste collection operation starts with the departure of compactor trucks from its base. They follow a collection route with a sequence of clients, which represent the demand points. The operation ends when the vehicle discharges the waste on the final destination, which is a transfer station. Nowadays, the company does not implement the selective collection on this operation.

To improve a waste management system aiming at a bigger recycling rate, it is key to implement an extensive selective collection. As PC does not operate selective collection routes today, a piece of software was used to optimize route planning if a selective collection were to be implemented.

The software chosen was ORTEC Routing and Dispatch (ORD) (ORTEC, 2016), a piece of software of vehicle routing and dispatch. It provides planning and optimization solutions and services, such as optimize fleet routing and dispatch, vehicle and pallet loading, workforce scheduling, delivery forecasting, supply chain network design, and warehouse control. To simulate the routes, some data need to be provided, such as location of each client and every possible final destination; the time window to collect each client; the duration time of the waste collection; quantity of waste of each client generates; and the vehicle fleet and the capacity of each vehicle. Most of the information demanded was supplied by average data from PC daily operation, which was observed in a period of 6 months.

4.4.2 Quantity of Waste Collected and Its Composition

Currently, PC collects 193,260.40 tons/day on average. Although there is data of the average waste weight collected in each client per day, this waste is not separated by the clients, so it is not possible to know how much of each type of waste is generated daily. To estimate the weight of the dry waste (recyclables) and wet waste (mainly organic), we used Eqs. (4.1) and (4.2), which are based on Asbea and Comcap (2014).

$$Q_{dry} = n \times k_1 \quad (4.1)$$

$$Q_{wet} = n \times k_2 \quad (4.2)$$

where Q_{dry} (kg/day): quantity of dry waste generated per day; Q_{wet} (kg/day): quantity of wet waste generated per day; n (kg/day): total of waste generated by the establishment; k_1 (%): percentual of dry waste generated; and k_2 (%): percentual of wet waste.

Thus, the authors classified each establishment according to the classification described by Asbea and Comcap (2014) classification. Then, the coefficients k_1 and k_2 were determined and are described on Annex 1.

Eqs. (4.1) and (4.2) indicate the variable n, which describes the total waste generated by each commercial establishment. Eq. (4.3) describes how to obtain this value.

$$n = v \times \gamma \times p \times m \quad (4.3)$$

Table 4.1 Types of waste disposal recipients and their volume

Type of waste disposal
Container of 1200 L
Container of 1000 L
Container of 360 L
Container of 240 L
Container of 120 L
Bag of 200 L
Bag of 100 L

where n (kg/day): total of waste generated by each establishment; v (m^3): volume of each type of waste disposal; γ (kg/m^3): specific weight; p (%): percentual of the waste disposal, which is not occupied; and m (un/dia): quantity of each unit of waste disposal, which is collected on each commercial establishment.

Table 4.1 shows each type of waste disposal recipients and the specific weight considered on this study is 360 kg/m^3. The waste occupied 50% of the total volume of the waste disposal unit and the quantity of units by each establishment was informed by PC. Both the specific weight and the fraction of the volume of waste occupied are data from field observation by the company.

4.4.3 Planned Scenarios

Because of the difficulty of the population awareness in relation to the environmental impacts of the waste management and the need of recycle the waste, it is considered that the collection will be implemented gradually by city area.

As there are five planning areas in Rio de Janeiro, this study will be based on six scenarios. Scenario 1 is the BAU scenario and represents the reality of waste collection in 2016 where no selective collection existed. The selective collection starts to be implemented from Scenario 2 and goes until Scenario 6, when it reaches all clients, considering a spatial planning, which is indicated on Fig. 4.7. The scenarios are cumulative, in other words, each scenario will consider that the collective selection has been implemented in all areas contemplated in the scenarios defined before.

From Scenario 2, when the recyclable waste starts to be segregated by the clients, the selective collection will be done by another type of truck, which does not have a compaction compartment. So, each client will be

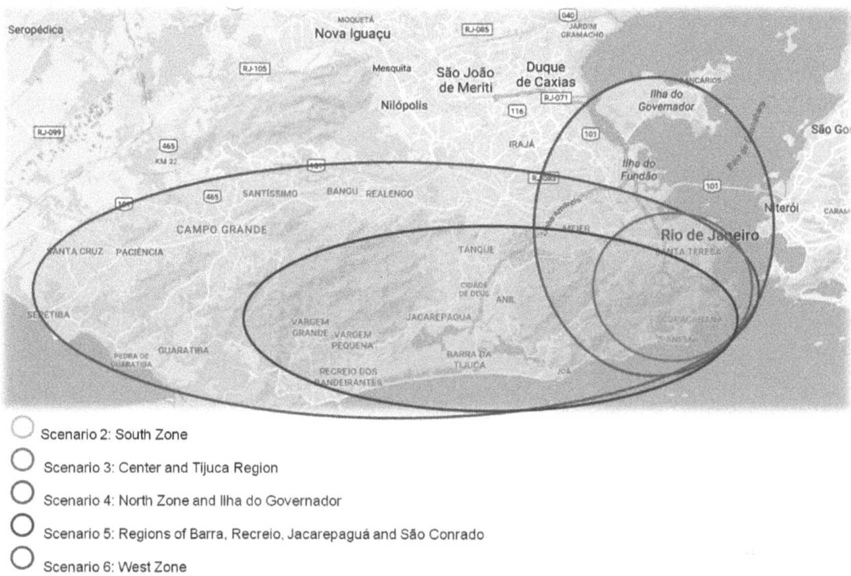

Figure 4.7 Regions of selective collection implemented by each scenario.

collected twice a day, one for the recyclable waste and one for the other fraction. It is considered that the selective routes for recyclable waste end on a sorting unit, which is established at the base of PC. On the other hand, the routes that collect the other fraction of the waste end on a composting facility.

The base, which is also the sorting unit, is located in Bonsucesso district; the transfer station considered on this work is located on the Jardim Gramacho region, and the composting unit is located on Caju district. These locations are illustrated by Fig. 4.8.

In addition, when the client segregates 100% of the recyclable waste, the wet fraction could be considered 100% organic. For simplicity, it was considered that there was no other type of waste rather than organic or recyclable waste coming from PC clients. So, in this case, it is assumed that the wet fraction is sent to a composting unit.

With that estimation, the routes were created using the software ORD. The routes are simulated for just 1 day of the week, Monday, because this is the day with the most number of collections points for the company.

To operate all routes, 44 trucks are needed. Most of those trucks of the company fleet followed EURO3 specifications, considering the limits of atmospheric pollutant emissions. The software used to draw the scenarios

Figure 4.8 Location of the facilities, which are considered on the scenarios. *(From Google Earth, 2017. Location of Facilities in Rio de Janeiro. https://www.google.com.br/googleearth.)*

distributed 44 trucks to operate these routes and specified how many kilometers and how many waste tons they would collect per day. Besides that, PC had average data for the diesel consumption of each truck as well as their specificities regarding capacity and truck types. The data for the routes of each scenario can be found in Annex II.

It is relevant to highlight that in Brazil diesel vehicles use a mixture of diesel and biodiesel as a fuel in increasing proportions because the bill was passed into the law 11.097/2005. From November 1st 2014 until the end of 2015, the mixture sold in Brazil was composed of 93% of diesel from petroleum and 7% of biodiesel B100 in terms of volume (Brasil, 2014). This diesel is called diesel B7.

Primary data for the transfer station, to where the total amount of waste in the BAU scenario (Scenario 1) goes, were obtained from its operator. The trucks from PC that arrive there leave the waste on the patio by lifting hydraulically their buckets. There are three shovel excavators that transfer the waste from the patio to bigger trucks that will drive to the landfill. The capacity of each shovel excavator is 1,229,863 kg of waste per liter of diesel B7.

Table 4.2 Amounts of waste going daily to transfer station, sorting center, or directly to composting facility

Scenario	Transfer station (kg)	Sorting center (kg)	Composting facility (kg)
1	193.2604	–	–
2	–	23.7759	169.4845
3	–	38.9209	154.3395
4	–	60.4685	132.7918
5	–	78.1611	115.0993
6	–	84.5744	108.6860

As the sorting center does not exist yet, data from a cooperative located in Riberão Pires, Brazil, were considered to describe the operation of such facilities. This cooperative is called Cooperpires, and its sorting operation was described by King and Gutberlet (2013). In Cooperpires facility, collected recyclables are manually transferred from trucks to a stationary triage table and then manually separated and pressed in a hydraulic press into 150 kg bales. The bales are then transported by electric forklift to be weighted and stocked. This routine consumed an average of 8.2 kWh/ton of waste.

Finally, the amount of waste that goes to the transfer station, sorting center, or directly to the composting facility is described in Table 4.2.

4.5 APPLIED CARBON FOOTPRINT METHODOLOGY

CF is a subcomponent of ecological footprint that is estimated by calculating the embodied life cycle GHG emissions associated with a consumption activity and aims at making such activity more sustainable (Cifrian et al., 2013).

The methodology adopted in this study to calculate the CF was based on ISO 14.040 and 14.046 standards from 2006. In other terms, likewise the work developed by Friedrich and Trois (2016), this study should be seen not as a detailed LCA but as CF analysis following a simplified LCA approach. So, only climate change indicator, more specifically Global Warming Potential (GWP), expressed in kg of CO_2eq, was considered in the analysis. The conversion factors to GWP for CH_4 and N_2O adopted were 28 CO_{2e} and 265 CO_{2e}, respectively (IPCC, 2014).

The waste management system operated by PC, and the processes that it includes, is presented in Fig. 4.9 inside the box for system boundaries. Fig. 4.10 presents the system boundaries for the modeled scenarios

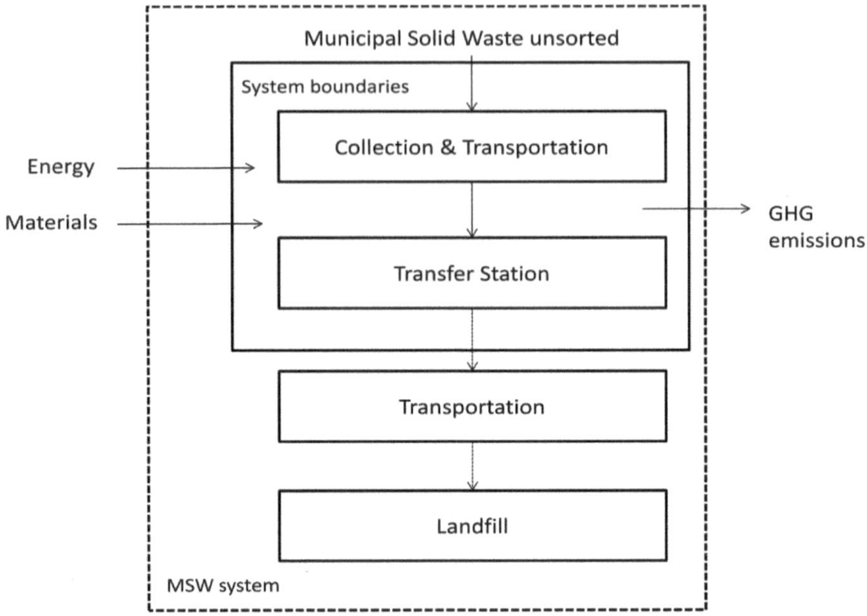

Figure 4.9 Carbon footprint system boundaries for Scenario 1 (business-as-usual). *GHG*, greenhouse gases.

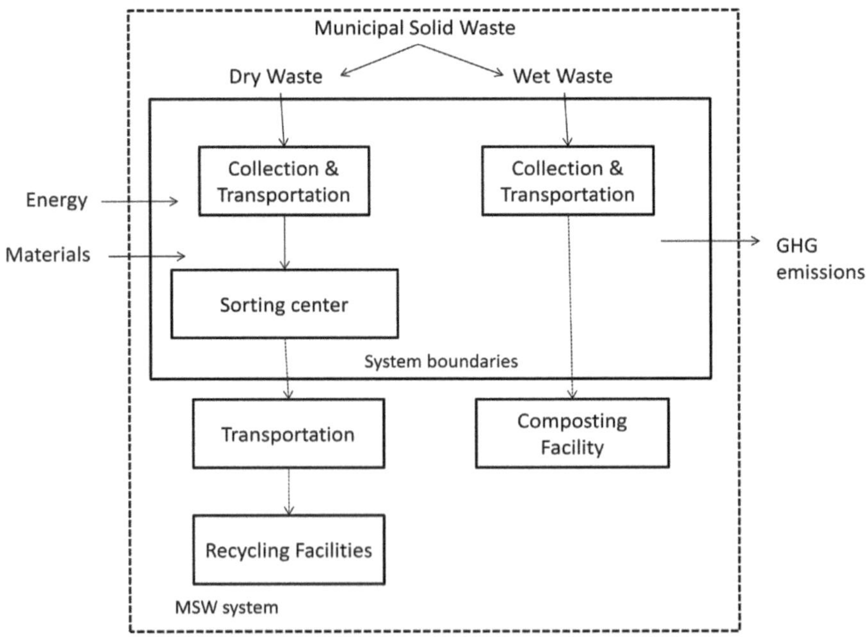

Figure 4.10 Carbon footprint system boundaries for Scenario 2–6. *GHG*, greenhouse gases.

(Scenario 2 to Scenario 6) where the selective collective is being gradually implemented.

As shown in Figs. 4.9 and 4.10, the other steps from waste management, like transportation of the sorted or unsorted materials to the final treatment or disposal will not be contemplated in this case study. This is because of the lack of primary data about the material composition of the waste collected currently by PC.

It is relevant to highlight that it was considered that the waste entering the system boundaries does not carry any of its product CF. In other words, a cut-off approach was used, which means that the recycled material will have only the environmental burden of the recycling efforts to turn waste into recycled material.

For the collection and transportation step, the GHG emissions were estimated combining the analysis of the direct emissions from the diesel B7 burned by the trucks, and indirect (upstream) emissions caused by the production and distribution of diesel and biodiesel, truck production and maintenance and road emissions.

The conversion factors for diesel upstream emissions came from datasets from the latest version of ecoinvent database and from a specific study for the Brazilian biodiesel (DeltaCO2 Sustentabilidade Ambiental, 2013).

For the direct emissions, as defined by the IPCC (2006), emission factors can be classified into three levels of detail, known as tiers. Tier 1 equals the least accurate level, and it is given by a default factor defined by the IPCC. Tier 2 and Tier 3 represent more specific factors for each national reality, requiring increasingly detailed data on the technologies used. Rio de Janeiro's city inventory adopted Tier 2 to CO_2 emissions from diesel and Tier 1 emissions for methane and N_2O (Coppe and Prefeitura do Rio de Janeiro, 2013), so these were the factors adopted in the present work.

In the case of biodiesel, the same factors used by the GHG Protocol, whose reference is the IPCC (2006), and the conversion factors (from TJ to liters) of the National Energy Balance (EPE, 2016) were adopted. It is relevant to point out that only CO_2, CH_4, and N_2O were considered in the direct emissions from the diesel B7 burning because of limited data.

DeltaCO2 Sustentabilidade Ambiental (2013) conducted a study about the CF of B100 in Brazil. For the present study, the maximum value for the CF of B100 distributed in Brazil was adopted.

For the Brazilian electric energy mix data from the latest version of ecoinvent database were used. It should be point out that GHG emissions generated in the production of collection bags, bins, and containers are not included in the scope of this study.

The metric used to calculate the amount of CO_2eq emitted was the GWP given by the IPCC (2014). GWP100 was chosen because it is the most frequently used metric (Goedkoop et al., 2013). Following this logic, the ReCiPe H Midpoint characterization method was chosen in the software SimaPro 8 to generate the emission factors of upstream processes used in this study.

All CO_2eq emission factors and their sources are described in Table 4.3.

4.6 RESULTS AND DISCUSSION

Based on the information regarding the scenarios and the emission factors, the CF of each scenario was calculated, and it is resumed in Table 4.4. Fig. 4.11 illustrates the difference of CF with each scenario with the previous one.

From the results, it becomes clear that the process that contributes the most for CF of each scenario is the collection & transportation, which highlights the importance on a good route planning for the selective collection. It is interested to notice by Fig. 4.11 that the CF grows in different proportion from one scenario to the other, increasing by 1.5% from Scenario 1 to Scenario 2 and by 8.3%, which is the highest increase, from Scenario 4 to Scenario 5. This is because of the fact that Scenario 5 considers a greater number of clients, which is collected, in the region of Barra, Recreio, Jacarepaguá, and São Conrado.

Counterintuitively, the CF of a scenario with more areas with selective collection might decrease compared with one with less areas covered. This is because of the increase in route efficiency as the same truck will be able to collect more tons/km. This situation can be acknowledge from Scenario 5 to Scenario 6, where a decrease of 1.5% can be noticed.

Similarly of what was developed in the present work, Eisted et al. (2009) have analyzed the CF of the collection, transfer, and transport of 1 ton of wet waste. If we divided the total CF per the total amount of waste for each scenario, we have an estimation of GHG emissions for 1 ton of wet waste generated. In this sense, the case studied would present a range from 40.63 (BAU scenario) to 47.29 kg CO_2eq/ton of wet waste in scenario 5. Eisted et al. (2009) findings follow the same growth trend of the present study. When considering residual waste, their results ranged from 9 to 57 kg CO_2eq/ton of waste. A substantial difference is perceived when the authors

Table 4.3 CO_2eq emission factors

	Life cycle stage	Emission factor	Unit	Source
Direct emissions	Diesel	2.7116	kg CO_2eq/L	Adapted from MMA (2011) and IPCC (2006)
	Biodiesel	2.4447	kg CO_2eq/L	Adapted from EPE (2016) and IPCC (2006)
Indirect emissions	Truck production (truck capacity from 7.5 to 16.0 metric tons)	0.0146	kg CO_2eq/tkm	Adapted from ecoinvent's dataset "Lorry, 16 metric ton {RoW}| production | Alloc Def, U"
	Truck Production (truck capacity from 3.5 to 7.5 metric tons)	0.0487	kg CO_2eq/tkm	Adapted from ecoinvent's dataset "Lorry, 16 metric ton {RoW}| production | Alloc Def, U"
	Light commercial vehicle production	0.1820	kg CO_2eq/tkm	Adapted from ecoinvent's dataset "Light commercial vehicle {GLO}| market for | Alloc Def, U"
	Truck maintenance (truck capacity from 7.5 to 16.0 metric tons)	0.0092	kg CO_2eq/tkm	Adapted from ecoinvent's dataset "Maintenance, lorry 16 metric ton {RoW}| processing | Alloc Def, U"
	Truck maintenance (truck capacity from 3.5 to 7.5 metric tons)	0.0308	kg CO_2eq/tkm	Adapted from ecoinvent's dataset "Maintenance, lorry 16 metric ton {RoW}| processing | Alloc Def, U"
	Light commercial vehicle maintenance	0.1721	kg CO_2eq/tkm	Adapted from ecoinvent's dataset "Maintenance, light commercial vehicle {GLO}| market for | Alloc Def, U"
	Road (truck capacity from 7.5 to 16.0 metric tons)	0.0182	kg CO_2eq/tkm	Adapted from ecoinvent's dataset "Road {GLO}| market for | Alloc Def, U"

Continued

Table 4.3 CO_2eq emission factors—cont'd

<table>
<tr><th></th><th>Life cycle stage</th><th>Emission factor</th><th>Unit</th><th>Source</th></tr>
<tr><td></td><td>Road (truck capacity from 3.5 to 7.5 metric tons)</td><td>0.0338</td><td>kg CO_2eq/tkm</td><td>Adapted from ecoinvent's dataset "Road {GLO}| market for | Alloc Def, U"</td></tr>
<tr><td></td><td>Road (light commercial vehicle)</td><td>0.0787</td><td>kg CO_2eq/tkm</td><td>Adapted from ecoinvent's dataset "Road {GLO}| market for | Alloc Def, U"</td></tr>
<tr><td></td><td>Diesel production and distribution (truck capacity from 7.5 to 16.0 metric tons)</td><td>0.0340</td><td>kg CO_2eq/tkm</td><td>Adapted from ecoinvent's dataset "Diesel, low-sulfur {RoW}| market for | Alloc Def, U"</td></tr>
<tr><td></td><td>Diesel production and distribution (truck capacity from 3.5 to 7.5 metric tons)</td><td>0.0642</td><td>kg CO_2eq/tkm</td><td>Adapted from ecoinvent's dataset "Diesel, low-sulfur {RoW}| market for | Alloc Def, U"</td></tr>
<tr><td></td><td>Diesel production and distribution (light commercial vehicle)</td><td>0.2056</td><td>kg CO_2eq/tkm</td><td>Adapted from ecoinvent's dataset "Diesel, low-sulfur {RoW}| market for | Alloc Def, U"</td></tr>
<tr><td></td><td>Diesel production and distribution (shovel excavator)</td><td>0.5436</td><td>kg CO_2eq/L</td><td>Adapted from ecoinvent's dataset "Diesel, low-sulfur {RoW}| market for | Alloc Def, U" and EPE (2016)</td></tr>
<tr><td></td><td>Biodiesel production and distribution</td><td>0.0009</td><td>Kg CO_2eq/L</td><td>Adapted from DeltaCO2 Sustentabilidade Ambiental (2013) and EPE (2016)</td></tr>
<tr><td></td><td>Brazilian electricity mix</td><td>0.0031</td><td>Kg CO_2eq/kWh</td><td>Adapted from ecoinvent's dataset "Electricity, low voltage {BR}| market for | Alloc Def, U"</td></tr>
</table>

Table 4.4 Carbon footprint of each scenario

	Carbon footprint (kg CO_2eq/day)			
Scenario	**Collection & transportation**	**Transfer station**	**Sorting center**	**Total**
1	7850.87	0.50	—	7851.37
2	7965.49	—	0.60	7966.09
3	8198.53	—	0.99	8199.51
4	8441.48	—	1.53	8443.02
5	9137.78	—	1.98	9139.76
6	8996.15	—	2.14	8998.29

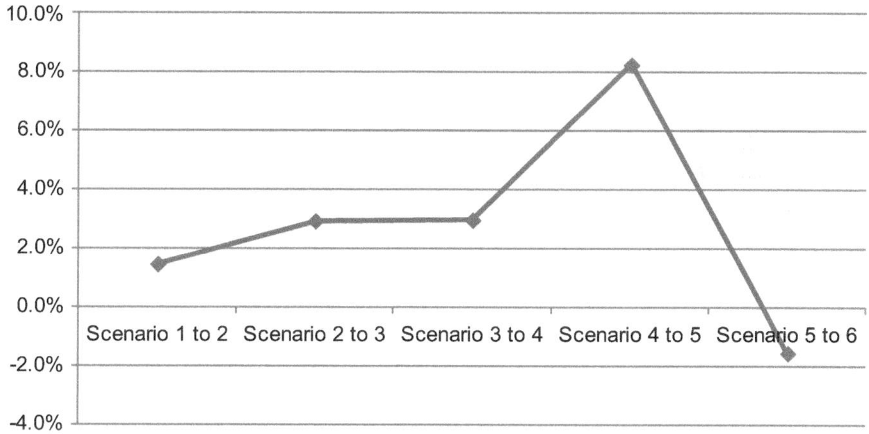

Figure 4.11 Comparison of carbon footprint of each scenario with the previous one.

analyzed examples where only recyclable paper of recyclable material, in general, with CFs ranging from 13 to 398 kg CO_2eq/ton of waste. However, the authors considered examples of developed countries, mainly, based on data originating from Europe and North America.

From a developing countries point of view, Friedrich and Trois (2013) elaborated a study to determine the GHG emission factors for the collection, transport, and landfilling of municipal waste in South African municipalities. They found that the emissions were 14.6 kg CO_2eq/ton of waste for collection and transport. That difference compared with the present study is probably because of the extent of the routes done by waste trucks, as the authors considered average data for a variety of municipalities in South Africa, none of them as big as Rio de Janeiro.

In addition, Fitzgerald et al. (2012) analyzed two different strategies to recyclable waste management. One considered no segregation of recyclables, and the other considered separation of paper and fiber from the rest of materials. The authors concluded that the introduction of more trucks in the second strategy did not significantly influenced the GHG emissions of an MSW management system as the avoid emissions from recycling are much larger.

Despite the effort to compare different studies, this is not an easy task. There is no standard definition of collection, transfer, and transport GHG emissions in the literature (Eisted et al., 2009).

Finally, as described in the case study, the Brazilian commercialized diesel has the characteristic of being 7% biodiesel. Because of that, we can consider that a percentage of the GHG emitted in the scenarios will be captured again by the plants used in the production of biodiesel. This is known as the biogenic carbon and would represent about 4% of the CF of each scenario.

4.7 CONCLUSIONS

This chapter evaluates the CF of selective collection of recyclable waste of a case study of a PC responsible for the MSW collection from commercial establishments in Rio de Janeiro, Brazil. A BAU scenario was modeled, and five scenarios were developed to program the selective collection by regions of the city as the segregation at source starts to exist.

From the results, it is possible to conclude that Scenario 6, which considers the selective collection implemented in all the regions of the city, emits a quantity of GHG emissions 13% bigger than the BAU scenario, which do not have any selective collection. This is because of the addition of more truck on the streets. In case of selective collection, each client is collected twice a day, one for the collection of the dry waste and another for the wet waste. This tendency could be observed in other studies on the literature, as shown in Friedrich and Trois (2013) and Fitzgerald et al. (2012).

On the other hand, the use of optimization route software could reduce the GHG emissions of the collection & transportation with the reduction of the total of kilometers traveled. So, the selective collection could have an efficient implementation, with a smaller environmental impact.

It is a limitation of the present study not to consider the disposal or end of life treatment of the waste collected by PC. As pointed out in the study of Fehr (2014), it is not safe to rely on general percentage waste composition because it may vary from one municipality to the other, depending even on the level of knowledge of separation practice by the citizens. So, the GHG emissions could be compensated by the recycling and composting processes.

For Hupponen et al. (2015) the significance of collection and transportation is higher from the costs perspective than from the point of view of GHG emissions. Therefore, it is a recommendation for future studies to analyze the scenarios considered here from a cost point of view.

ANNEX I

Coefficients of the gravimetric composition of each commercial establishment are presented in Table I.1.

Table I.1 Coefficients of the gravimetric composition

Type of commercial establishment	k_1	k_2	References
Gym	0.20	0.80	Asbea and Comcap (2014)
Fishmonger/butchery	0.37	0.63	de Campos and de Lima (2014)
Commercial association	0.70	0.30	Asbea and Comcap (2014)
Office	0.40	0.60	Asbea and Comcap (2014)
Bank	0.70	0.30	Asbea and Comcap (2014)
Bar/Restaurant	0.40	0.60	Asbea and Comcap (2014)
Coffee shop	0.40	0.60	Asbea and Comcap (2014)
Party house	0.20	0.80	Asbea and Comcap (2014)
Rest house	0.20	0.80	de Conto et al. (2004)
Commercial center	0.70	0.30	Asbea and Comcap (2014)
Cultural center	0.70	0.30	Asbea and Comcap (2014)
Cinema	0.20	0.80	Asbea and Comcap (2014)
Club	0.20	0.80	Asbea and Comcap (2014)
Car concessionaire	0.50	0.50	Asbea and Comcap (2014)
Breeder	0.40	0.60	Asbea and Comcap (2014)
Distribution center	0.50	0.50	Asbea and Comcap (2014)
Fuel distribution	0.50	0.50	Asbea and Comcap (2014)
Event company	0.20	0.80	Asbea and Comcap (2014)
Cleaning company	0.70	0.30	Asbea and Comcap (2014)
Transportation company	0.50	0.50	Asbea and Comcap (2014)
Tourism company	0.70	0.30	Asbea and Comcap (2014)
School/University	0.45	0.55	Araújo and Viana (2012)
Food industry	0.40	0.60	Asbea and Comcap (2014)
Bag industry	0.60	0.40	Asbea and Comcap (2014)
Communication company	0.60	0.40	Asbea and Comcap (2014)
Equipment industry	0.60	0.40	Asbea and Comcap (2014)
Tape industry	0.60	0.40	Asbea and Comcap (2014)

Continued

Table I.1 Coefficients of the gravimetric composition—cont'd

Type of commercial establishment	k_1	k_2	References
Glass industry	0.60	0.40	Asbea and Comcap (2014)
Mechanical peaces industry	0.60	0.40	Asbea and Comcap (2014)
Plastic industry	0.60	0.40	Asbea and Comcap (2014)
Chemical good industry	0.60	0.40	Asbea and Comcap (2014)
Plastic bowl industry	0.60	0.40	Asbea and Comcap (2014)
Clothes industry	0.60	0.40	Asbea and Comcap (2014)
Pharmaceutical industry	0.30	0.70	Asbea and Comcap (2014)
Pharmacy	0.20	0.80	Asbea and Comcap (2014)
Graphic	0.80	0.20	Asbea and Comcap (2014)
Hospital/Clinic	0.60	0.40	Asbea and Comcap (2014)
Hotel	0.20	0.80	de Conto et al. (2004)
Church	0.70	0.30	Asbea and Comcap (2014)
Real state company	0.70	0.30	Asbea and Comcap (2014)
Rent shop	0.70	0.30	Asbea and Comcap (2014)
Shop	0.70	0.30	Asbea and Comcap (2014)
Supermarket	0.37	0.63	de Campos and de Lima (2014)
Civil construction area	0.20	0.80	de Araújo and Carnaúba (2010)
Mechanical workshop	0.50	0.50	Asbea and Comcap (2014)
Bakery/Snack bar	0.40	0.60	Asbea and Comcap (2014)
Gas station	0.50	0.50	Asbea and Comcap (2014)
Film production company	0.70	0.30	Asbea and Comcap (2014)
Truck seller	0.70	0.30	Asbea and Comcap (2014)
Saloon	0.60	0.40	Asbea and Comcap (2014)
Security company	0.70	0.30	Asbea and Comcap (2014)
Hydraulic services	0.70	0.30	Asbea and Comcap (2014)

ANNEX II

This annex indicates the characteristics of the routes created by the ORD (ORTEC, 2016), which are used in each planned scenario. Tables II.1–II.6 shows the distance traveled, the total waste weight carried, and the diesel efficiency of each route. The name of route, which started with "A" indicates that this route collects the waste without separation and sends it to the transfer station. In case of the name of the route starts with "O" indicates that this route collects organic waste and sends it to the composting unit. Finally, when the name of the route starts with "R" indicates that this route collects recyclable waste and sends it to the sorting unit.

Table II.1 Transport data from Scenario 1 (business-as-usual)

Route	Distance traveled (km)	Total waste weight carried (tons)	Diesel B7 efficiency (km/L)
A1	147	8.61	1.96
A2	169	11.96	1.96
A3	132	6.45	1.96
A4	128	12.49	1.91
A5	113	7.97	1.96
A6	69	0.04	1.89
A7	135	6.79	1.96
A8	227	6.98	1.63
A9	227	11.29	1.96
A10	70	1.89	1.63
A11	253	12.40	1.91
A12	161	5.99	1.99
A13	181	7.84	1.89
A14	180	11.98	1.96
A15	201	12.49	1.91
A16	140	12.49	1.91
A17	170	9.61	1.96
A18	249	11.55	1.96
A19	169	12.49	1.91
A20	181	10.86	1.91
A21	136	4.99	1.99
A22	133	6.00	1.99

Table II.2 Transport data from Scenario 2

Route	Distance traveled (km)	Total waste weight carried (tons)	Diesel B7 efficiency (km/L)
O1	208	11.79	1.96
O2	107	3.98	1.96
O3	218	12.00	1.96
O4	129	12.49	1.91
O5	204	12.00	1.96
O6	168	11.99	1.96
O7	147	8.99	1.63
O8	178	1.86	1.63
O9	83	4.34	1.96
O10	180	9.00	1.63
O11	181	12.49	1.91
O12	294	4.46	1.65
O13	293	7.80	1.89

Continued

Table II.2 Transport data from Scenario 2—cont'd

Route	Distance traveled (km)	Total waste weight carried (tons)	Diesel B7 efficiency (km/L)
O14	93	1.77	1.96
O15	100	10.37	1.91
O16	104	9.74	1.91
O17	188	9.48	1.96
O18	70	1.10	2.17
O19	189	11.58	1.96
O20	62	1.34	1.91
O21	173	8.25	1.96
O22	51	2.67	1.91
R1	94	2.39	4.03
R2	89	5.29	4.03
R3	87	5.30	4.03
R4	103	5.23	4.03
R5	69	1.05	4.03
R6	134	4.28	4.03
R7	36	0.23	9.20

Table II.3 Transport data from Scenario 3

Route	Distance traveled(km)	Total waste weight carried (tons)	Diesel B7 efficiency (km/L)
O1	192	11.66	1.96
O2	188	8.93	1.96
O3	221	11.10	1.96
O4	115	0.20	1.91
O5	227	11.29	1.96
O6	289	5.50	1.96
O7	183	7.56	1.63
O8	78	8.21	1.63
O9	195	11.10	1.96
O10	271	2.98	1.63
O11	181	12.49	1.91
O12	149	11.10	1.96
O13	90	9.83	1.91
O14	130	10.10	1.91
O15	218	11.99	1.96
O16	213	8.12	1.96
O17	149	8.58	1.91
R1	155	5.30	4.03

Table II.3 Transport data from Scenario 3—cont'd

Route	Distance traveled(km)	Total waste weight carried (tons)	Diesel B7 efficiency (km/L)
R2	70	5.30	4.03
R3	102	5.29	4.03
R4	99	4.63	4.03
R5	72	5.30	4.03
R6	73	0.43	9.37
R7	76	1.10	9.37
R8	107	1.98	9.37
R9	65	0.25	9.37
R10	76	0.81	9.37
R11	84	2.98	9.20
R12	182	4.66	4.03

Table II.4 Transport data from Scenario 4

Route	Distance traveled (km)	Total waste weight carried (tons)	Diesel B7 efficiency (km/L)
O1	163	7.22	1.96
O2	190	9.61	1.96
O3	225	11.24	1.96
O4	77	8.37	1.91
O5	292	3.29	1.96
O6	191	2.51	1.96
O7	197	6.31	1.63
O8	113	2.51	1.63
O9	187	10.49	1.96
O10	181	12.49	1.91
O11	195	11.10	1.96
O12	93	9.96	1.91
O13	113	10.26	1.91
O14	195	8.21	1.96
O15	225	11.10	1.96
O16	147	8.57	1.91
R1	164	5.30	4.03
R2	83	5.30	4.03
R3	104	5.30	4.03
R4	89	5.30	4.03
R5	72	5.30	4.03
R6	78	1.10	9.37

Continued

Table II.4 Transport data from Scenario 4—cont'd

Route	Distance traveled (km)	Total waste weight carried (tons)	Diesel B7 efficiency (km/L)
R7	98	1.10	9.37
R8	103	1.10	9.37
R9	65	1.10	9.37
R10	65	1.10	9.37
R11	155	1.10	9.37
R12	126	5.30	4.03
R13	73	1.10	9.37
R14	82	1.75	9.37
R15	90	1.57	9.37
R16	88	0.74	9.37
R17	84	2.96	9.20
R18	66	5.29	4.03
R19	131	2.39	4.03

Table II.5 Transport data from Scenario 5

Route	Distance traveled (km)	Total waste weight carried (tons)	Diesel B7 efficiency (km/L)
O1	163	7.1	1.96
O2	187	8.31	1.96
O3	306	6.72	1.96
O4	77	8.37	1.91
O5	267	1.59	1.96
O6	197	6.30	1.63
O7	80	0.09	1.63
O8	186	10.12	1.96
O9	256	12.50	1.91
O10	177	8.62	1.96
O11	105	9.85	1.91
O12	143	10.07	1.91
O13	230	11.67	1.96
O14	264	4.87	1.96
O15	125	8.91	1.91
R1	119	5.30	4.03
R2	96	5.29	4.03
R3	98	4.34	4.03
R4	84	5.30	4.03
R5	72	5.30	4.03

Table II.5 Transport data from Scenario 5—cont'd

Route	Distance traveled (km)	Total waste weight carried (tons)	Diesel B7 efficiency (km/L)
R6	129	1.04	9.37
R7	166	1.98	9.37
R8	187	5.30	4.03
R9	72	1.10	9.37
R10	81	1.57	9.37
R11	138	1.88	9.37
R12	66	1.10	9.37
R13	84	0.32	9.37
R14	83	2.34	9.20
R15	137	5.30	4.03
R16	115	5.30	4.03
R17	94	5.27	4.03
R18	171	5.30	4.03
R19	175	5.14	4.03
R20	104	5.30	4.03
R21	190	2.62	4.03

Table II.6 Transport data from Scenario 6

Route	Distance traveled (km)	Total waste weight carried (tons)	Diesel B7 efficiency (km/L)
O1	163	7.11	1.96
O2	187	8.29	1.96
O3	291	3.74	1.96
O4	77	8.37	1.91
O5	197	6.30	1.63
O6	80	0.09	1.63
O7	186	10.12	1.96
O8	209	12.50	1.91
O9	177	8.62	1.96
O10	105	9.85	1.91
O11	143	10.07	1.91
O12	199	8.95	1.96
O13	251	5.76	1.96
O14	125	8.91	1.91
R1	179	5.30	4.03
R2	96	5.29	4.03
R3	98	4.34	4.03

Continued

Table II.6 Transport data from Scenario 6—cont'd

Route	Distance traveled (km)	Total waste weight carried (tons)	Diesel B7 efficiency (km/L)
R4	84	5.30	4.03
R5	72	5.30	4.03
R6	80	1.10	9.37
R7	41	1.10	9.37
R8	69	1.10	9.37
R9	94	1.10	9.37
R10	60	1.10	9.37
R11	124	1.10	9.37
R12	167	5.30	4.03
R13	72	1.99	9.37
R14	81	1.57	9.37
R15	138	1.88	9.37
R16	66	1.10	9.37
R17	84	0.32	9.37
R18	83	2.34	9.20
R19	170	5.30	4.03
R20	137	5.27	4.03
R21	137	5.26	4.03
R22	221	5.21	4.03
R23	215	4.53	4.03
R24	185	4.56	4.03

REFERENCES

Abrelpe, 2014. Solid Waste Panorama in Brazil. Brazilian Association of Public Cleaning Companies (ABRELPE). Brasil.

Araújo, R.S., Viana, E., 2012. Diagnóstico Dos Resíduos Sólidos Gerados Na Escola De Artes, Ciências E Humanidades (Each) Como Instrumento Para a Elaboração De Um Plano De Gestão Na Unidade. Revista Eletrônica em Gestão Educação e Tecnologia Ambiental 8 (8), 1805—1817.

de Araújo, N.M.C., Carnaúba, T.M.G.V., 2010. Composição Gravimétrica E Massa Específica Dos Rcd Oriundos De Obras De Edificações Verticais De Maceió. Canela, Rio Grande do Sul: XIII Encontro Nacional de Tecnologia do Ambiente Construído.

Asbea and Comcap, 2014. Manejo de Resíduos Sólidos — Manual para Edificações Multifamiliares e de Uso Misto. Associação Brasileira dos Escritórios de Arquitetura (ASBEA) e Companhia Melhoramentos da Capital (COMCAP). Florianópolis, SC, Brasil.

Bing, X., Bloemhof, J., Ramos, T., Barbosa-Povoa, A.P., Wongd, C., van der Vorst, J., 2016. Research challenges in municipal solid waste logistics management. Waste Management 48, 584—592.

Brasil, 2014. Lei n. 13.033 de 24 de setembro de 2014. [S.l.].

de Campos, R.F.F., de Lima, C., 2014. Sustentabilidade através de remanejo de resíduos com prática de gestão ambiental implantado no Supermercado Cereal. Ignis 3, 25—44.

Cempre, 2013. Cempre Review 2013. Business Commitment for Recycling. Brazil.

Cifrian, E., Andres, A., Viguri, J.R., 2013. Estimating Monitoring Indicators and the Carbon Footprint of Municipal Solid Waste Management in the Region of Cantabria, Northern Spain, pp. 271–285.
City Hall of Rio de Janeiro, 2012. Municipal Plan for Integrated Management of Solid Waste of Rio de Janeiro. Rio de Janeiro, Brazil.
City Hall of Rio de Janeiro, 2015. Preliminary Diagnostico f Solid Waste of Rio de Janeiro. Rio de Janeiro, Brazil.
de Conto, S.M., et al., 2004. Geração de resíduos sólidos em um meio de hospedagem: Um estudo de caso. Florianópolis, Santa Catarina: ICTR – Instituto de Ciência e Tecnologia em Resíduos e Desenvolvimento Sustentável.
Coppe and Prefeitura do Rio de Janeiro, 2013. Inventário de Gases de Efeito Estufa da Cidade do Rio de Janeiro em 2012 e Atualização do Plano Municipal para Redução das Emissões. Rio de Janeiro.
DeltaCO2 Sustentabilidade Ambiental, January 2013. Carbon Footprints in the Production of Biodiesel from Soybeans. Reference Period: 2008/09.
Eisted, R., Larsen, A.W., Christensen, T.H., March 2009. Collection, transfer and transport of waste: accounting of greenhouse gases and global warming contribution. 738–745.
EPE, 2016. Balanço Energético Nacional 2016: Ano base 2015. Empresa de Planejamento Energético. Rio de Janeiro, p. 292.
Fehr, M., 2014. The management challenge for household waste in emerging economies like Brazil: realistic source separation and activation of reverse logistics.
Ferri, G.L., Chaves, G.L.D., Ribeiro, G.M., 2015. Reverse logistics network for municipal solid waste management: the inclusion of waste pickers as a Brazilian legal requirement. Waste Management 40, 173–191.
Fitzgerald, G.C., Krones, J.S., Themelis, N.J., 2012. Resources, conservation and recycling greenhouse gas impact of dual stream and single stream collection and separation of recyclables. Resources, Conservation and Recycling 69, 50–56.
Friedrich, E., Trois, C., 2013. GHG emission factors developed for the collection, transport and landfilling of municipal waste in South African municipalities. Waste Management 33 (4), 1013–1026.
Friedrich, E., Trois, C., 2016. Current and future greenhouse gas (GHG) emissions from the management of municipal solid waste in the eThekwini municipality – South Africa. Journal of Cleaner Production 112.
Goedkoop, M., Heijungs, R., Huijbregts, M., Schryver, A., de Struijs, J., Van Zelm, R., 2013. Report I: Characterisation. ReCiPe a Life Cycle Impact Assess. Method Which Comprises Harmonised Category Indicators at the Midpoint and the Endpoint Level. version 1.08.
Google Earth, 2017. Location of Facilities in Rio de Janeiro. https://www.google.com.br/googleearth.
Hupponen, M., Grönman, K., Horttanainen, M., 2015. How should greenhouse gas emissions be taken into account in the decision making of municipal solid waste management procurements? A case study of the South Karelia region, Finland. Waste Management 42.
IBAM, 2001. Manual for Integrated Solid Waste Management. Brazilian Institute of Municipal Management (IBAM) and Special Secretary of Brazilian Urban Development (SEDU/PR). Rio de Janeiro, Brazil.
IPCC, 2006. IPCC Guidelines for National Greenhouse Gas Inventories. National Greenhouse Gas Inventories Programme/IGES. Interngovernmental Panel on Climate Change [S.l.].
IPCC, 2014. Climate Change 2014: Synthesis Report. Contribution of Working Groups I, II and III to the Fifth Assessment Report of the Intergovernmental Panel on Climate Change. In: Core Writing Team, Pachauri, R.K., Meyer, L.A. (Eds.). IPCC, Geneva, Switzerland, 151 pp.

IPEA, 2010. Research Report – Research of Payment Over Urban Environmental Services for Waste Management. Research Institute of Applied Economic Research, Brazil.

King, M.F., Gutberlet, J., 2013. Contribution of cooperative sector recycling to greenhouse gas emissions reduction: a case study of Ribeirão Pires, Brazil. Waste Management 33 (12), 2771–2780.

Law no. 12.305/2010, 2010. Institution of National Policy for Solid Waste. National Government of Brazil.

Loureiro, S.M., Rovere, E.L.L., Mahler, C.F., 2013. Analysis of potential for reducing emissions of greenhouse gases in municipal solid waste in Brazil, in the state and city of Rio de Janeiro. Waste Management 33 (5), 1302–1312.

MMA, 2011. 1° Inventário Nacional de Emissões Atmosféricas por Veículos Automotores Rodoviários. Ministério do Meio Ambiente, Brasília.

MMA, 2015. Data about the National Law no. 12.305/10. http://www.mma.gov.br/pol%C3%ADtica-de-res%C3%ADduos-s%C3%B3lidos.

ORTEC, 2016. Optimized Vehicle Routing – Key Ingredient for Successful Last Mile Fulfillment. Last Mile Logistics, Singapura.

Programme, U.N.E. Global Waste Management Outlook. [s.l: s.n.].

SNIS, 2016. Diagnóstico do Manejo de Resíduos Sólidos Urbanos – 2014. Sistema Nacional de Informações sobre Saneamento (SNIS) – Ministério das Cidades. Brasil.

UNEP, 2015. Global Waste Management Outlook. United Nations Environment Programme, London, UK.

UNEP and ISWA, 2015. Global Waste Management Outlook. Viena, Áustria. United Nations Environment Programme (UNEP) e International Solid Waste Association (ISWA). International Solid Waste Association General Secretariat.

Weidema, B.P., Thrane, M., Christensen, P., 2008. Carbon footprint a catalyst for life cycle assessment, 12 (1).

World Bank, 2016. Data About Urban Solid Waste Management. http://go.worldbank.org/A5TFX56L50.

CHAPTER 5

Carbon Footprint Analysis of Personal Electronic Product—Induction Cooker

Winco K.C. Yung[1], Subramanian Senthilkannan Muthu[2], Karpagam Subramanian[1]
[1]The Hong Kong Polytechnic University, Kowloon, Hong Kong SAR; [2]Bestseller, Kowloon, Hong Kong SAR

5.1 INTRODUCTION

Induction cooker is one of the most common and popular household electronic product used in recent decades. However, similar to all other electronic products, the various steps or stages in the production, usage, and disposal of these products also lead to substantial amount of greenhouse gas (GHG) emissions. Sustainability and the interest of consumers to opt for eco-friendly and low-carbon products are also simultaneously increasing in recent years. To achieve sustainable consumption and production, the stages in the life cycle of a product that create these carbon emissions have to be identified, measured, and then mitigated. Hence calculating carbon emission and carrying out a detailed analysis of electronic products such as induction cookers will help manufacturers achieve sustainable production and get an edge over other similar products in the market. Product carbon footprint (PCF) analysis of an induction cooker performed and reported here is the first of its kind in this area of research. The results will help the manufacturer identify the hotspots and reduce carbon emissions accordingly in various life cycle phases of the studied product.

The goal in this PCF study is to calculate the potential contribution of the chosen product, induction Cooker (hereafter named as "the cooker"), to global warming expressed as grams or kilograms of CO_{2eq}, by quantifying all significant GHG emissions and removals over the product life cycle (only "cradle to gate" stages are considered in this study). This report follows the standard of ISO/TS 14067:2013 "Greenhouse gases—Carbon footprint of

Environmental Carbon Footprints
ISBN 978-0-12-812849-7
http://dx.doi.org/10.1016/B978-0-12-812849-7.00005-2

products—Requirements and guidelines for quantification and communication" (International Organization for Standardization (ISO), 2013). The results, data, methods, assumptions, and life cycle interpretations are presented in sufficient detail to allow the reader to comprehend the complexities and trade-offs inherent in the PCF study. The PCF value is measured by quantifying the GHG emissions for the entire product life cycle expressed in carbon dioxide equivalents value (kg CO_{2eq}), as per ISO/TS 14067.

$$CF_{total} = CF_{RM} + CF_{M} + CF_{D} + CF_{U} + CF_{EOL} \quad (5.1)$$

where CF_{RM} is the carbon footprint of the raw material stage (kg CO_{2eq}); CF_{M} is the carbon footprint in manufacturing stage (kg CO_{2eq}); CFD is the carbon footprint in distribution stage (kg CO_{2eq}); CFU is the carbon footprint during use stage (kg CO_{2eq}); and CFEOL is the carbon footprint at the end-of-life stage (kg CO_{2eq}).

5.2 METHODOLOGY

5.2.1 Scope

5.2.1.1 Product System and Its Function(s)

The product selected for this study is an induction cooker, which is a home appliance for cooking. The cooker is mainly made from ceramic, plastic, and metals. It provides three types of cooking methods—heating with set temperatures (60, 90, 120, 160, and 200°C), boiling, and steaming/stewing. Electricity is used to power this cooker. Fig. 5.1 shows the outlook of the induction cooker.

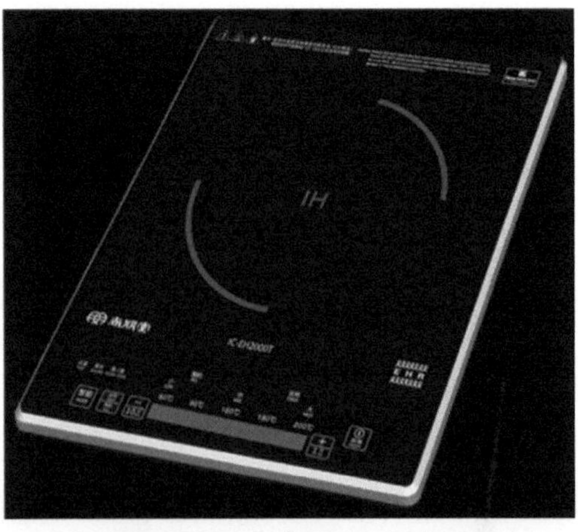

Figure 5.1 Induction cooker.

5.2.2 Functional Unit

As there is no specific product category rule (PCR) for an induction cooker, the functional unit (f.u.) of the cooker is referenced to the PCRs of similar electronic products in Korea (microwave ovens), Sweden (electrical machinery and apparatus), and Taiwan (printed circuit board assembly). The f.u. of this study is defined as one unit of induction cooker, including one electronic induction printed circuit board assembly, along with primary packaging (PE bag and packaging foam) and secondary packaging (printed carton box). The reference flow is defined as one unit of induction cooker in this study.

5.2.2.1 Data and Data Quality

5.2.2.1.1 Description of Data

This chapter covers "cradle to gate" stages of product life cycle of the induction cooker. Data of the unit process obtained for the raw material, manufacturing, and transportation were primarily collected by Sunpentown (Guangzhou) Appliance Co., Ltd., which had financial and operational control of the unit processes. The emission factors used were as local as possible, with geographic, technology, and time-related considerations. Table 5.1 shows details of the data.

5.2.2.1.2 Characterization of Data Quality

The best quality data were used to reduce bias and uncertainty as far as practically possible. Primary and secondary data were selected to meet the goal and scope of the PCF study.

Table 5.1 Description of data

Life cycle stage	Type of data	Data
Raw material stage	Primary data (site specific)	Amount of materials and components used
	Secondary data	Emission factors of material and components
Manufacturing stage	Primary data (site specific)	Electricity consumption of manufacturing process
	Secondary data	Emission factors of electricity
Transportation stage	Primary data	Transportation tools used
	Secondary data	Emission factors of transportation route
		Distance of transportation routes

5.2.2.2 Cut-Off Criteria and Cut-Offs

No materials or processes have been omitted in this study; thus, no cut-off criteria were established.

5.2.2.3 Allocation Procedures

Allocation of electricity consumption is made according to production volumes of domestic and commercial models of induction cookers in the Guangzhou factory, from August 2013 to July 2014. Only domestic models of induction cooker production were considered in this study. The allocation was applied to the electricity used in the manufacturing stage of domestic induction cookers (Section 5.4.1) based on their total production volume.

5.2.2.4 Geographical and Time Boundary of Data

The induction cookers were manufactured in Guangzhou, China, from August 2013 to July 2014 and are sold in mainland China and Hong Kong.

5.2.2.5 System Boundary

As a product manufacturer, Sunpentown (Guangzhou) Appliance Co., Ltd. provides service and products to customers (B2C business model).

Regarding the determination of the system boundary of this study, the "use" stage and the "end-of-life" stage were not included. This was because the factory could not provide any testing report of product endurance or specific product end-of-life treatment policy. Also there is a huge variation in users' usage patterns in terms of different temperature selection and cooking methods and in geographical differences regarding electricity production in different regions. As such, it is difficult to build reliable use stage and end-of-life stage scenarios (life cycle assessment (LCA)) because of these practical issues. To reduce the uncertainty of the end result (PCF) of the cooker, the system boundary for this study is confined to "cradle to gate" of the product life cycle stages, including raw material extraction, manufacturing, and product distribution stages. However, in this study, the calculation of emission in distribution stage is not included in the final PCF results and is reported for reference only. Fig. 5.2 shows the system boundary in this study.

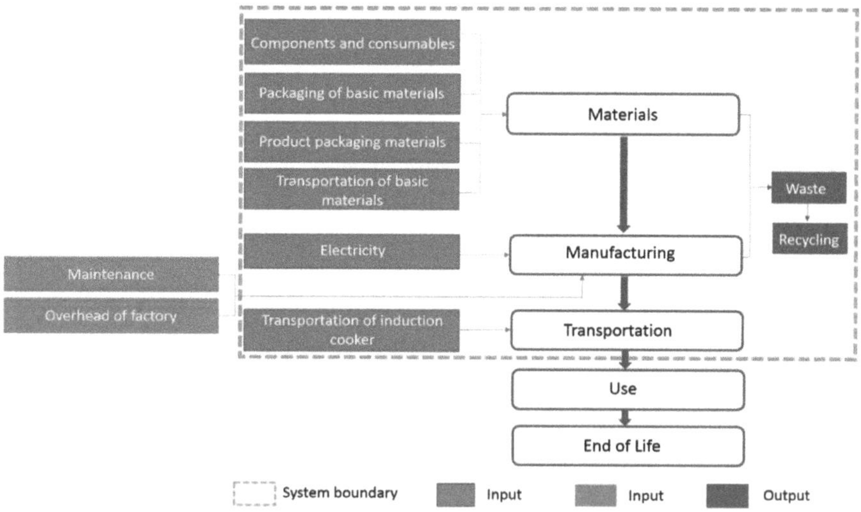

Figure 5.2 System boundary of carbon footprint analysis of "the cooker."

All unit processes included in this study directly contributed to the end result (GHG emissions) from "cradle to gate" life cycle stages. The overheads of the factory and equipment maintenance in all life cycle stages were excluded in this study because they were not significant and not directly related to the GHG emissions and removal of the cookers. Regarding the waste generated from the material and manufacturing stages, the transportation of the waste from the factory to recycling site was included in this report, but the environmental burdens from the recycling processes were not included in the calculation (Fig. 5.4).

5.2.2.6 Relevant Assumptions in This Study

5.2.2.6.1 Assumptions in "Raw Material Stage"

- Packaging of materials and packaging of packaging materials for each type of material or component were assumed to be the same in all the orders, although packaging methods or packaging materials might vary from batch to batch because of variation in the production quantities.
- The transportation of material and packaging material was grouped by suppliers, and transportation methods were the same from the same supplier.

- Inner packaging of certain components (diodes, triodes, and transistors) was left out because of the small contribution to carbon emissions and the difficulty in measuring.
- Similar components (for example, different types of diodes) were grouped as one (diodes) when conducting their life cycle inventory and assessment for PCFs.
- The packaging of materials was sorted into two main categories—plastic materials and paper materials. All the packaging was assumed to be recycled in the local recycling plant.
- All distances in each waste treatment transportation routes were estimated based on Google Map.
- The closest and most similar material emission factor was chosen for the emission factor selection; for example, emission factors for polybutylene terephthalate (Wikipedia, 2014), tie wrap (Cable Tie Express, 2014), jump wire (Taitron, n.d.), constantan wire (Rapidonline, n.d.), plastic housing connector (Alibaba, 2014), and power plug (Blackshaw Inc, n.d.) in the material stage.
- There were differences among the weighting data for some of the components and consumables; the variances were due to different suppliers for the components and consumables.

5.2.2.6.2 Assumptions in the "Manufacturing Stage"

- The energy consumption of each cooker (domestic model) manufactured in the factory was assumed to be the same. The percentage of commercial (large working load) products out of total production was assumed to be 10% (data come from factory sources).
- The consumables used for each cooker manufactured in the factory were assumed to be the same.
- Electricity consumption in the office area was excluded from the electricity consumption of the induction cooker production.
- Electricity consumption of the lift was included in the production.
- The annual energy consumption was averaged over 12 months of data as the cooker was only manufactured from August 2013 to July 2014. No seasonal production factors were included in this study.
- The amount of evaporation and metal cut-offs were estimated based on the calculation of material loss during the manufacturing process.
- All the waste produced (offcuts from metal components) from manufacturing processes were recycled to the local recycling plants.

- Data regarding electricity usage and recycling details were provided by factory personnel.

5.2.2.6.3 Assumptions in "Transportation Stage"
- All distances in each transportation activities routes were obtained from Google Map.
- The distribution of products in this study was assumed to be in China because more production orders come from China (66.13%) than from Hong Kong (33.87%).
- Data regarding product order information were provided by factory staff.

5.2.2.7 *Treatment of Electricity*

The electricity emission factor for the manufacturing stage was obtained from the Chinese Government National Development and Reform Commission (NDRC). The emission factor for southern China in 2013 was 0.9223 kg CO_{2eq}/kWh (National Development and Reform Commission, 2013).

5.3 CARBON FOOTPRINT ANALYSIS

5.3.1 Carbon Emission Calculation

According to the data provided by Sunpentown (Guangzhou) Appliance Co., Ltd., calculation of the "cradle to gate" assessment of PCF is based on the LCA approach, which includes the raw material stage and the manufacturing stage, when the distribution stage is listed for reference. The analysis is based on the concept of LCA and ISO/TS 14067:2013 (International Organization for Standardization (ISO), 2013) and allows computation of the total amount of carbon emission within the selected system boundary. This PCF study, according to ISO/TS 14067:2013, includes four phases of LCA, i.e., goal and scope definition, life cycle inventory (LCI), life cycle impact assessment, and life cycle interpretation (LCI). Fig. 5.3 shows the framework of LCA from ISO 14040 (International Organization for Standardization (ISO), 2006).

The unit processes comprising the product system are grouped into selected life cycle stages in this study, i.e., raw material acquisition and manufacturing. GHG emissions and removals from the product's life cycle are assigned to the life cycle stage in which the GHG emissions and removals occur. Partial PCFs may be added together to quantify the PCF,

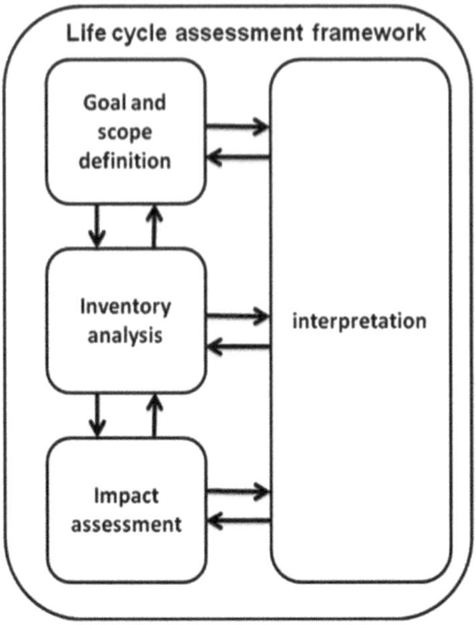

Figure 5.3 Framework of life cycle assessment (Taitron, n.d.).

provided that the same methodology is used and there are no gaps and overlaps. The PCF is calculated and presented as a carbon dioxide equivalent (kg CO_{2eq}) using the relevant 100-year global warming potential. Findings in the impact assessment and the sensitivity and contribution analysis, performed as part of the interpretation, help identify the most relevant and contributory ("key") processes and basic flows of the system.

$$\text{Carbon Emission} = \text{Activity Data} \times \text{Emission Factor} \times \text{Global Warming Potential} \quad (5.2)$$

The "global warming potential" should be excluded, where the unit of the "emission factor" was "kg CO_{2eq}/unit." The activity data collected from each unit process were inputted into an online PCF analyzer (Green Bill of Materials (G-BOM) analyzer provided by The Hong Kong Polytechnic University) and computed as in European Commission (2012). The working principle, the methodology behind the software, and access can be found in the following link: http://www.pctech.ise.polyu.edu.hk/ecodesign/gbom_analyzer.html.

The G-BOM analyzer is online software based on the structure of the bill of material. The G-BOM analyzer allows manufacturers to input their product information/data according to the product life cycle, which includes the raw materials, manufacturing, distribution, usage, and end-of-life stages and simply calculates the PCF using the GHG emissions on the basis of the database provided. Users can define the material names, emission factors, and quantities according to their life cycle inventory. The GHG emission database is pregrouped into five life cycle stages. The emission factor can be selected manually in life cycle groups or by a keyword search engine.

5.4 LIFE CYCLE INVENTORY FOR PRODUCT CARBON FOOTPRINT

5.4.1 Life Cycle Inventory of "Raw Material" Stage

5.4.1.1 Sources of Data

1. NDRC—Chinese Government National Development and Reform Commission
2. Ecoinvent—Swiss Ecoinvent data version 2.2

Database selection was based on credibility and geographical, technological, and time-related considerations. Emission factor for electricity consumption is obtained from NDRC 2013, and the data used for components, consumables, packaging materials, and transportation are obtained from Ecoinvent version 2.2.

5.4.2 Life Cycle Inventory of "Manufacturing" Stage

5.4.2.1 Process Flow in the "Manufacturing" Stage

- All activity data of the resources and energy consumption in the "manufacturing" stage were measured onsite and provided by Sunpentown (Guangzhou) Appliance Co., Ltd. in August 2014.

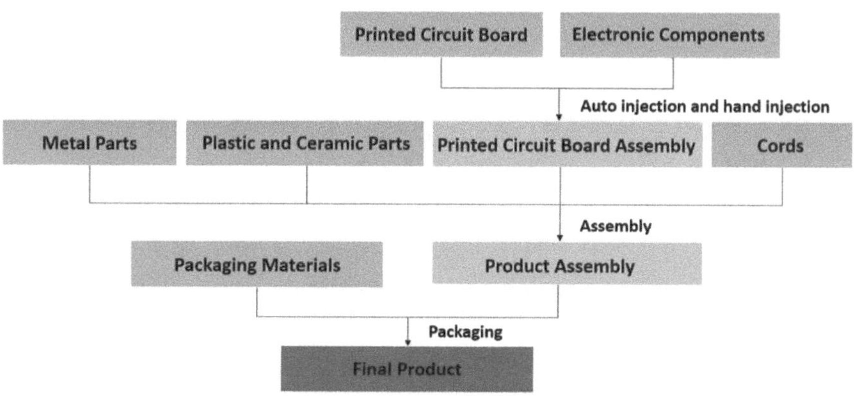

Figure 5.4 Process flow in the "manufacturing" stage.

- The emission factor of the electricity used was obtained from the NDRC. The emission factor for southern China for 2013 was 0.9223 kg CO_{2eq}/kWh (Cable Tie Express, 2014).
- The energy consumption in the manufacturing stage was provided by Sunpentown (Guangzhou) Appliance Co., Ltd. It was based on the allocation of energy consumption according to the production volume of domestic models and commercial models of induction cookers in the factory from August 2013 to July 2014. Only domestic models of induction cooker production were considered in this study.
- The energy consumption for manufacturing each induction cooker was calculated to be 1.1618 kWh.
- The wastes (offcuts of metal components) produced during manufacturing processes and the packaging of basic materials (components, consumables, and product packaging) are assumed to be recycled in local recycling plants. The emission from transportation of waste from the factory to recycling site was included, but the environmental burdens from the recycling processes were excluded from the calculation.
- The defect rate of products within the factory was assumed to be 0%. According to the factory staff, all the default products found in the quality control section are returned to the manufacturing line for fixing and retesting until qualified.

5.4.3 Life Cycle Inventory of "Transportation" Stage

- Transportation destinations and tools for the finished product were provided by Sunpentown (Guangzhou) Appliance Co., Ltd.
- The distance traveled for each delivery was estimated based on Google Map.

5.5 LIFE CYCLE IMPACT ASSESSMENT OF PRODUCT CARBON FOOTPRINT ANALYSIS

The calculation of the carbon footprint of the cooker was based on a comprehensive coverage of raw material stage, manufacturing stage, and transportation stage. By assigning quantities for the input and output in each process, the carbon footprint of each input, output, and process were obtained and are shown in Table 5.2.

Table 5.2 Carbon footprint results from "cradle to gate" life cycle stages

No.	Item	Carbon footprint (kg CO_{2eq}/unit)	Percentage of total carbon footprint (%)
Components, consumables, product packaging, and corresponding packaging			
Components			
1	Bottom cover	2.5745	7.65
2	Upper lid	0.5279	1.57
3	Aluminum plate (round shape)	0.9834	2.92
4	IGBT (insulated gate bipolar transistor) heat sink	1.7021	5.06
5	Heating coil plate	0.7191	2.14
6	Copper wire	0.6188	1.84
7	Tin	0.2840	0.84
8	Tie wrap	0.0019	0.01
9	Bridge diode	1.4700	4.37
10	Diode	0.7006	2.08
11	IGBT	0.8983	2.67
12	Triode	0.4122	1.23
13	Transistor	0.0872	0.26
14	Integrated circuit	4.1951	12.47
15	Carbon film resistor	0.2871	0.85
16	Metal film resistor	0.0616	0.18
17	Metal oxide film resistor	0.0525	0.16
18	Wirewound resistor (high-pressure metal glass glaze)	0.0274	0.08
19	Rheostat (variable resistor)	0.0156	0.05
20	Ceramic capacitor	0.1490	0.44
21	Plastic capacitor	1.1054	3.29
22	Electrolytic capacitor	0.3136	0.93
23	X capacitor	1.4978	4.45
24	Iron core	0.6816	2.03
25	Ring core choke coil inductor	2.0203	6.01
26	Transformer	0.0386	0.11
27	Wire set	0.0122	0.04
28	Wire set shielded line	0.0249	0.07
29	Copper terminal	0.0005	0.00
30	Connector terminal	0.0277	0.08
31	Jump wire	0.0061	0.02
32	Constantan wire	0.0017	0.00
33	Tin-coated copper wire jumper	0.0016	0.00
34	Plastic housing connector	0.0201	0.06
35	Power cord	0.4263	1.27

Continued

Table 5.2 Carbon footprint results from "cradle to gate" life cycle stages—cont'd

No.	Item	Carbon footprint (kg CO_{2eq}/unit)	Percentage of total carbon footprint (%)
36	Wire	0.0034	0.01
37	Rubber pad	0.0144	0.04
38	Power plug	0.2631	0.78
39	Silicone fibreglass sleeving	0.0031	0.01
40	PVC tube	0.0001	0.00
41	Glass ceramic board	1.1029	3.28
42	Touch spring	0.0181	0.05
43	U shape terminal	0.0224	0.07
44	Silicone pad	0.0062	0.02
45	Thermal fuse	0.0443	0.13
46	Silicone mushroom head (round)	0.0106	0.03
47	Sticker	0.0019	0.01
48	Fuse	0.0788	0.23
49	Two digit segment LED display	0.3972	1.18
50	One digit segment LED display	0.2135	0.63
51	Colored LED	1.0538	3.13
52	DC brushless fan	1.0679	3.17
53	PCB	3.0969	9.21
54	Thermistor	0.0156	0.05
55	Buzzer	0.9128	2.71
56	Tapping screw	0.0479	0.14
57	Machine screw	0.0189	0.06
58	Pan head machine screw with spring washer	0.0171	0.05
59	Pad	0.0009	0.00
60	Thermal grease	0.0011	0.00
61	Silicone (black) for upper cover	0.0049	0.01
62	Silicone (transparent) for silicone pad	0.0022	0.01
Product packaging			
1	User manual (primary packaging)	0.0706	0.21
2	Colored carton (primary packaging)	0.7411	2.20
3	Outer box (secondary packaging)	0.4010	1.19
4	User manual packaging bag (primary packaging)	0.0101	0.03
5	Product packaging bag (primary packaging)	0.0518	0.15
6	Packaging foam (primary packaging)	0.1047	0.31
7	Polystyrene styrofoam (primary packaging)	0.0370	0.11

Table 5.2 Carbon footprint results from "cradle to gate" life cycle stages—cont'd

No.	Item	Carbon footprint (kg CO_{2eq}/unit)	Percentage of total carbon footprint (%)
8	Sealing tape (secondary packaging)	0.0017	0.01
9	Plastic handle (primary packaging)	0.0210	0.06
Consumables			
1	Soldering flux (consumable)	0.0001	0.00
2	Polishing detergent (consumable)	0.0001	0.00
3	Cleansing cotton cloth	0.0596	0.18
4	Water (for testing)	0.0001	0.00
5	Solid waste and evaporation loss from manufacturing	0.0370	0.11
Packaging of components and consumables			
1	Corrugated carton	0.3222	0.96
2	Plastic woven bag	0.0084	0.02
3	Poly straps	0.0015	0.00
4	Kraft paper box	0.0206	0.06
5	White cardboard box	0.0046	0.01
6	PE plastic bag	0.0428	0.13
7	Plastic bucket	0.0065	0.02
8	Kraft paper board	0.0019	0.01
9	Plastic stretch film	0.0003	0.00
10	Newspaper	0.0032	0.01
11	Plastic roll core	0.0177	0.05
12	IC plastic tube	0.0141	0.04
13	Polystyrene styrofoam	0.0209	0.06
14	Expandable polyethylene	0.0053	0.02
15	PVC plastic wrapping straps	0.0041	0.01
16	Nylon cord	0.0000	0.00
17	Aluminum tube	0.0029	0.01
18	Plastic partition board	0.0002	0.00
Packaging of product packaging materials			
1	Wrapping paper	0.0125	0.04
2	Poly straps	0.0073	0.02
3	PE plastic bag	0.0043	0.01
4	Plastic woven bag	0.0010	0.00
5	Paper roll core	0.0003	0.00
6	Rubber band	0.0000	0.00
7	Corrugated carton	0.0001	0.00
Transportation of components and consumables			
1	Bottom cover Upper lid (silver)	0.0003	0.00
2	Aluminum plate (round shape)	0.0010	0.00

Continued

Table 5.2 Carbon footprint results from "cradle to gate" life cycle stages—cont'd

No.	Item	Carbon footprint (kg CO_{2eq}/unit)	Percentage of total carbon footprint (%)
3	IGBT heat sink	0.0057	0.02
	Ring core choke coil inductor		
4	Heating coil plate	0.0002	0.00
5	Copper wire	0.0045	0.01
6	Solder	0.0019	0.01
7	Bridge diode	0.0145	0.04
8	Diode	0.0002	0.00
	Zener diode		
9	Diode	0.0000	0.00
10	IGBT	0.0013	0.00
11	Triode	0.0001	0.00
	Integrated circuit		
12	Transistor	0.0001	0.00
13	Integrated circuit	0.0005	0.00
14	Integrated circuit	0.0001	0.00
15	Carbon film resistor	0.0016	0.00
	Metal film resistor		
	Metal oxide film resistor		
	Wirewound resistor (high pressure metal glass glaze)		
16	Rheostat	0.0001	0.00
17	Ceramic capacitor	0.0008	0.00
	Plastic capacitor		
	Electrolytic capacitor		
18	Electrolytic capacitor	0.0004	0.00
19	Plastic capacitor	0.0082	0.02
	X capacitor		
	Iron core (high conductivity)		
	Transformer (horizontal)		
20	Wire set	0.0013	0.00
	Plastic housing connector		
	Wire		
	Copper terminal		
	Connector terminal		
21	Wire set shielded line	0.0006	0.00
22	R-type connector terminal	0.0044	0.01
23	Jump wire	0.0002	0.00
24	Constantan wire	0.0001	0.00
	Tin coated copper jump wire		
25	Power cord	0.0075	0.02
26	Rubber pad	0.0004	0.00
	Silicone mushroom head (tall)		

Table 5.2 Carbon footprint results from "cradle to gate" life cycle stages—cont'd

No.	Item	Carbon footprint (kg CO_{2eq}/unit)	Percentage of total carbon footprint (%)
27	Power plug	0.0037	0.01
28	Silicone fibreglass sleeving	0.0000	0.00
29	PVC tube	0.0000	0.00
30	Tie wrap	0.0000	0.00
31	Manufacturing sticker Serial number sticker	0.0000	0.00
32	Glass ceramic board	0.0784	0.23
33	Touch spring U shape terminal	0.0012	0.00
34	Silicone pad	0.0031	0.01
35	Thermal fuse	0.0017	0.01
36	Silicone mushroom head round pad	0.0001	0.00
37	Silicone	0.0001	0.00
38	Soldering flux Polishing detergent	0.0000	0.00
39	Silicone (transparent)	0.0001	0.00
40	Fuse	0.0002	0.00
41	Two digit segment LED display One digit segment LED display	0.0004	0.00
42	Colored LED high brightness	0.0010	0.00
43	DC brushless fan	0.0034	0.01
44	Single side PCB PCB PCB board	0.026	0.01
45	Thermistor	0.0000	0.00
46	Buzzer	0.0002	0.00
47	H word tapping screw Tapping screw Pan head tapping screw Machine screw (round head)	0.0028	0.01
48	Cleansing cotton cloth	0.0001	0.00
49	Thermal Grease	0.0001	0.00
Transportation of product packaging			
1	User manual	0.0013	0.00
2	Printed carton box	0.0034	0.01
3	Outer box	0.0012	0.00
4	User manual packaging bag	0.0002	0.00

Continued

Table 5.2 Carbon footprint results from "cradle to gate" life cycle stages—cont'd

No.	Item	Carbon footprint (kg CO_{2eq}/unit)	Percentage of total carbon footprint (%)
5	Product packaging bag		
6	Packaging foam	0.0001	0.00
7	Polystyrene styrofoam		
8	Sealing tape		
9	Plastic handle	0.0001	0.00
	Subtotal	32.5671	96.81
Manufacturing stage			
Electricity consumption			
1	Electricity	1.0715	3.19
Transportation of waste			
1	Metal offcuts	0.0027	0.01
2	Plastic materials		
3	Others (cardboard, boxes, paper, etc.)		
	Subtotal	1.0742	3.19
Transportation stage (for reference only)			
1	Transportation of final product	0.4474	NA
	Subtotal	0.4474	NA
Total product carbon footprint (excluding distribution)		33.6413	100

5.6 LIFE CYCLE INTERPRETATION

5.6.1 Results of Life Cycle Interpretation

1. The carbon footprint per f.u. (one unit of induction cooker with packaging) is 33.6413 kg CO_{2eq}. The summary of the cooker's carbon footprint for each life cycle is presented in Table 5.3.
2. The "raw material" stage is the most significant life cycle stage and contributes to a significant amount of carbon emission (96.81%) in the whole life cycle. In the "raw material" stage, the integrated circuit (12.47%) has a significant contribution, as shown in Table 5.2 and Fig. 5.5.
3. The "manufacturing" stage contributes 3.19% of total product carbon emission.

5.6.2 Specific Greenhouse Gas Emissions and Removals

Table 5.4 separately shows the specific GHG emissions and removals arising from this study.

Table 5.3 Contribution of carbon footprint for each product life cycle stage

Life cycle stage		Carbon footprint (kg CO_{2eq}/unit)		Contribution to total product carbon footprint (%)	
Raw materials	Components and consumables	30.4633	32.5671	90.55	96.81
	Product packaging	1.4390		4.28	
	Packaging of components, consumables, and product packaging	0.5030		1.50	
	Transportation of components, consumables, product packaging, and corresponding packaging materials	0.1618		0.48	
Manufacturing process		1.0742		3.19	
Transportation (for reference only)		0.4474		NA	
Total product carbon footprint (excluding transportation)		33.6413		100.00	

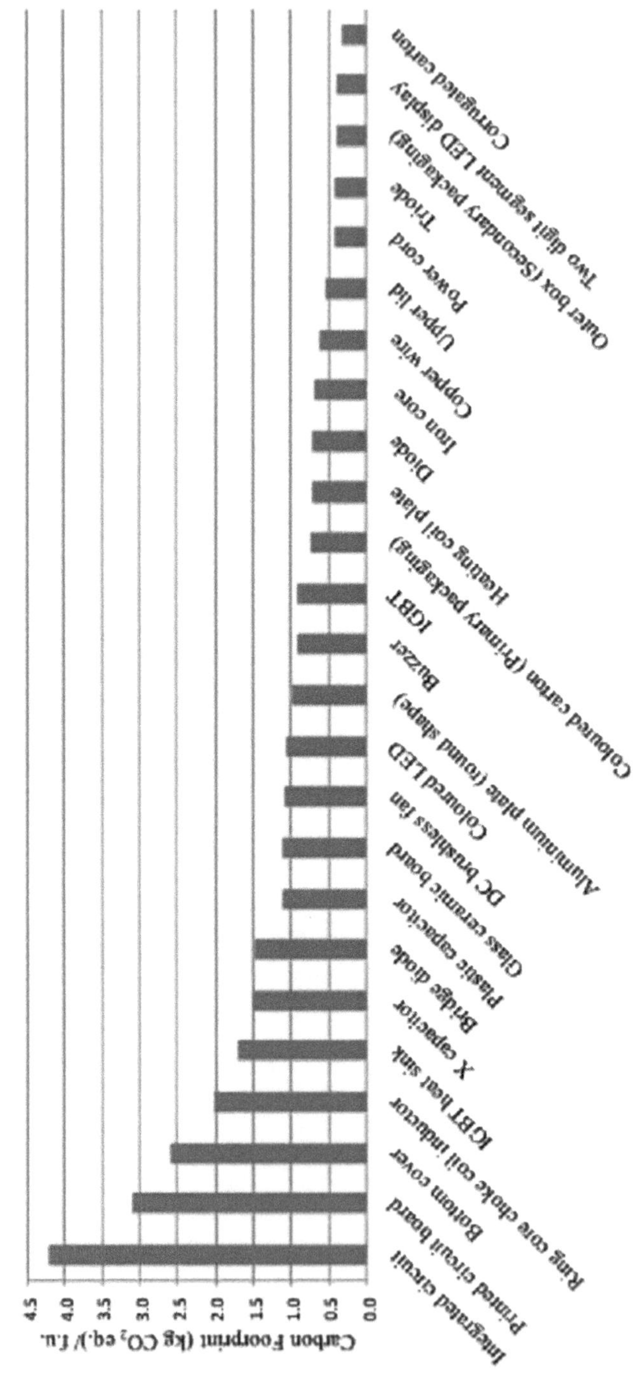

Figure 5.5 Contribution of carbon footprint by components in the raw material stage. *IGBT*, insulated gate bipolar transistor.

5.6.3 Limitations

5.6.3.1 Focus on a Single Environmental Issue

The PCF only reflects GHG emissions and removals of a product system, expressed as CO_{2eq}, associated with raw material acquisition, manufacturing, and transportation of a product. Although PCF can be an important environmental aspect of the life cycle of a product, affecting "climate," a product's life cycle may have other environmental impacts that are also of concern (e.g., resource depletion, air, water, soil, and ecosystems).

5.6.3.2 Limitations Related to the Assumptions

Assumptions were made in each stage of this study that may affect the absolute accuracy of the final result.

5.6.3.3 Limitations Related to the Methodology

The PCF was calculated based on LCA methodology. ISO 14040 and ISO 14044 address the inherent limitations and trade-offs in LCA methodology. These include the establishment of an f.u. and the system boundary, the availability and selection of appropriate data sources, allocation rules, and assumptions regarding the transport. Some of the chosen data may be limited to a specific geographical area (e.g., national electricity grid) and/or may vary in time (e.g., seasonal variations). The choice of the f.u. and allocation rules may also be important in more accurately modeling the life cycle.

5.6.4 Disclosure and Justification of Value Choices

One of the important value choices is the database selection. The priority for the database selection was based on credibility, geographical, technological,

Table 5.4 Specific greenhouse gas emissions and removals arising from this study

Source	Emission and removals
Fossil carbon sources and sinks	NA
Biogenic carbon sources and sinks	NA
Direct land use change	NA
Aircraft transportation	NA

and time-related considerations. In this report, emission factor for electricity consumption is obtained from NDRC 2013, and the data used for components, consumables, packaging materials, and transportation are obtained from Ecoinvent version 2.2.

5.6.5 Sensitivity Analysis

This section describes the sensitivity analysis regarding the influence of choice of emission factor database on the elements with most significant contributions to the GHG emissions, and electricity, on the final PCF result.

5.6.5.1 Sensitivity Analysis of Significant Input Data in Material Stage

As shown in Table 5.3, the integrated circuits (IC) and printed circuit boards (PCB) used in the material stage and the electricity in manufacturing stage contribute large amounts of GHG emissions, and their emission factors are secondary data, which are not localized. Sensitivity analysis was conducted for these three inventories. Different emission factors were used to analyze the sensitivity. For IC, Table 5.5 and Fig. 5.6 show the different carbon footprint results when changing the emission factors of the IC from Ecoinvent to JEMAI.

For PCB, Table 5.6 and Fig. 5.7 show the different carbon footprint results when changing the emission factors of the PCB from Ecoinvent to Korean LCI 2008.

5.6.5.2 Sensitivity Analysis of Significant Input Data in Manufacturing Stage

Table 5.7 and Fig. 5.8 show different carbon footprint results when changing the electricity emission factor provided by NDRC to the emission factors obtained from Ecoinvent v2.2 and from a Chinese research journal (Hou et al., 2012), respectively.

This section showed how the changes of emission factor database selection affect the end result (PCF) from selected product life cycle stages in this study.

5.6.6 Uncertainty Analysis

In this study, the data includes primary data and secondary data. Some uncertainties may exist. Table 5.8 summarizes the type of data and the possible errors that may contribute to the uncertainty of the results.

Table 5.5 Result of changing emission factor of integrated circuits

Material	Amount (g)	Database	Name in database	Emission factor (kg CO_{2eq}/kg)	Carbon emission (kg CO_{2eq})/f.u.	Final carbon footprint result (kg CO_{2eq}/f.u.)
Integrated circuit	4.14	Ecoinvent v2.2	Integrated circuit, IC, logic type, at plant [global]	1013.3	4.1951	33.6413
		JEMAI	Standard logic integrated circuit	0.1003	0.5015	29.9477

Table 5.6 Result of changing emission factor of printed circuit boards

Material	Amount (g)	Database	Name in database	Emission factor (kg CO_{2eq}/kg)	Carbon emission (kg CO_{2eq})/f.u.	Final carbon footprint result (kg CO_{2eq}/f.u.)
Printed circuit boards	74.28	Ecoinvent v2.2	Printed wiring board, through-hole, at plant [global]	41.69	3.0969	33.6413
		Korean LCI 2008	Printed circuit board	10.2	0.7576	31.3020

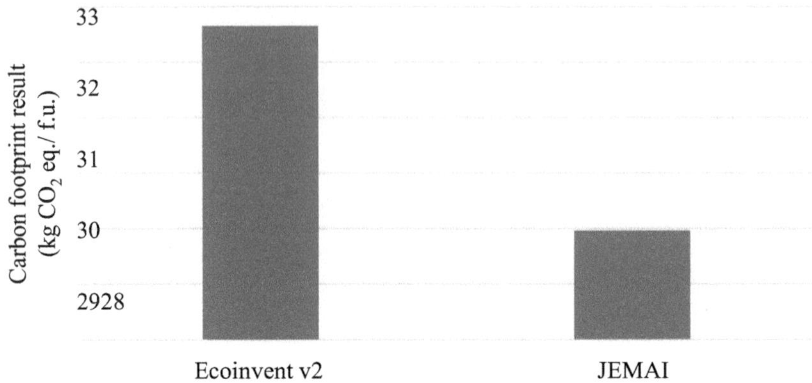

Figure 5.6 Result of changing emission factor of integrated circuits.

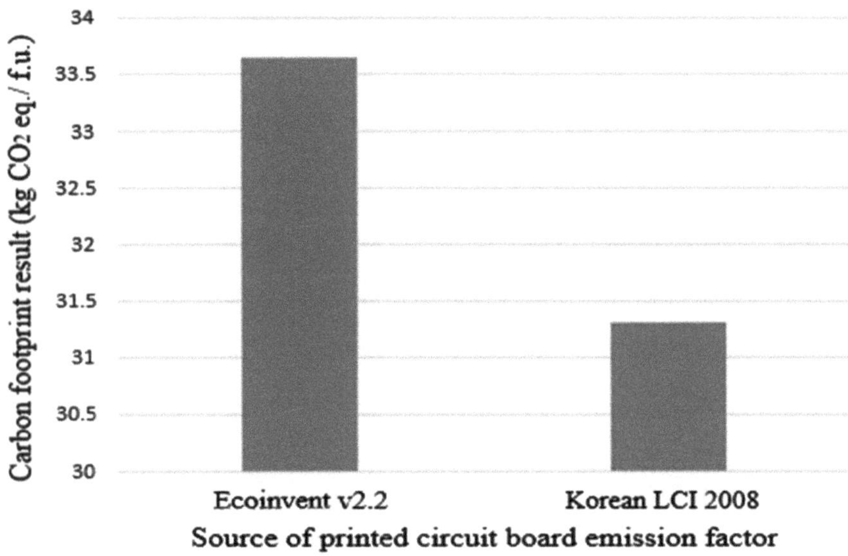

Figure 5.7 Result of changing emission factor of printed circuit boards.

Table 5.7 Result of changing emission factor of electricity used in domestic induction cooker manufacturing processes

Electricity consumption	Database	Emission factor (kg CO_{2eq}/kWh)	Carbon emission (kg CO_{2eq}/f.u.)	Final carbon footprint result (kg CO_{2eq}/f.u.)
1.1618	NDRC 2013	0.9223	1.0715	33.6413
	Ecoinvent v2.2	1.51	1.7543	34.3241
	Hou Ping et al. (2012)	0.853	0.9910	33.5608

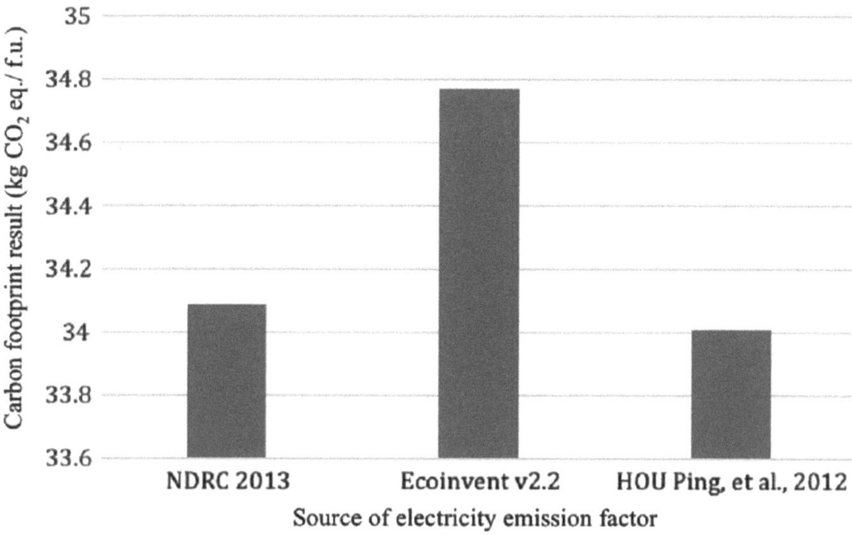

Figure 5.8 Result of changing emission factor of electricity used in manufacturing.

Table 5.8 Summary of type of data and the possible error that may contribute to the uncertainty of the results

Type of data	Type of error contributing uncertainty to results
Process data	Allocation of electricity for manufacturing one unit of scale
Process data	Seasonal factor of production
Process data	Scenarios adopted in use stage and end-of-life stage may be different with the reality
Emission factor	Localized emission factor is the first choice and only adopts other emission factors if localized emission factor is not applicable or available

5.7 CONCLUSION

According to the results, IC (12.47%) and PCB (9.219%) have large contributions to the total GHG emissions from "cradle to gate" life cycle stages. A better design of product with IC and PCB usage might be adopted. Also, alternative materials may be employed in substation for components such as bottom cover (7.65%), and glass ceramic board (3.28%) could reduce the GHG emissions because of its comparatively high emission value. The carbon footprint results reflect that any improvement in the raw material stage (96.81%) may effectively reduce the carbon footprint of the product. Thus, the design and the usage of material are critical in eco-product design to reduce the PCF. Hence, we hope that this PCF case study of personal electronic product such as induction cooker will help many manufacturers who are keen in delivering green products to their consumers in right direction, facilitating green production and consumption. Also these PCF studies act as an inventory source for future LCA studies covering more impact categories including a comprehensive evaluation.

APPENDIX

Photos taken onsite in Sunpentown (Guangzhou) Appliance Co., Ltd. (Photos 1–8).

Photo 1 Manufacturing line of domestic models of induction cookers.

Photo 2 Factory warehouse.

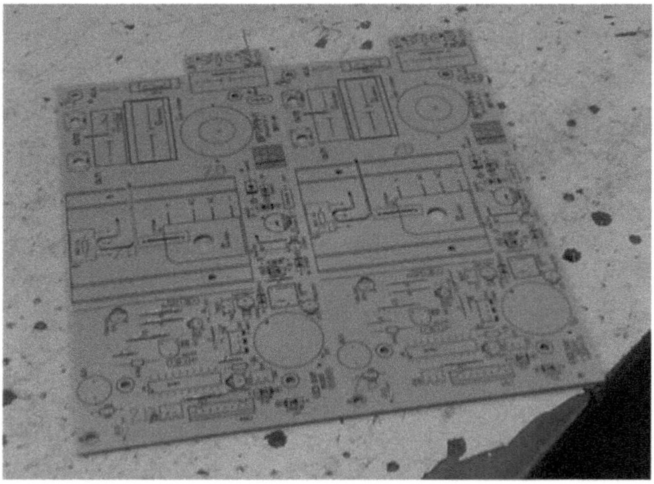

Photo 3 Component (printed circuit boards) of induction cooker.

Photo 4 Packaging materials of components.

Photo 5 Manufacturing process of induction cooker.

Photo 6 Manufacturing process of induction cooker.

Photo 7 Quality testing process.

Photo 8 Packaging materials of components.

ACKNOWLEDGMENTS

This work is supported by the Trade and Industry Department of HKSAR Government under **SME Development Fund**, "Development of an embedded Greenhouse Gas (GHG) emissions database with a G-BOM analyzer and a SME advisory kit for electrical and electronic industries to respond to the implementation and compliance of ISO 14067 (carbon footprint of products)."

We would like to express our thanks to SGS Hong Kong Limited for their support of the Project "*Development of an embedded Greenhouse Gas (GHG) emissions database with a G-BOM analyzer and a SME advisory kit for electrical and electronic industries to respond to the implementation and compliance of ISO 14067 (carbon footprint of products).*"

This case study report (Brand: Sunpentown, Model: EH2000T) of an induction cooker's product carbon footprint was made possible by the technical and professional contribution of our collaborating organization: **SGS Hong Kong Limited**. We would like to thank the above organization who has given valuable technical support and professional advices and has participated on site assessment.

REFERENCES

Alibaba, 2014. New Plastic Nylon 66 Housing Electrical Connector. Available from: http://www.alibaba.com/product-detail/Sineyi-093-Yuyao-2014-New-plastic_257523226.html.

Blackshaw Inc., (n.d.). 3 Pin Plug Manufacture. Available from: http://blackshawinc.weebly.com/plugs.html.

Cable Tie Express, 2014. Quality Cable Ties, Tie Wraps and Zip Ties. Available from: http://www.cabletieexpress.com/cable-ties.

European Commission, 2012. Directive 2012/19/EU of the European Parliament and of the Council on Waste Electrical and Electronic Equipment (WEEE) Recast.

International Organization for Standardization (ISO), 2006. ISO 14040:2006 Environmental Management – Life Cycle Assessment – Principles and Framework.
International Organization for Standardization (ISO), 2013. ISO/TS 14067:2013 Greenhouse Gases – Carbon Footprint of Products – Requirements and Guidelines for Quantification and Communication.
National Development and Reform Commission, 2013. 2013 Baseline Emission Factors for Regional Power Grids in China. Available from: http://www.ccchina.gov.cn/archiver/cdmcn/UpFile/Files/Default/20130917081426863466.pdf.
Hou, P., Wang, H.-T., Zhang, H., Fan, C.-D., Huang, N., 2012. Greenhouse gas emission factors of Chinese power grids for organization and product carbon footprint. China Environmental Science 32, 961–967.
Rapidonline, (n.d.). Constantan Wire – An Introduction to Constantan and its Applications. Available from: http://www.rapidonline.com/pdf/05-0400e.pdf.
Taitron, (n.d.). Tinned-Copper Jumper Wire. Available from: http://www.taitroncomponents.com/Catalog/Datasheet/JPW06T7526.pdf.
Wikipedia, 2014. Polybutylene Terephthalate. Available from: http://en.wikipedia.org/wiki/Polybutylene_terephthalate.

FURTHER READING

British Standards Institution (BSI), 2011. PAS 2050:2011 Specification for the Assessment of the Life Cycle Greenhouse Gas Emission of Goods and Services.

CHAPTER 6

Carbon Footprint Analysis of a Selected Indian Power Plant

Debrupa Chakraborty
Netaji Nagar College, Kolkata, India

6.1 INTRODUCTION

The concept of carbon footprint (CF) was introduced over more than a decade ago though the concept has become popular since 2005. CF has become one of the more important environmental protection indicators over the past few years (Wiedmann and Minx, 2008) and commonly signifies a certain amount of gaseous emissions that are relevant to climate change. It is associated with human production and consumption activities years (Wiedmann and Minx, 2008). It is a measure of the total amount of carbon dioxide (CO_2) emissions that is directly or indirectly caused by an activity or is accumulated over the life cycles stages of product. This includes activities of individuals, population, governments, companies, organizations, processes, industry, sectors, etc. In all of the above cases all direct (on-site, internal) and indirect (off-site, external, embodied, upstream, downstream) emissions are needed to be taken into account (Wiedmann et al., 2007b).

Currently the CF concept is being widely used as a way of reducing greenhouse gas (GHG) emissions by governments, companies, nongovernment organizations, and private authority's response to the increasing concern about global warming and climate change. This is all more so as CO_2 is the single most human—emitted GHG in the atmosphere contributing around 63.5% to the total global radioactive forcing circulating within the earth's surface(World Meteorological Organization, 2009) The response strategies are mainly of two kinds (1) reduction and (2) offsetting. Reduction refers to undertaking activities in ales carbon intensive way. Offsetting refers to taking external actions to compensate for CF by means of adopting ways of emission reduction or carbon capture elsewhere by others. Both of these strategies are widely supported by governments and businesses across the globe. Businesses, governments, and policy makers

Environmental Carbon Footprints
ISBN 978-0-12-812849-7
http://dx.doi.org/10.1016/B978-0-12-812849-7.00006-4

across the globe have started giving attention to the need for taking actions to mitigate global warming and climate change.

It is important to know the impact on ecological resources and nonrenewable energy consumed by the industries in their production process and CO_2 emissions resulting from such processes. Electricity production contributes toward 37% of the total global emissions according to the Intergovernmental Panel on Climate Change (www.world.nuclear.org), and electricity demand is expected to increase by 43% over the next 20 years (WNA, 2011). There are a number of different electrical generation methods, each one of which produces GHG in varying quantities through construction, operation, and decommissioning. Coal-based generation power plants release the majority of GHGs during operation, whereas wind and nuclear power plants release the majority of emissions during construction and decommissioning. Life cycle approach encompasses accounting for emission from all the above three phases of electricity generation (www.world.nuclear.org).

Indian power sector has recorded an impressive growth over the last few decades from 1713 MW in 1950 to 199,877.03 MW as on March 31, 2012. Out of the total capacity generation of 199,877.03 MW in 2012 sector wise distribution of installed capacity reveals that thermal power plant occupies a lion's share (131,603.18 MW) followed by hydro (38,990.40 MW) taking the second position and renewable (24,503.45 MW) and nuclear (4780 MW) occupying the third and fourth positions(CEA, 2013). Again, installed capacity is predominantly coal based and therefore is a major source of CO_2 emissions in India. Hence, there exists scope for reducing the CO_2 emissions in the country by way of fuel substitution, increased use of renewable energy sources, and also by improving the thermal efficiency of power generation (CEA, 2013). The Indian power system has divided into five independent regional grids, namely Northern, Eastern, Western, Southern, and North-Eastern. Each grid covered several states. The case study thermal power generation unit belongs to the eastern grid.

A number of methodologies for calculating CF developed by government or businesses, varying in their approach and mode of functioning are in operation or in the course of being developed. Of the various models for calculating the CF of a product or process, the life cycle assessment (LCA)—Principles and framework (International Standard Organization (ISO) 14040/14044: 2006), ISO 14067, and British Publicly Available Specification (PAS) 2050 framework are the most significant ones. ISO 14067 is the most robust and the more general standard, with some specific requirements e.g., requirements on the use of green electricity. On the other hand, PAS 2050 and LCA model provide more detailed requirements and guidance with less space for interpretation. CF is calculated using LCA or input—output analysis (Hertwich and Peters, 2009). LCA is a bottom-up method, which has been developed to understand the environmental impacts of individual products from cradle to grave process based (LCA): from resource extraction and processing, through construction, manufacturing and retail, distribution and use, repair and maintenance, disposal and reuse/recycling. LCA approach suffers from difficulties when used for deriving CF of larger entities such as household, industrial sectors, or governments. So this method is suitable for assessment of CF at the micro level (i.e., an individual, product, a particular process, individual businesses, a small group of individual products) because they suffer from a system boundary problem (Weidmann and Minx, 2008). CF of an organization or a company can be quantified by adopting LCA described in ISO standard.

Environmental Input—Output (EIO) analysis provides an alternative top-down approach to CF taking into account the entire supply chain. Input—output tables are economic accounts providing a picture of all economic activities at the meso (sector) level. The EIO analysis is superior for establishment of CF's of industrial sectors, larger product groups, household, government, etc (Foran et al., 2005a; SEI et al., 2006; Wiedmann et al., 2007a,b). Whatever method is used to calculate CF it is important to avoid double counting along supply chain and life cycles. Because they will have significant impacts on the practices of carbon trading and carbon offsetting (Lenzen, 2007; Lenzen et al., 2007). A big advantage of input—output based approach is that it requires small amount of time and manpower once proper data are fitted the model but it lack the precision at the process level.

The remaining sections of the paper are organized as follows: Section 6.2 discusses prior research on CF accounting; Section 6.3 explains the goals and objectives of the study. Data sources and research methodology used for the study have been discussed in Section 6.4, Section 6.5 presents the results and analysis, conclusion are drawn in the final Section 6.6.

6.2 REVIEW OF LITERATURE

In literature CF has been estimated at national level, (Lenzen et al., 2004; Peters and Hertwich, 2006, 2008; Wiedmann et al., 2007a,b, 2010) sub-national level, municipalities, and cities (Kennedy et al., 2009); at the sector level (Lenzen, 2003; Foran et al., 2005b; Peters and Hertwich, 2006; Peters et al., 2009) for lifestyle groups (Duchin, 1998; Minx and Baiocchi, 2009) and for businesses (Weidmann and Lenzen, 2006; Wiedmann et al., 2007a,b; Cen, 2009; Penela et al., 2009; Wood and Dey, 2009; Eckel, 2007). CF has been quantified by adopting LCA approach—at the national level for electricity generation in Ontario (CERI, 2008); in Sweden (Vattenfall, 2005); biomass-based energy system in Denmark (Tonini and Astrup, 2012). At the industrial level LCA methodology has been used for estimating GHG emissions of—coal fired electricity generation (Whitaker et al., 2012), wind electricity (Pandey et al., 2012); various sources of electricity generation (Cameo Corporation, 2010). LCA study has been made on personal computers(Atlantic Consulting and IPU, 1998); for identifying environmental impacts of color TV sets in China LCA approach has been adopted(Feng and Ma, 2009); LCA of personal computers and its effective recycling rate have been studied (Choi et al., 2006) and (Lu et al., 2006). Measurement of direct and indirect GHG emissions has been done for coal-based Indian thermal power plants (Chakraborty et al., 2008). The potential of using sawdust or its low temperature char as cofuel with coal for generation of electricity in Indian power plants as a part of renewable energy development program in India has been studied (Sarkar et al., 2014). Impact of using of different biomass feedstocks in the gasifier as fuel in thermal power plants in India and bioenergy maximum generation rate, and the minimum specific fuel cost has been estimated (Roy et al., 2013). Literature survey in the Indian context reveals that considerable gap in knowledge exists in the performance evaluation of the industries through resource use assessments. This paper makes an attempt toward estimation of CF for one thermal power production unit in India.

6.3 GOALS AND OBJECTIVES

This study aims to provide a comprehensive scope of analysis of the CO_2 emissions resulting from the case study unit, the results from which can help toward decision-making and identification of "hotspots" and sustainable options. The case study thermal power plant is situated in West Bengal having a total area of 750 acres and built up area of 600 acres. It is one of the most reliable and prestigious coal-fired power plants in West Bengal in India as well. In two stages the total-generation capacity of the plant is (5 units × 210 MW), i.e., 1050 MW. The unit has a total generating capacity of 7300 Million Unit (MU) of electricity. However during the year 2015—16 (taking 365 working days) the power plant generated 6669 MU of electricity. The study unit has been chosen for the present study is a good representation of Indian thermal power generating units and belongs to a company that is fore runner so far as modernization and sustainable development goals are concerned. This paper demonstrates the ways where CF can be estimated using unit level data collected from the specific industry. Final choice of the study unit has been determined by willingness of the unit to cooperate in data sharing and time commitment for verbal communication.

6.4 RESEARCH METHODOLOGY AND DATA SOURCES

6.4.1 Research Methodology

It was decided that LCA Approach to be adopted as this methodology allows quantification of CF of processes with great deal of precision. The leading standards for LCA are ISO 14040 and ISO 14044. These international standards focus mainly on the process of performing an LCA and its principles and framework is divided into in four following phases (Klemes, 2015):

Goal and scope definition; life cycle inventory (LCI); life cycle impact assessment (LCIA); and interpretation.

1. Goal and scope definition—During the first phase the goals and objectives of the LCA and the systems boundaries must be clearly defined. This step includes defining the functional unit, life cycle stages, assumptions and limitations, the methods that are chosen for allocating the impacts of processes and impact categories.
2. LCI—In this second phase data relating to input (raw material, energy, water, etc.) and output (product, emission to air, soil, and water) are collected, recorded, and quantified. All data should relate to the functional unit stated in the first phase and must be expressed on a common unit.
3. LCIA—The third phase involves evaluation of the significances of the impact of all the outputs of the LCI on the environment. This includes impact of using resources, GHG/CO_2 emissions, land use, etc.
4. Interpretation—In this last phase of LCA study the results of LCI and LCIA are to be summarized and evaluated in a systematic way taking into consideration the completeness, sensitivities, and consistencies of the study. Results should be interpreted in such a way so as to identify the areas that have potential for improvement, proposing recommendations in the process.

Selecting boundaries is very important for LCA of any process or product. The wider the boundary the greater will be the precision of obtained results depending on the availability of required detailed data. The possible range of boundaries is given as under (Klemes, 2015):

- Cradle to cradle: From resource extraction (cradle) to recycling of new product (100% utilization of wastes)
- Cradle to grave: From resource extraction (cradle) to disposal (grave)
- Cradle to gate: From extraction of resources (cradle) to factory gate before sending the final products to the customers (gate) excludes the use and disposal phases.
- Gate to gate: this option includes only one process within the entire production chain i.e., processing within a factory

However, the preferred options are "cradle to cradle" and "cradle to grave". However, cradle-to-grave approach has been used in this paper as the study boundary for calculation of emission.

6.4.2 Scope and Functional Unit

The process of electricity generation has been selected for this study. The functional unit is 6669 MU i.e., 6669 million kWh of electricity generated during the year 2015–16 (taking 365 days as working days). 1 MU = 10,00,000 kWh of generated electricity.

6.4.3 System Boundary

Organizational boundary of this study encompasses the GHG (only CO_2 emissions), emitted by the facilities while generating electricity. The operational boundary on the other hand includes the *Cradle-to-grave* approach. The complete life cycle comprising of

1. transportation of coal (hard grade C/F), fuel (light diesel oil or heavy fuel oil), and chemicals into the factory premises;
2. the power plant's operation and service facilities.

   For service facilities energy requirements related to

   a. power plants' administrative office
   b. the employees' business trips

   So emission from production process includes emission from coal, fuel used in the production process and emission from auxiliary operations related to production. Auxiliary operations include electricity consumed during production process, in sealing and cooling operations, electricity consumed in administrative office. Auxiliary operation constitutes 9.5% of total 6669 MUs of electricity generated by the plant.

   The amount of energy used for business travel by the employees required for the proper functioning of the power plant is also considered though less significant as compared with other items under study. The mode of transport availed by the employees are bikes, cars, and bicycles.

3. Transportation and disposal of the wastes generated was taken into account in calculating the CF of the unit. Only the CO_2 produced during the process has been taken into consideration. There are three types of waste generated by the case study power plant unit—wet ash, dry fly ash, and mill rejects.

   Once boundaries are selected, relevant primary data are collected from the factory managers of the case study unit using a preset questionnaire, and CO_2 emissions are quantified. Current paper uses activity data based on the primary data collected from the factory managers of the case study unit, and the emission factors provided by a number of secondary sources.

Emission factors have mainly being taken from secondary sources for e.g.,

- Emissions resulting from transportation of raw materials, waste, and employees business' travel are calculated by taking into account the emission factors published in India GHG Program, 2015, Version 1.0 by World Resource Institute (WRI) India, The Energy and Resources Institute (TERI), and Confederation of Indian Industry (CII).
- For calculating emission generated from coal, fuel used in production process emission factors published in Guidance for Voluntary Corporate Greenhouse Gas Reporting – 2015 by Ministry of Environment and GHG Program, 2015, Version 1.0 by WRI India, TERI, and CII as these are Indian industry—specific emission factors.
- Emissions from disposal of wastes have been calculated by using emission factors published by the PAS 2050: 2011 guidelines developed by British Standard Institute, cosponsored by the Carbon Trust and Department for Environment, Food and Rural Affairs and emission factor published by GHG Program, 2015, Version 1.0 by WRI India, TERI, and CII.

Emission factors quantify the emissions resulting from individual processes (as CO_2e). GHG emission for each processes are calculated from the product of the organization's activity data and the corresponding and relevant emission factors (expressed in units of measurement) are used. At this stage, all the data having relevance to this case study are collected. Using the following formula the CF has been calculated:

Activity data (kg/liters/kWh/tkm, etc.) × emission factor (kg CO_2e/kg/liters/kWh/tkm, etc.)

All these data are summed up so as to provide the total CF for each life cycle stage and for the total system.

6.4.4 Data Sources

Data for calculating CF in this study have been gathered verbally from the case study production unit using a preset questionnaire (questionnaire is available from the corresponding author on request). CF of the thermal power production unit for the year 2015–16 encompasses emission from both direct and indirect sources of energy and direct and indirect CO_2eq emissions.

- Direct emission (e.g., from the stack of power plant, operation of power plant)
- Indirect emission related
  - to upstream provisions of fuel, resources, goods, etc. (e.g., transportation of raw materials, the employees business trips, and energy requirement in administrative office)
  - downstream management of wastes and by-products

Required data, sources of collection and data sources for emission factors for estimating process CF using LCA has been shown in Table 6.1.

6.4.5 Assumptions

6.4.5.1 Assumptions in the Transportation of Raw Material Stage

- Average distance of raw material transportation provided by the managers was taken as the actual distance for calculation of CF.

Table 6.1 Required data, sources of collection, and emission factors for estimating process carbon footprint using life cycle assessment

Data required	Sources of collection	Sources of emission factor
Primary data		
Transportation of raw materials into the factory premises; **1.** Coal (hard grade C/F), **2.** Fuel (light diesel oil or heavy fuel oil)	Primary data from the case study unit	Secondary data source
3. Power plant's operation **4.** Service facilities energy requirements related to **a.** Power plants' administrative office **b.** Employees' business trips	Primary data from the case study unit	Secondary data source
5. Transportation and disposal of the wastes generated	Primary data from the case study unit	Secondary data source

- For calculating emission from transportation of fuel oil the emission factor of km was taken as a unit instead of tkm as density of oil could not be measured in km.
- For transportation of chemicals by road, it is assumed that chemicals have been transported by 100% laden heavy duty vehicles (HDV) of more than 12 MT of carrying capacity.
- Distances from which raw materials have been transported to the power plant were obtained from Google Maps.

6.4.5.2 Assumptions During the Power Plant Operation and Its Service Facilities

Power plant operation—Production stage

- Standard reheat regenerative ranking cycle has been considered.
- Standard mass balance with losses for 210 MW plant has been considered.
- Auxiliary consumption is considered to be 9.5% of generation.
- Data gathered mostly over verbal discussion, standard human error may prevail.
- For calculation emissions from employees' business trips, it is assumed that the scooters and bikes are of more than 1500 cc engine capacity, and cars belong to the category of small cars having less than 1000 cc engine capacity with an uplift factor.
- The estimations were made by the managers and engineers of the power plant.

6.4.5.3 Assumptions During the Transportation and Through Disposal of Waste Materials From the Power Plant

- Average distance of raw material transportation provided by the managers was taken as the actual distance for calculation of CF.
- It is assumed that both dry and wet ashes are transported in 100% laden dumper truck.
- Mill rejects constitute of a mixture of different types of wastes the default emission factor (also a national average in India) is taken.
- Distances to which wastes have been transported from the power plant were obtained from Google Maps.

6.5 RESULTS AND ANALYSIS

The detailed calculation of CF encompasses the components of production with corresponding units, emission factors used, sources of used emission factors, and the total emission in each stage of production for generating 6669 MUs i.e., 6669 million kWh of electricity, which has been presented in Table 6.2.

6.5.1 Carbon Dioxide Emissions Originating From Raw Material Transportation

The raw materials i.e., coal and fuel (light diesel oil) was transported to the power plant from various suppliers. Coal India Ltd. is the largest supplier of coal supplying 60% of the total coal requirement, 10% of coal is procured through e-auction, and the rest 30% provided by other suppliers namely Mahanadi Coalfields Limited, Bharat Coking Coal Limited, and Eastern Coalfields Limited. All of these coalfields and suppliers of coal are located in India. Emission from use of coal has been calculated based on the assumption that noncoking coal is used by the unit. In 2015−16 the quantity of coal and fuel was 46,23,048 Mt and 14,365 kL of light vehicle oil, respectively. Coal is transported from various locations (distance from different locations ranging from 20 to 250 km). So an average of 100−150 km has been taken for transportation of 46,23,048 Mt coal to the power plant from various locations by rail. Emissions from transportation of coal by rail have been calculated by using emission factor of 0.0095 kgCO_2e/tkm [India GHG Program, 2015, Version 1.0].

The total quantity of fuel (14,365 KL) is transported to the power plant from a distance of 600 kms on an average from the oil refineries by rail. This is done twice a year. So fuel is transported from a distance of 1200 km in 2015−16. It is assumed that fully loaded goods train run by electricity are used for transportation. So for calculating emissions from transportation of fuel (in liters) by rail, emission factor of 0.7353 kg CO_2e/km have been used.

Details of transportation of chemicals (quantity and distance from which they are transported) used in production process are given in Table 6.3. Chemicals have been transported from a distance of 50−250 km on an average, and approximately 1599 MT of chemicals were transported to the power plant during 2015−16 by road. It is assumed that chemicals have been transported by 100% laden HDV of more than 12 MT of carrying capacity. So emission from transportation of raw material by road has been calculated by using emission factor of freight vehicle, which is 0.0143 kg CO_2e/tkm [India GHG Program, 2015, Version 1.0].

Total annual emissions from transportation of raw materials (coal, fuel, and chemicals) in 2015−16 was **26,028 tCO_2e.**

Table 6.2 Detailed calculation of carbon footprint of power plant

Components of production	Quantity (total: 1050 MW electricity generation)	Unit	Emission factor	Reference/source	GHG emission (tCO_2e) for 6669 million unit (MU) = 6669 million kWh of electricity generation
Coal(by rail)	273,7500,000	tkm	0.0095 $kgCO_2e/tkm$	India GHG Program, 2015, Version 1.0	26,006
Fuel oil: light diesel oil (by rail)	1200	Km	0.7353 kg CO_2e/km	India GHG Program, 2015, Version 1.0	1
Transportation of chemicals used in production process (by road, 100% laden freight)	145,609	tkm	0.143 kg CO_2e/tkm	India GHG Program, 2015, Version 1.0	21
Subtotal (A)					**26,028**
Coal used in boiler (bituminous coal)	46,23,048	MT	2.65 $kgCO_2e/kg$	Guidance for Voluntary Corporate Greenhouse Gas Reporting – 2015	122,51,077
Fuel used	143,65,000	liters	2.95 $kgCO_2e/L$	Guidance for Voluntary Corporate Greenhouse Gas Reporting – 2015	42,377
Subtotal (B1)					**122,93,454**

Continued

Table 6.2 Detailed calculation of carbon footprint of power plant—cont'd

Components of production	Quantity (total: 1050 MW electricity generation)	Unit	Emission factor	Reference/source	GHG emission (tCO_2e) for 6669 million unit (MU) = 6669 million kWh of electricity generation
Employees business travel					
Bikes and scooter	11,52,000	Passenger kms	0.0387 $kgCO_2e$/pax-km	India GHG Program, 2015, Version 1.0	44.58
Cars	9,90,000	Passenger kms	0.111 $kgCO_2e$/pax-km	India GHG Program, 2015, Version 1.0	10.98
Subtotal (B2)					**56.00**
Subtotal (B1 + B2)					**122,93,510**
Transportation of wastes (by road, 100% laden freight)	543,45,600	tkm	0.143 $kgCO_2e$/tkm	India GHG Program, 2015, Version 1.0	7771
Wastes: disposal wet ash (landfill wastes)	7,30,000	MT	1.59 kg CO_2e/kg	PAS 2050:2011 Guidelines	1160
Wastes: disposal dry fly ash (sold off and assumed to have been recycled)	10,95,000	MT	0	PAS 2050:2011 Guidelines	Nil
Mill rejects	14,60,000	MT	1.13 kg CO_2e/kg	India GHG Program, 2015, Version 1.0	1650
Subtotal (C)					**10,581**

GHG, greenhouse gas.

Table 6.3 Details of transportation of chemicals used in production process

Material	Quantity per year (in MT) (2015–16)	Transportation distance (in km)	Transportation distance (in tkm)
Hydrochloric acid	170	58	9860
Caustic soda	25	219	5475
Hydrazine hydrate	2.2	219	482
Ammonia	1.18	219	258
Chlorine	1100	58	63,800
Alum	250	219	54,750
Polyformic chloride	50	219	10,950
Trisodium phosphate	0.3	112	34
Disodium phosphate	0.0015	112	0.168
	1599		**145,609**

6.5.2 Carbon Dioxide Produced During the Power Plant Operation and Its Service Facilities

The plant have five generating units of 210 MW capacity each that is a total generation capacity of 1050 MW using coal having 35% ash content. Depending on the quality of available coal, the feeding rate varied from unit to unit, as and when required. Noncoking bituminous coal is fed into the boiler. For efficient working of power plant and for carrying out auxiliary operations, a substantial quantity of electric power is required, estimated to be 9.5% of the total 6669 MUs or 6669 million kWh of generated electricity in 2015–16. However, these energy requirements are met by the electrical power that is generated by the power plant itself.

For service facilities energy requirement related to the following categories were taken into consideration:

- The power plant's administrative offices (obtained from electricity generated by the plant so not considered separately)
- Employees' business trips

From the employees travel register the following details of business travel were obtained:

1. 400 bikes and scooters ply on an average 288 days in 2015–16 per day (mainly diesel run bikes and scooters). A distance of 10 km/day is traveled by each vehicle.

2. 100 cars and bus ply on an average 330 days in 2015–16 per day (mainly diesel run bikes and scooters). A distance of 30 km/day is traveled by each vehicle.
3. 500 bicycles and scooters ply on an average a distance of 5 km/day for 365 days in 2015–16.

Based on the above information calculation of the distances traveled by the employees have been done. Emission factor for scooters and bikes is taken to be 0.0387 kg CO_2e/pax-km [India GHG Program, 2015, Version 1.0] based on the assumption that the scooters and bikes are of more than 1500 cc engine capacity [India GHG Program, 2015, Version 1.0]. Annual emission from employees commuting in bikes and scooters (owned vehicle) for 2015–16 is 44.58 tCO_2. In similar way emission factor for cars/bus is taken to be 0.111kgCO_2e/pax-km based on the assumption that the cars belong to the category of small cars having less than 1000 cc engine capacity with an uplift factor [India GHG Program, 2015, Version 1.0]. Emission from employees commuting in cars/bus (hired vehicles) for 2015–16 is 10.98 tCO_2. Emission from bicycle is taken to be zero. So the total emission from employee's business trip in 2015–16 was **55.56 tCO_2e** i.e., **56 tCO_2e.** So emission produced during the power plant operation and its service facilities totaled. **122,93,510 tCO_2e** in 2015–16. Out of which emission from production facilities is **122,93,454 tCO_2e** and from service facilities is **56 tCO_2e.**

6.5.3 Carbon DiOxide Produced During the Transportation and Through Disposal of Waste Materials From the Power Plant

There are three types of waste generated by the case study power plant unit—wet ash, dry fly ash, and mill rejects. Dry ash constitutes of fly ash generated in the production process. 3000 MT fly ash is transported per day in dampers/trucks over a distance of 30–40 km on an average and is sold off to the vendors or directly to the industries manufacturing cement from fly ash. 2000 MT of wet ash is transported per day to landfills at a distance of 30 kms/day on an average by trucks and disposed off as landfill waste. It is assumed that both dry and wet ashes are transported in 100% laden dumper truck. So emission from transportation of waste by road is considered, which 0.143 kg CO_2e is/tkm [India GHG Program, 2015, Version 1.0].

Annual emissions from transportation of three types of wastes—wet ash, dry ash, and mill rejects amounted to 7771 tCO_2e during 2015–16.

Emission from disposal of 7,30,000 MT wet ash as landfill waste amounted to 1160 tCO_2e taking 1.59 kg CO_2e/kg as the emission factor [PAS 20150:2011 Guidelines]. Dry ash (fly ash) of 10,95,000 MT is sold off to cement manufactures and local vendors for recycling and thus emission from dry/fly ash is taken as zero.

Mill rejects amounted to 100 MT/day that are transported by tractor to a distance of 2 km on an average every day before being disposed off as landfill waste. Because the mill rejects constitute of a mixture of different types of wastes the default emission factor (also a national average in India) is taken as 1.13 kg CO_2e/kg [India GHG Program, 2015, Version 1.0]. Emission from disposal of 14,60,000 MT of mill rejects disposed of as landfill waste amounted to 1650 tCO_2e for the year 2015–16.

Annual emission from disposal of wastes totaled to 2810 tCO_2e during 2015–16.

Total annual emissions from transportation and disposal of wastes amounted to **10,581 tCO_2e** for the year 2015–16.

6.5.4 Determination of the Total Carbon Footprint of the Thermal Power Plant

The GHG emissions in each phase of the case study power plant were ascertained considering its complete life cycle and summing up the emissions from each phase gave the CF of the power plant for the year 2015–16, which was **123,30,119 tCO_2e** (Fig. 6.1) for generating 6669 million kWh of electricity.

The CF of transportation of raw materials into the power plant is moderate compared with the quantity of raw material mainly coal that is being transported to the power plant. The large quantity of coal is carried to the plant from an average distance of 100–150 km, which accounts for a

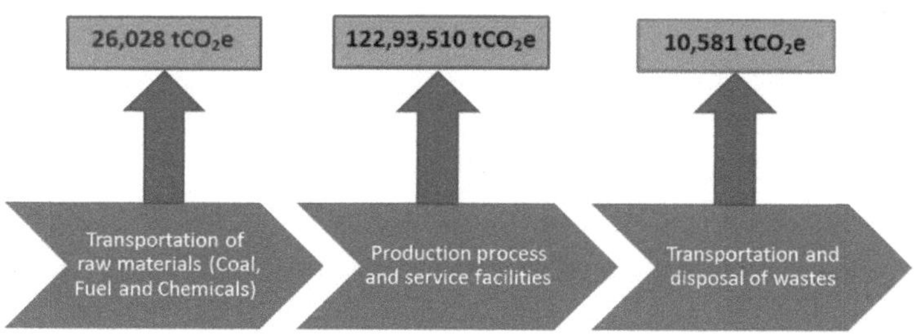

Figure 6.1 Greenhouse gases emissions of the thermal power plant complete life cycle (cradle to grave) expressed in tCO_2e (2015–16).

major share of emission during raw material transportation. This is because raw material is transported by rail and electricity consumed by rail is very low as compared with other mode of transportation. So mission from transportation of raw materials during the year 2015–16 was **26,028 tCO_2e.**

The CF of power plants operation and service facilities is **122,93,510 tCO_2e**. Out of which emission from production facilities is **122,93,454 tCO_2e** and from service facilities is **56 tCO_2e**. The emissions related to production process of the power plant constitute the most dominant part (largest contributor) of the CF. This is because of the large quantity of coal used in the electricity generation process. Annual emission from service facilities is quite negligible.

Emissions from transportation and disposal of wastes amounted to **10,581 tCO_2e** for the year 2015–16, the third largest contributor in CF of the plant. Out of which emission from waste transportation was 7771 tCO_2e, and that from disposal of generated waste being 2810 tCO_2e.

6.5.5 Direct and Indirect Emission

1. The sources for direct (or on-site) emissions of the production unit for the year 2015–16 are from operation of power plant amounting to **122,93,510 tCO_2e** or 99.7% of unit's total CF.
2. The least portion of the unit's CF for the year 2015–16 consisted of indirect emissions constituting of
   a. upstream emissions mainly from transportation of raw materials, the employees' business trips, energy requirement in administrative office adding up to **26,084 tCO_2e**
3. Downstream management of wastes and by-products totaling to **10,581 tCO_2e** in 2015–16.

Indirect emission (sum total of upstream and downstream) of 36,665 **tCO_2e** constitutes only 0.3% of the total CF of the case study power plant unit.

6.5.6 Limitations of the Study

- Sum total of emissions from only direct GHGs (CO_2) expressed in tCO_2e associated with raw material transportation, power plant operations and service facilities, and transportation and disposal of end wastes have been calculated. Indirect emissions (CO, NO, and SO_2) of GHGs have not been considered and calculated.

- Also requirements on specific issues relevant for CFs, including land-use change, carbon uptake, biogenic carbon emissions, and soil carbon change (ISO 14067) have not been considered in this study.
- Assumptions have been made at all stages of this study, absence of which could give a more accurate final result.

6.6 CONCLUSIONS

In this study the GHG emissions during the full life cycle of a thermal power generation unit in West Bengal, India has been estimated. The GHG emissions projected in the full life cycle was **123,30,119 tCO_2e** in 2015–16 for generating 6669 million kWh of electricity. The studied power plant's production results in considerably high GHG emissions as the production process is based on coal: a nonrenewable source of energy. The most significant GHG emission factor contributing toward CF of the plant is the production process, followed by emissions from transportation of raw materials and emissions from transportation and disposal of wastes being the least contributor. The auxiliary operation, an inseparable part of the production process, of the power plant consumes 9.5% of the total 6669 MUs of generated electricity resulting in an emission of 8,17,286 tCO_2e annually in 2015–16. The CF of the power plant can be reduced by using solar energy for carrying out the auxiliary operations. The wastes are transported by the plant to landfills or vendors situated at a distance of 30–40 km on an average every day. Emissions resulting from transportation of wastes can be reduced if wastes are transported to landfills or vendors located in nearby vicinity of the power plants. Combustion of coal is the major contributor toward CF of the power plant. Because of this reason emission from direct sources constituted 99.7% of total CF and emission from indirect sources being only as insignificant as 0.3%. So it is important that power generation from nonrenewable sources such as coal is substituted by some renewable sources of energy to moderate extent for reducing the carbon footprint of the power plant and contribute in its own way toward sustainability.

ACKNOWLEDGMENTS

I extend my heartfelt thanks to Dr. Niladri Chakraborty, Professor, Department of Power Engineering, Jadavpur University, Kolkata, India for providing useful guidance and suggestions. I also acknowledge with thanks the anonymous referee(s) for their encouraging and valuable observations on the earlier versions of the paper.

REFERENCES

Atlantic Consulting and IPU, 1998. LCA Study of the Product Group Personal Computers in the EU Ecolabel Scheme. European Commission.

Cameo Corporation, 2010. Comparison of Lifecycle Greenhouse Gas Emissions of Various Electricity Generation Sources.

Canadian Energy Research Institute (CERI), 2008. Comparative Life Cycle Assessment (LCA) of Base Load Electricity Generation in Ontario. A report authored by Jazayeri S, Kralovic P, Honarvar A, Naini A, Rozhon J, Shabaneh R, Walden T and prepared for the Canadian Nuclear Association (CNA).

Cen, S.A., 2009. Final report on the Carbon Footprint and Climate Footprint of Highlands and Islands Enterprise 2007–08. Centre for Sustainability Accounting, Research Report 09–01.

Central Electricity Authority, January 2013. CO_2 Baseline Database for the Indian Power Sector: User Guide. Version 8.0. published by Government of India, Ministry of Power, Central Electricity Authority.

Chakraborty, N., Mukherjee, I., Santra, A.K., Chowdhury, S., Chakraborty, S., Bhattacharya, S., Mitra, A.P., Sharma, C., 2008. Measurement of CO_2, CO, SO_2, and NO emissions from coal-based thermal power plants in India. Atmospheric Environment 42, 1073–1082.

Choi, B.C., Shin, H.S., Lee, S.Y., Hur, T., 2006. Life cycle assessment of a personal computer and its effective recycling rate. International Journal of Life Cycle Assessment 11 (2), 122–128.

Duchin, F., 1998. Structural Economics: Measuring Change in Technology, Lifestyle and Environment. Washington DC, Island.

Eckel, A., 2007. Gravure 21 (2), 35–36.

Feng, C., Ma, X.Q., 2009. The energy consumption and environmental impacts of a color TV set in China. Journal of Cleaner Production 17 (1), 13–25.

Foran, B.M., Lenzen, M., Dey, C., 2005a. Balancing Act – A Triple Bottom Line Analysis of the Australian Economy, vols. 1–3. The University of Sydney and CSIRO, Sydney.

Foran, B.M., Lenzen, M., Dey, C., Bilek, M., 2005b. Integrating sustainable chain management with triple bottom line accounting. Ecological Economics 52, 143–157.

Guidance for Voluntary Corporate Greenhouse Gas Reporting – 2015: Using Data and Methods from the 2013 Calendar Year Published in April 2015 by Ministry of Environment, INFO 734.

Hertwich, E.G., Peters, G.P., 2009. Carbon footprint of nations: a global, trade – linked analysis. Environmental Science & Technology 43, 6414–6420.

India GHG Program, 2015. India Specific Road Transport Emission Factors. Version 1.0. Published by WRI India, TERI and CII.

Kennedy, C., Rawaswami, A., Carney, S., Dhakal, S., 2009. Greenhouse gas emissions baselines for global cities and metropolitan regions. In: Paper Presented at the Fifth Urban Symposium, 27 June – 1 July, 2009, in Marseille, France.

Klemes, J.J., 2015. Overview of environmental footprints. In: Assessing and Measuring Environmental Impact and Sustainability. Elsevier, pp. 131–193.

Lenzen, M., 2007. Aggregation (in-) variance of shared responsibility: a case study of Australia. Ecological Economics 64 (1), 19–24.

Lenzen, M., 2003. Environmentally important paths, linkages and key sectors in the Australian economy. Structural Change and Economic Dynamics 14, 1–34.

Lenzen, M., Dey, C., Foran, B., 2004. Energy requirements of sydney households. Ecological Economics 49, 375–399.

Lenzen, M., Murray, J., Sack, F., Wiedmann, T., 2007. Shared producer and consumer responsibility - theory and practice. Ecological Economics 61, 27–42.

Lu, L.T., Wernick, I.K., Hsiao, T.Y., Yu, Y.H., Yang, Y.M., Ma, H.W., 2006. Balancing the life cycle impacts of notebook computers: Taiwan's experience. Resources, Conservation and Recycling 48 (1), 13–25.
Minx, J.C., Baiocchi, G., 2009. Time use and sustainability. In: Suh, S. (Ed.), Handbook of Input-Output Methods for Industrial Ecology. Springer, Dordrecht.
Pandey, P., Blanc, I., Le Boulch, D., Xiusheng, Z.A., 2012. Simplified life cycle approach for assessing greenhouse gas emissions of wind electricity. Journal of Industrial Ecology 16, S28–S38.
Penela, A.C., Negro, M.D.C.G.C., Quesda, J.L.D., 2009. A methodological proposal for corporate carbon footprint and its application to a wine producing company in Galicia, Spain. Sustainability 1, 302–318. http: doi: 10.339/su1020302.
Peters, G., Hertwich, G., 2006. Structural analysis of international trade: environmental impacts of Norway. Economic Systems Research 18, 155–181.
Peters, G., Hertwich, E., 2008. CO_2 embodied in international with implications for global climate policy. Environmental Science and Technology 42, 1401–1407.
Peters, G., Marland, G., Hertwich, E., Saikku, L., Rauatianinen, A., Kauppi, P., 2009. Trade, transport, and sink extend the carbon dioxide: responsibilities of the countries. Climatic Change.
Roy, P.C., Dutta, A., Chakrabort, N., 2013. An assessment of different biomass feedstocks in a downdraft gasifier for engine application. Fuel 104, 864–868.
Sarkar, S., Sahu, S.G., Mukherjee, A., Kumar, M., Adak, A.K., Chakraborty, N., Biswas, S., 2014. Co-combustion studies for potential application of sawdust or its low temperature char as co-fuel with coal. Applied Thermal Engineering 63 (2), 616–623.
SEI, WWF, CURE, 2006. Counting Consumption – CO_2 Emissions, Material Flows and Ecological Footprint of the UK by Region and Devolved Country. Published by WWF-UK, Godalming, Surrey, UK.
Tonini, D., Astrup, T., 2012. LCA of biomass-based energy systems: a case study for Denmark. Applied Energy 99, 234–246.
Vattenfall, 2005. Life-Cycle Assessment Vattenfall's Electricity in Sweden.
Weidmann, T., Lenzen, M., 2006. Tripple bottom line accounting of social, economic and environmental indicators: a new life – cycle software tool for U.K. business. In: Third Annual International Sustainable Development Conference on Sustainability – Creating the Culture. Perth, Scotland, United Kingdom.
Whitaker, M., Heath, G.A., Donoughe, P., Vorum, M., 2012. Life cycle greenhouse gas emissions of coal fired electricity generation: systematic review and harmonization. Journal of Industrial Ecology 16.
Wiedmann, T., Minx, J., 2008. A definition of "carbon footprint". In: Pertsova, C.C. (Ed.), Ecological Economics Research Trends. Nova Science Publishers, Hauppauge, NY, USA, pp. 1–11 (Chapter 1).
Wiedmann, T., Barret, J., Lenzen, 2007a. Companies on the scale: comparing and benchmarking the footprints of Businesses. In: Paper Prepared for the International Ecological Footprint Conference. Cardiff, Australia.
Wiedmann, T., Lenzen, M., Turner, K., Minx, J., Barrett, J., 2007b. Multiregional input-output modelling opens new opportunities for the estimation of ecological footprints embedded in international trade. In: Paper Prepared for the International Ecological Footprint Conference May 8–10, 2007, Cardiff, and UK.
Wiedmann, T., Wood, R., Minx, J., Lenzen, M., Guan, D., Harris, R., 2010. Carbon footprint time series of the UK – results from a multi-region input–output model. Economic Research 22 (1), 19–42.
Wood, R., Dey, C., 2009. Australia's carbon footprint. Economic Research 21, 243–266.
World Meteorological Organization, 2009. WMO Gas Bulletin. The State of Greenhouse Gases in the Atmosphere Using Global Observations Through 2008.

World Nuclear Association (WNA) Report, 2011. Comparison of Lifecycle Greenhouse Gas Emissions of Various Electricity Generation Sources. http://www.world-nuclear.org/uploadedFiles/org/WNA/Publications/Working_Group_Reports/comparison of_lifecycle.pdf.

FURTHER READING

Odeh, N., Cockerill, T., 2008. Life Cycle analysis of UK coal fired power plants. Energy Conversion and Management 49, 212–220.

Pablo, C.M., Quesada, J.L.D., Mantecon, I.M., 2008. Corporate Ecological Footprint: New Conversion Factors. Hindawi Publishing Corporation, Research Letters in Ecology. http://dx.doi.org/10.1155/2008/415934.

CHAPTER 7

Carbon Footprint in the Wine Industry

Flavio Scrucca, Emanuele Bonamente, Sara Rinaldi
University of Perugia, Perugia, Italy

7.1 INTRODUCTION

Anthropogenic activities are widely recognized as a driver for the impact on the environment because of the significant pressures they are recognized to generate. Among all the production sectors, the food sector represents one of the major impacting ones because of its major contribution to land occupation and use (about 38% of Earth's terrestrial surface (Foley et al., 2011)), greenhouse gases (GHGs) emissions (up to 29% of all anthropogenic emissions (Vermeulen et al., 2012)), and freshwater consumption (about 70% of all the human uses (FAO, 1996; UN – Sustainable Development Solutions Network, 2013)). Within all the products, wine emerges as one of the most analyzed because it has a great relevance both in the economic production and in the world distribution market.

The generally used approach for the evaluation and communication of the total direct and indirect environmental impact of a product or activity is based on the footprint concept, with the carbon footprint (CF) being one of the most widespread indicator for the assessment of the environmental effects related to food production and consumption.

This chapter is aimed at giving a comprehensive view of the environmental pressures related to the life cycle of a wine bottle, thus allowing the identification of environmental hot spots of the production process and providing support for strategic decisions aimed at its improvement and optimization.

A brief analysis of the wine sector worldwide, in terms of production volumes and economic relevance, is carried out in Section 7.2. In Section 7.3 a review of the scientific literature is reported together with a review of different experiences regarding CF evaluation in the wine sector focused on nationally/internationally valid protocols. Section 7.4 describes the

Environmental Carbon Footprints
ISBN 978-0-12-812849-7
http://dx.doi.org/10.1016/B978-0-12-812849-7.00007-6

methodology used for the evaluation of the CF of the different case studies considered, and Section 7.5 reports the results related to those case studies regarding typical Italian wineries.

7.2 THE WINE SECTOR WORLDWIDE

Agriculture accounts globally for more than one-third of the land area. In particular, with a percentage in the order of 35%–36% (Fig. 7.1), the agricultural land is estimated in about 4.5 billion hectares, 1.5 of which—i.e., about 12% of the world's land area—are used for crop production (FAO, 2015). Vines represent a small part of the global crop production. The global area under vines in 2015 was 7.5 million hectares (about 0.5% of the total cultivated area), with a world production of grapes equal to 75.7 million tons. Anyway, despite this small percentage contribution in terms of land area, the wine sector has a great relevance in the economies of the countries worldwide. In 2015 the global market, considered here as the total exports of all countries, reached 104.3 million hl in terms of volume (with an increase of almost 2% compared with 2014) and 28.3 billion € in terms of value (a rise of 10% compared with 2014) (OIV Statistical Report on World Vitiviniculture, 2016).

Looking at the wine production, it emerges that it has been in the range of 250–300 million hl in the last 15 years, as shown in Fig. 7.2. In 2015, in particular, it was observed an increase of 1.1% compared with 2014, and the world wine production was 274 million hl (OIV Statistical Report on World Vitiviniculture, 2016).

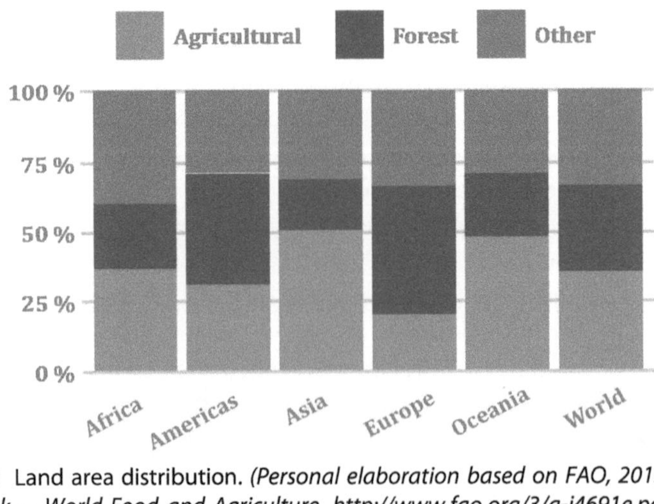

Figure 7.1 Land area distribution. *(Personal elaboration based on FAO, 2015. Statistical Pocketbook – World Food and Agriculture. http://www.fao.org/3/a-i4691e.pdf.)*

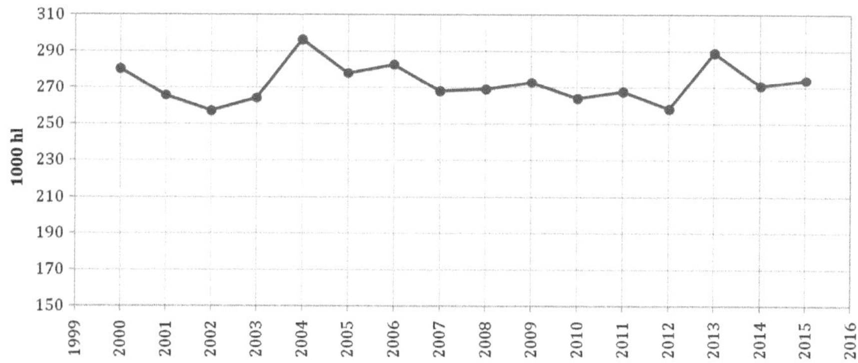

Figure 7.2 World wine production trend (excluding juice and musts). *(Personal elaboration based on OIV Statistical Report on World Vitiviniculture, 2016. http://www.oiv.int/public/medias/5029/world-vitiviniculture-situation-2016.pdf and OIV Databases and Statistics. http://www.oiv.int/en/databases-and-statistics.)*

The European Union (EU) remains the world's biggest wine producer, with a share of about 60% on the world total wine production (European Commission, 2015). Italy remains the first producer, with a share of 18.2%, followed by France and Spain with a share, respectively, of 17.3% and 13.6%. Within the so-called "New World countries," i.e., within the countries outside the traditional wine-growing areas of Europe and the Middle East, the United States represented the top producer in 2015 (8.1% of the world total production), followed by Argentina (4.9% of the world total production), Chile (4.7% of the world total production), Australia, China, and South Africa (each one about 4% of the world total production) (OIV Statistical Report on World Vitiviniculture, 2016) (Table 7.1).

7.3 AVAILABLE LITERATURE AND EXISTING EXPERIENCES REGARDING CARBON FOOTPRINT CALCULATION

7.3.1 Scientific Literature

Available scientific literature deals with the analysis of the wine production from different points of view, and therefore, there are studies that analyze the whole life cycle of the product, as well as studies regarding specific phases of the production process. In particular, available studies quantify the CF value both from a *cradle to grave* and a *cradle to gate* perspective and, in some cases, also model the transportation and distribution phase without considering the end of life. The differences on system boundaries definition mostly depend on data availability. In fact, although data are typically

Table 7.1 Wine production data (OIV Statistical Report on World Vitiviniculture, 2016)

mhl	2011	2012	2013	2014	2015
Italy	42.8	45.6	54.0	44.2	50.0
France	50.8	41.5	42.1	46.5	47.4
Spain	33.4	31.1	45.3	39.5	37.3
United States	19.1	21.7	23.6	23.7	22.1
Argentina	15.5	11.8	15.0	15.2	13.4
Australia	11.2	12.3	12.3	11.9	11.9
China	13.2	13.8	11.1	11.1	11.5
South Africa	9.7	10.6	11.0	11.5	11.2
Chile	10.5	12.6	12.8	10.5	12.9
Germany	9.1	9	8.4	9.2	8.8
Portugal	5.6	6.3	6.2	6.2	7.0
Russian Federation	7.0	6.2	5.3	4.9	4.9
Romania	4.1	3.3	5.1	3.7	3.5
New Zealand	2.4	1.9	2.5	3.2	2.3
Hungary	2.8	1.8	2.6	2.6	2.9
Greece	2.8	3.1	3.3	2.8	2.5
Serbia	2.2	2.2	2.3	2.3	2.3
Austria	2.8	2.1	2.4	2.0	2.3
Georgia	1.1	0.8	1.0	1.1	1.7
Moldova	1.5	1.5	2.6	1.6	1.7
Brazil	3.5	3.0	2.7	2.7	2.8
Bulgaria	1.1	1.3	1.8	0.8	1.3
OIV world total	268	258	289	271	274

OIV, Organization of Vine and Wine.

available for representative phases of the production process, such as viticulture and wine making, reliable information may not always be accessible to consistently model transportation and/or end-of-life phases. A gate-to-gate approach is instead used when there is the need a to give decisional support in order to compare different scenarios.

Most of the reviewed studies refer to a 0.75 L bottle as functional unit, whereas a few refer to other functional units (1 L of wine, 1 t of wine/grapes, etc.). According to Rugani et al. (2013)—given the variability in technological, geographical, and viticulture conditions, and the uncertainty they imply—the CF for a generic bottle of wine is on average $2.2 \pm 1.3\ kgCO_{2eq}$, with viticulture activities, packaging processes, and end-of-life being the most significant processes. The average value becomes $1.91 \pm 1.3\ kgCO_{2eq}$ if critical values (two landfill scenarios that greatly

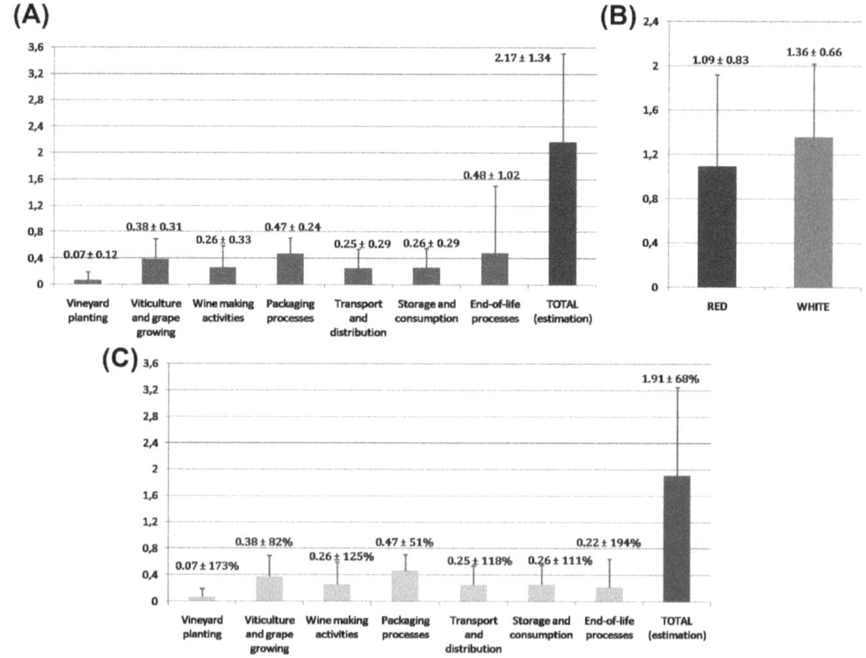

Figure 7.3 Average carbon footprint value of a generic wine bottle in $kgCO_{2eq}$/bottle: (A) cradle to grave, (B) cradle to gate, (C) cradle to grave excluding critical values for the end of life phase (Rugani et al., 2013).

influence the variability of results) are excluded from the analysis. Moreover, considering a "cradle-to-gate" approach, i.e., the results related to the wine bottled at winery gate, the average CF is equal to 1.09 ± 0.83 $kgCO_{2eq}$ for a red wine and 1.36 ± 0.66 $kgCO_{2eq}$ for a white wine (Rugani et al., 2013) (Fig. 7.3).

Looking at the different life cycle stages, the vineyard-planting phase emerged as a significant contributor to the wine CF according to Bosco et al. (2011), while based on the results reported in Fig. 7.3A on average its share on the CF value of a generic bottle of wine could be considered in the order of 3%–5%.

Viticulture activities contribution is estimated to be, on average, equal to 17%, even if its share on the total CF value could reach values round to 50% (Gazulla et al., 2010). The use of pesticides and fertilizers is identified as an issue of this life cycle stage in most of the reviewed papers (Aranda et al., 2005; Vázquez-Rowe et al., 2012) because of their production/transportation and application. In fact, resources consumption (e.g., energy, fuels, etc.) in the production process, coupled with the transportation activities and the large release of N compounds during the treatment of the grapevines, can affect significantly the overall CF results. Another significant contributor to the CF value included in the viticulture stage is the diesel consumption for field activities. In the Marras et al. (2015) study, which evaluate the CF of a mature vineyard in Southern Italy, diesel consumption and treatments accounted, respectively, for 44% and 50% of the overall impact. In this regard Villanueva-Rey et al. (2014) compared conventional to biodynamic viticulture activities. Diesel production and consumption used in field operations emerged as the main source of environmental impacts in all the different agricultural management techniques considered. Anyway, results reported in gCO_{2eq}/bottle show a value in the order of 70–90 for the biodynamic wine and of 280–370 for the conventional one, according to the characteristics of different harvest years. This significant reduction is the consequence, within all the differences, also of the absence of fertilization, the characteristics of the pesticides (only copper-based), and the noteworthy reduction in their use (about 75% less in terms of mass).

Wine-making activities, on average, contribute to the total CF of a generic wine bottle for a 10%–15% even if are available studies that found a contribution of this phase on the total CF round to 27% (Fusi et al., 2014). Literature focused on this life cycle stage (Iannone et al., 2016) highlighted that about the 90% of the total CO_2 emissions associated with this phase are because of energy consumption at the winery, i.e., electricity consumption in the different industrial phases including the use of air-conditioning systems.

Packaging processes is another key issue identified by the literature, mainly because of the glass bottle production. On average the contribution of this component of the life cycle could be considered in the order of 20%–22% of the total CF value of a generic wine bottle, but also greater values are reported (Carballo-Penela et al., 2009; Meneses et al., 2016). According to Bonamente et al. (2016), for instance, the CF of a protected designation of origin Italian red wine is estimated to be 1.068 $kgCO_{2eq}$, with a contribution of packaging of 0.599 $kgCO_{2eq}$ (about 56%). A study focused on the analysis of different wine and spirit packages (Cleary, 2013) seems to confirm the literature results, bringing out that a single use glass bottle (volume of 1 L) represents the most impacting packaging solution in terms of CO_2eq emissions.

Few studies (Point et al., 2012; Comandaru et al., 2012; Greenhaigh et al., 2011; Soosay et al., 2012; WRAP-Waste Resources Action

Programme, 2007) evaluated the contribution of the storage and consumption phase, which on average results in the order of 10%–15% for a generic wine bottle, whereas large variability in results was observed for transportation and end of life. These phases, in fact, are influenced by assumptions and choices made when modeling the life cycle of the product.

In this regard, the end of life is the phase characterized by the largest variability in results because the quality of the data used (primary or national/international average), the nature of the selected processes (landfill disposal, reuse, recycling), and the allocation choices (recycled materials used or not as "secondary" raw materials) play a key role on the final CF value. Results presented in Fig. 7.3 can be considered representative of this issue; in fact they show how the contribution of the end of life can vary from about a 10% to about a 42%.

Regarding transportation, literature focused on the evaluation of the influence of different transport modes/choices and are available studies that analyze alternative scenarios (Cholette and Kumar, 2009; Harris et al., 2016). For instance, analyzed different scenarios regarding how wines (produced in California) may be routed to consumers, calculating the energy and carbon emissions associated with each one of them. The least efficient scenario, which considers the consumers driving to the winery in a typical passenger car, results in 80 times the emissions that would occur if that local delivery was shipped to customers through a third-party logistic company. In a study by Amienyo et al. (2014) focused on the environmental impacts of consumption of Australian red wine in the United Kingdom, moreover, it also emerges that shipping in bulk rather than bottled wine would appreciably reduce (13%) the total impact of global warming potential (GWP).

7.3.2 Protocols and Guidelines for Carbon Footprint Calculation

Beyond the scientific studies, the wine industry has developed an environmental awareness in recent years and is becoming more and more committed to sustainability. It is possible to identify different initiatives (programs, EU-funded, or National projects, etc.) aimed at implementing sustainability principles, as reported, for instance, Corbo et al. (2014) for the Italian wine sector. In this section, the attention is focused on nationally/internationally valid protocols/guidelines the assessment of CF in the wine sector.

A collaboration of several national agencies, i.e., the Wine Institute of California, New Zealand Winegrowers, Integrated Production of Wine South Africa, and the Winemakers Federation of Australia, led to the development of Version 1.0 of the International Wine Carbon Calculator (IWCC) by the end of 2007. Version 1.1 of the document was released by Provisor Pty Ltd on February 2008, and Version 1.2 was then released in July 2008 by the International Federation of Wines and Spirits (FIVS). The IWCC is tailored for use at the international level by all segments of the wine industry (e.g., growers, wineries, contract bottlers, etc.) and is aimed at offering a common framework for calculating, reporting, and communicating about CO_2 emissions in compliance with international standards (FIVS, 2008). The calculator considers both emissions related to company activities (Scope 1 and Scope 2) and emissions that occur as a consequence of producing a finished salable vitivinicultural product (Scope 3), and the calculation methodology uses an activity-based approach. Therefore, corporate activity data like fuel use or glass purchased are accounted and then input into the calculator, where they are either multiplied by an emission factor or input into an emission model. Recently, FIVS has revised and updated the Protocol and in February 2016 it has been released the IWCC—Version 2.0 (FIVS, 2016).

Following this experience, an Italian company in collaboration with the Winemakers Federation of Australia, under the scientific supervision of the University of Milan (Department of Agricultural and Environmental Sciences), developed the Italian Wine Carbon Calculator (Ita.Ca), which represents the adaption of the IWCC to the Italian wine sector (Agronomic Study SATA). Moreover, as a national application of the IWCC, in 2009 it was also launched the Australian Wine Carbon Calculator, which was built on the IWCC and included components specific to Australian needs (Australian Wine Research Institute, 2009).

In 2008 it was published the first version of the Publicly Available Specification (PAS) 2050 (BSI, 2011) for the assessment of the life cycle GHG emissions of goods and services, which was then updated in 2011. The PAS 2050, prepared by British Standards Institute—and cosponsored by the Carbon Trust and the Department for Environment, Food and Rural Affairs (Defra)—was not specific to the wine sector, but it was one of the first documents tailored on the CF assessment of a "product" and following its specifications for calculation a New Zealand wine obtained the first certification in the sector (New Zealand Wine Company).

In 2011, the International Organization of Vine and Wine (OIV) set up the general principles of the OIV GHG accounting protocol, based on the "International Wine Carbon Protocol," on the ISO 14040, 14044, and 14064 Standards and on relevant international and national standards (OIV Resolution OIV-CST 431-2011, 2011). Then in 2015, OIV approved the requirements related to the calculation of GHGs emission balance and published its GHG accounting protocol, aimed at providing provide specific information on the items to be considered when accounting for GHG emissions of an enterprise or of a specific product (OIV Resolution: OIV-CST 503AB-2015, 2015). The document, in fact, includes two different elements. The first one is the Enterprise Protocol that includes the evaluation both of direct and indirect emissions and is intended to provide a clear and consistent method for the complete assessment of the GHG emissions associated with vine and wine companies' activities. The second one is the Product Protocol, which is intended to be used for business-to-consumer assessments, thus including all the emissions arising from the full life cycle of the final product in compliance with international standards for lifecycle analysis. Another relevant document is the GHG Protocol Agricultural Guidance, released by the World Resources Institute in 2014.

The GHG Protocol Agricultural Guidance, representing the first global guidance to measure GHG emissions for the agriculture sector, is intended to promote sector-wide adoption of emissions management practices by harmonizing how agricultural companies worldwide measure and report their emissions. The guidance is a supplement to GHG Protocol's Corporate Standard and covers all agricultural subsectors, including livestock, crop production, and land use change, and the calculation approach is activity data-based.

7.4 CARBON FOOTPRINT METHODOLOGY IN THE WINE INDUSTRY

Because of their success in reaching a great audience and to their ease of understanding, environmental footprint indicators such as CF and water footprint (WF) have become largely widespread for the evaluation of the total direct and indirect environmental impacts related to food production and consumption. A reliable and consistent CF analysis of a generic process, in particular, has to be carried out following a life cycle assessment (LCA) approach (ISO 14040:2006; ISO 14044:2006), which allows the evaluation of impacts from a cradle-to-grave perspective, following the requirements of the reference international standards, respectively, ISO 14067 (ISO (International Standard Organization), 2013) for a product-oriented study and ISO 14064 for an organization-oriented study (ISO 14064—1:2012).

The methodology should be preferably based also on Environmental Labels and Declarations (ISO 14020:2000; ISO 14024:1999; ISO 14025:2006), to guarantee that the relevant outputs are comparable with different results worldwide. According to the standardized procedure, methodological choices are made, if present, following the prescriptions provided by the Product Category Rules (PCR) for wine. Because groups of products usually differ in their inherent environmental performance, they require specific rules that can be applied to similar products, the so-called PCR. Such documents are, in general, made available by the International EPD System, as it is the case of wine products.

PCRs detail the definition of the product, declared unit, system boundaries, allocation, and cutoff rules. For the wine sector, two PCRs are available: sparkling wine of fresh grapes (Environdec, 2014) and wine of fresh grapes, except sparkling wine; wine wust (Environdec, 2010). In the system definition, three different life cycle stages are considered as follows:

- Upstream processes that include the following:
  - production of grapes in vineyard including all operation for the production of grapes;
  - generation of energy wares used in agriculture;
  - production of other ingredients used in sparkling wine, or in wine of fresh grapes, or grape must;
  - production of primary, secondary, and third tier packaging materials;
  - production and use of fertilizers, pesticides, insecticides, water, posts trellis, yarns.
- Core processes that include the following:
  - external transportation to the core processes (i.e., from the vineyard to the cellar);
  - production and packaging of the final product;
  - maintenance activities more frequent than 3 years;
  - waste treatment if waste generated during manufacturing;
  - electricity production;
  - wastewater treatment of the wastewater generated during vinification.

- Downstream processes that include:
  - transportation from production to an average distribution platform;
  - end-of-life processes of packaging waste.

CF methodologies describe the procedure to evaluate a single issue indicator that is commonly used to quantify the pressure of human activities on the environment. Equivalent carbon dioxide (CO_2eq) emissions are evaluated for any activity, process, or product as the sum of the total amount of direct and indirect GHGs emissions related to any stage of the life cycle according to the system boundaries defined for the specific case study.

ISO 14064 is the international standard for the evaluation, monitoring, communication, and verification of the GHG emissions produced by an organization. To ensure a widespread and flexible application, the standard describes the general principles and the guidelines rather than the specific procedure to be used. The aim of this standard is to provide organizations with an instrument to compare their environmental performance with other organizations worldwide and develop a set of common best practices. Particular care has to be paid to fulfill basic principles such as follows:

- relevance—selection of appropriate GHG sources, sinks, reservoir, data, and methodologies;
- completeness—all relevant GHG emissions and removals must be included;
- consistency—results related to GHG emissions must allow a reliable comparison;
- accuracy—uncertainties and bias must be contained and properly treated;
- transparency—the communication of results needs to be sufficient and appropriate.

Furthermore, GHG emissions must be divided into three scopes, as defined in the GHG Protocol (World Resources Institute and World Business Council for Sustainable Development, 2004):

- Scope 1: direct emissions;
- Scope 2: indirect emissions from energy consumption—electric and thermal energy produced outside the organization boundaries;
- Scope 3: other indirect emissions—different from Scope 2 emissions.

The standard requires to explicitly explain the methodological choices adopted for the emission quantification and the system boundaries. Scope 3

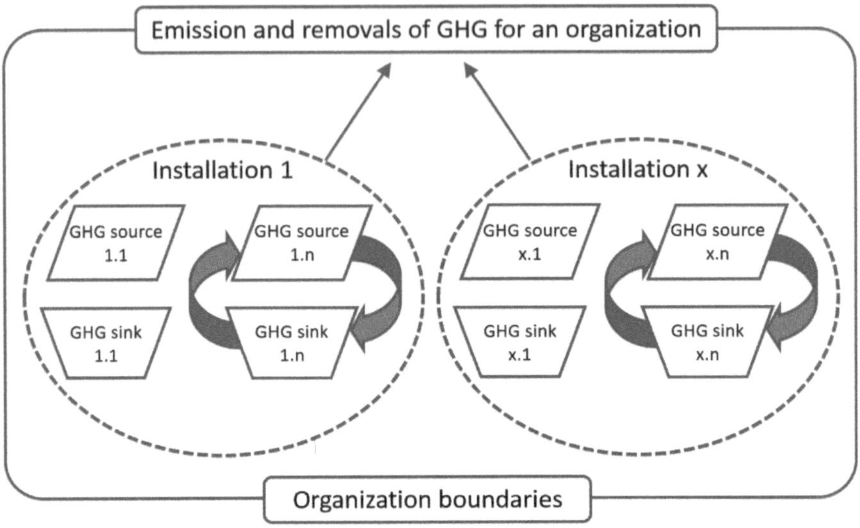

Figure 7.4 Relationship between greenhouse gas (GHG) sources, sinks, and facilities (ISO 14064–1:2012).

emissions may be not included or partially included as long as it is clearly stated. Data used for the emission inventory are divided into the following:

- Primary data, deriving from direct measurements;
- Secondary data, obtained from databases, simulation, or theoretical models.

The system boundaries (e.g., the organization limits) must be defined to include all relevant emissions, and they can be obtained dividing the organization into subentities (i.e., facilities), as shown in Fig. 7.4.

The more recent international standard ISO 14067 provides a description of principles, requirements, and guidance for the quantification and communication of the carbon footprint of a product, which includes any study centered on a given good or service. The methodology is based on the use of specific emission and removal factor that multiplied activity data to compute the emissions of GHG associated to each process or subprocess contributing to the different phases of the life cycle. Therefore, the GHG emission related to i-th process included in the lifecycle of the product (CF_i) is computed using the following equation:

$$CF_i = EF_i \cdot A_i$$

where A_i is the activity data and EF_i is the emission factor of the i-th process (e.g., $kgCO_{2eq}$ per unit of A_i).

Table 7.2 Global warming potential of relevant greenhouse gases

Name	Formula	GWP	
Carbon dioxide	CO_2	1	$kgCO_{2eq}/kgCO_2$
Methane	CH_4	28	$kgCO_{2eq}/kgCH_4$
Nitrous oxide	N_2O	265	$kgCO_{2eq}/kgN_2O$

GWP, global warming potential.

The emission factors to be used must be in compliance with the definition reported by Intergovernmental Panel on Climate Change (Climate Change, 2013) and computed considering each GHG emission generated by the process. Emission factors for unit processes are obtained considering the GWP of each GHGs emitted according to the following equation:

$$EF_i = \sum_j GWP_j \cdot e_{j,i}$$

where $e_{j,i}$ is the emission (in mass unit) of the j-th GHG associated to the i-th process per unitary amount. In Table 7.2, an example is given for the GWP of most relevant GHGs for a 100-year time horizon.

7.4.1 System Boundaries and Functional Unit

The system boundaries represent the interface between the product system and the environment and their definition determines, which unit processes shall be included within the assessment. Consistently with the goal of the studies presented later in this chapter, the system boundaries include grapes production, vinification, and marketing of the final product, whereas the final transportation of the product from the retailer to the end consumer is not included. The product life cycle is modeled considering three main modules: upstream, core, and downstream. All the inflow of raw materials and energywares required for the wine production is included within the upstream module. The core module includes the production and the packaging of the final product (including internal transportation and external transportation of raw materials and energywares), whereas the downstream module comprises the transportation to a distribution platform and the handling (recycling or disposal) of packaging materials (Fig. 7.5).

The functional unit (FU) is defined to quantify the performance of the product system, and its primary purpose is to provide a reference to which the inputs and outputs are related. The FU used in this study is a 0.75 L wine bottle.

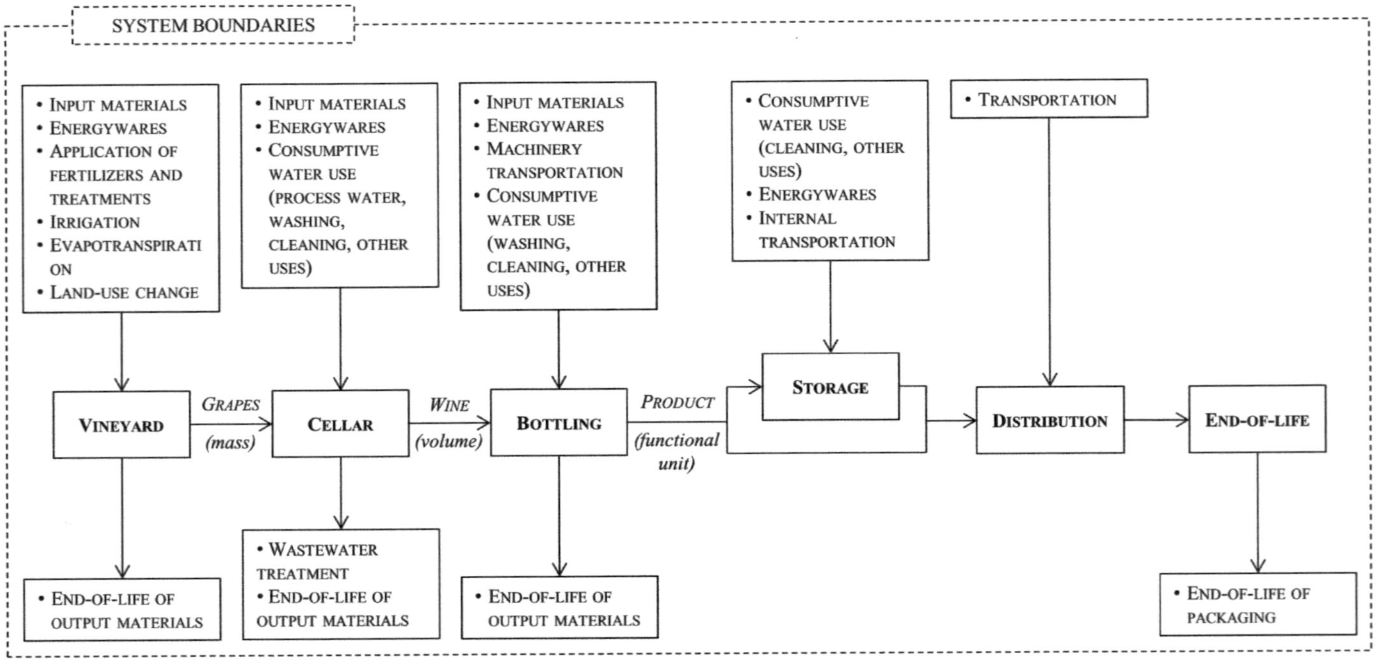

Figure 7.5 System boundaries and flow diagram.

7.5 CASE STUDIES AND RESULTS

In this section, an overview of the environmental footprint associated to the wine sector, from the first formulation of the CF concept to the state-of-the-art approach, is presented. Results from case studies from the Umbria region in Italy are introduced and discussed to discover how the environmental performance of wine sector is improved during the years and how the continuous optimization of the assessment methodologies represents a strong support for companies.

7.5.1 Case Study #1: ISO 14064

The very first application of the ISO 14064 standard back in 2009 was realized for an Umbrian agricultural company active in the wine and olive production (Rinaldi, 2010; Torquati et al., 2010).

The company decided to account and report GHG emissions from 2003 (base year) to evaluate the environmental impact of their own activities and processes, monitoring the trend in emissions and reductions as a results of organization strategies and policies over the years. Starting from 2003, in fact, the company had begun a crop reconversion process adopting less energy-intensive cultures instead of more impacting ones (e.g., tobacco). Over the years, the company took care of planting over 10,000 trees within its premises in an effort to boost overall GHG absorption. In addition, 250 kW photovoltaic panels have been installed on the rooftops of existing buildings, and the use of first-generation biodiesel was tested as an effective solution toward environmental sustainability of products.

Five facilities, under which GHG sources and sinks were aggregated, have been identified. The organization boundaries, defined according to the control approach, overlap with the operational boundaries as 100% of the GHG emissions come from operations under direct control of the company itself.

The system boundaries included Scope 1 and Scope 2 only, as required by the standard. Other indirect emissions, included in the following case studies, were not taken into account in this study. This has to be considered when interpreting present results.

In particular, direct emissions accounted in Scope 1 were because of:

- mobile combustion, deriving from the use of diesel and biodiesel fuels in agricultural equipment;
- stationary combustion of natural gas for heating system in winery and farm offices, including tobacco drying;
- use of nitrogen fertilizers;
- fermentation process.

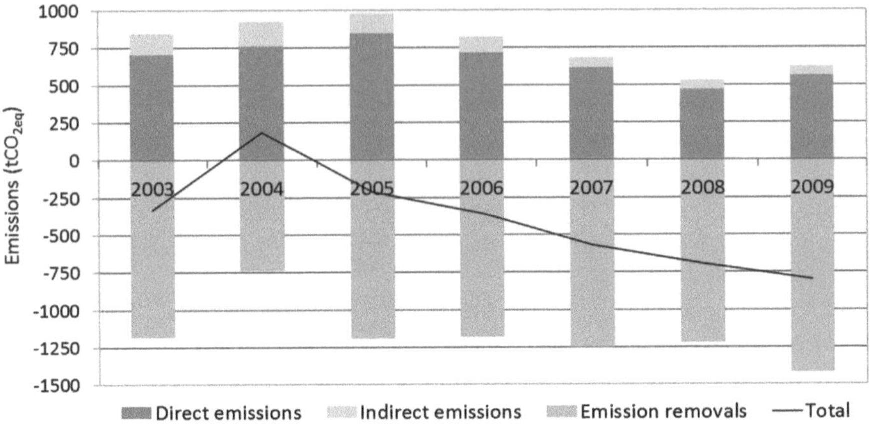

Figure 7.6 Greenhouse gas emissions for organization.

Emissions accounted for in Scope 2, derived from the generation of electric energy purchased and consumed by the company.

Unlike the following case studies, only GHG emissions from the combustion of natural gas, biodiesel, and diesel were evaluated. Emissions from the production of energywares should be included in Scope 3, which is not mandatory according to ISO 14064. An uncertainty assessment was also computed, considering the uncertainties associated to activity data (e.g., diesel consumption), emission factors, and methodology.

Results were certified by a third-party entity, and they were relative to the entire company as shown in Fig. 7.6.

Results are a balance between emissions and removals, and the overall trend is dominated by the latter. To compare such results to product-oriented studies, which do not include offsetting, it is necessary to consider emissions only (direct and indirect).

Analyzed in detail, there is a net decrease of indirect emissions since 2004, primarily because of the transition from tobacco to olive and wine production and the consequent decommissioning of tobacco ovens, leading to a strong reduction electric energy consumption (Fig. 7.7). Tobacco production ceased in 2007.

Direct emissions (Fig. 7.8) show a less regular trend, having a maximum in 2005.

Emissions form stationary combustion follow the same behavior of Fig. 7.7 because natural gas was used for tobacco drying.

An approximate allocation of the impacts between wine and other products was performed on a land surface basis (Fig. 7.9). At the current

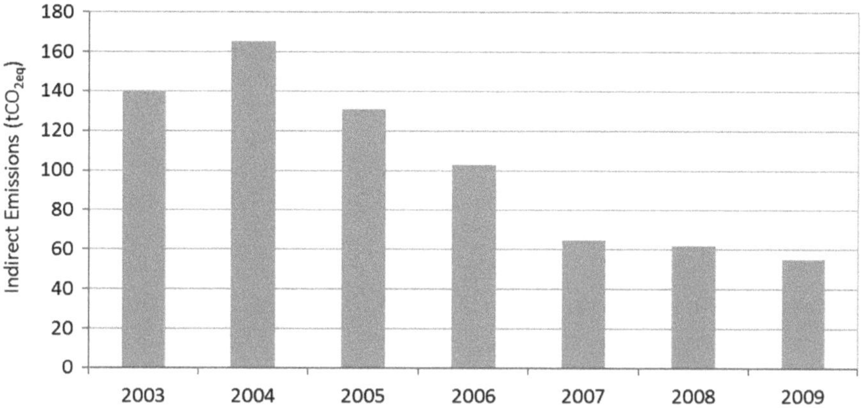

Figure 7.7 Indirect emissions trend.

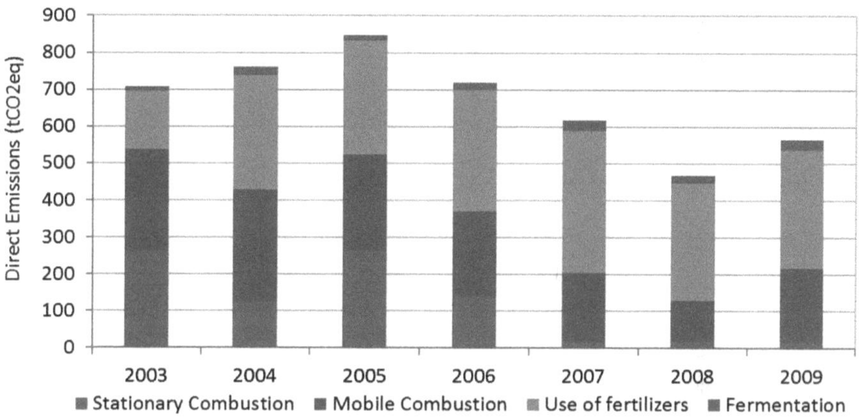

Figure 7.8 Direct emissions trend.

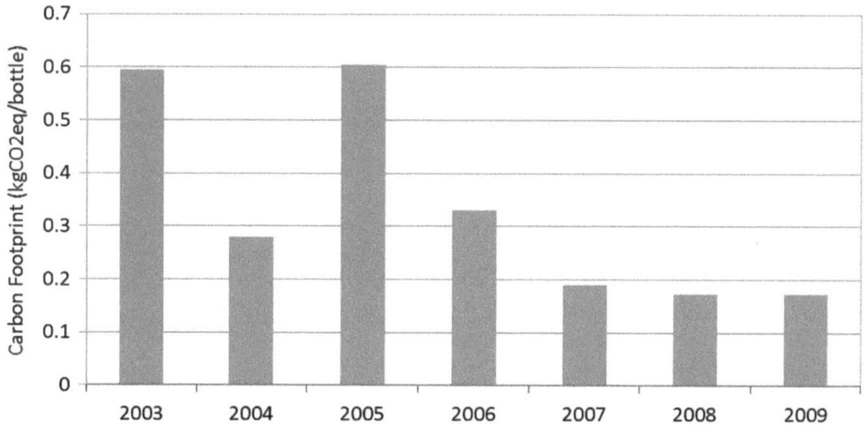

Figure 7.9 Carbon footprint of one bottle of wine, including only Scope 1 and 2 emissions.

state, this is the only way of comparing those results with other studies realized using a product-oriented approach. The CF of wine, considering Scope 1 and Scope 2 only and neglecting emission removals (mostly due to CO_2 capture from forests), is between 0.17 and 0.60 $kgCO_{2eq}$/bottle. There is a general decrease in the CF of the product, from 2003 to 2009, which is mostly related to the increase in productivity (ton of grapes per hectare). The anomalous behavior if 2005 is due a very low productivity in this year. The almost constant value during the last 3 years can be explained by the fact that the overall company production was almost the same, and there was no fundamental change in tillage techniques.

7.5.2 Case Study #2: VIVA Sustainable Wine

One of the first effort for the evaluation of the CF of the wine sector from a product perspective is represented by the VIVA Sustainable Wine project (VIVA Sustainable Wine; Lamastra et al., 2014). The goal of this pilot project, sponsored by the Italian Ministry for the Environment and realized in collaboration with research institutions and wine producers, was to develop an innovative methodology for the quantification of performance of wineries and wine products and support the organization choices for a transition to a more sustainable wine production process.

Along with CF, also other indicators are computed to estimate the comprehensive impact in air, water, territory, and vineyard, identifying and measuring the environmental and socioeconomic performance of wine and wineries.

The four indicators are as follows:

- Air—direct and indirect GHG emissions are evaluated for the entire life cycle of the product and the organization;
- Water—total volume of freshwater consumed and polluted because of vineyard and cellar activities;
- Vineyard—an indicator taking into account the agronomic management practices in the vineyards, including pest management practices, fertilization, soil fertility, soil erosion, and landscape;
- Territory—an indicator for the evaluation of the impact and preservation of protected areas and heritage sites including biodiversity, landscape, society, and collectivity.

Also, a label and a smartphone application enabling the final consumers to recognize producers committed to the project were developed (Fig. 7.10).

Figure 7.10 Label and quick response-code of the VIVA project.

The methodology for CF evaluation was originally designed starting from an organization perspective (i.e., based on an ISO 14064) and subsequently modified and adapted to focus on the product as long as on the corporate performance. This hybrid approach represents the first attempt to produce an LCA assessment for the evaluation of impacts related the global warming that was able to overcome the limits of organization-oriented studies. In this case, some of the controversial sections of the ISO 14064, such as CO_2 offsetting, were still preserved.

However, results shown in Fig. 7.11 were obtained neglecting the CO_2 offset contribution because it was subsequently excluded starting from the release of the ISO 14067. This is a crucial improvement with respect to ISO 14064 in scientific reliability of the LCA results and in clarity and acceptability of the effectiveness of company strategies. More in detail, as of ISO 14064, the acquisition of forests or other surfaces, where CO_2 capture and

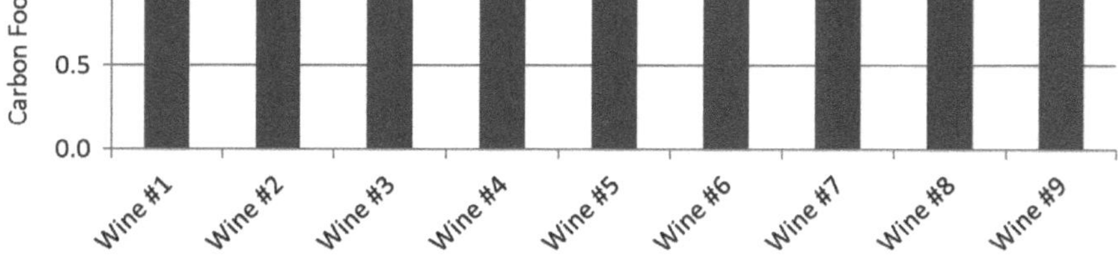

Figure 7.11 Carbon footprint of nine red wines computed using the VIVA methodology.

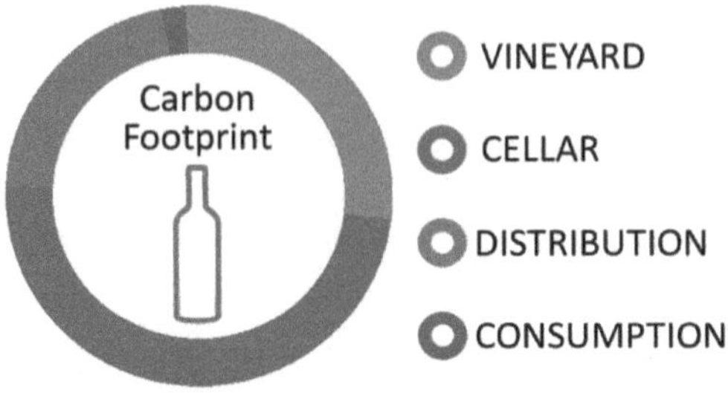

Figure 7.12 Details of carbon footprint results for Red #6.

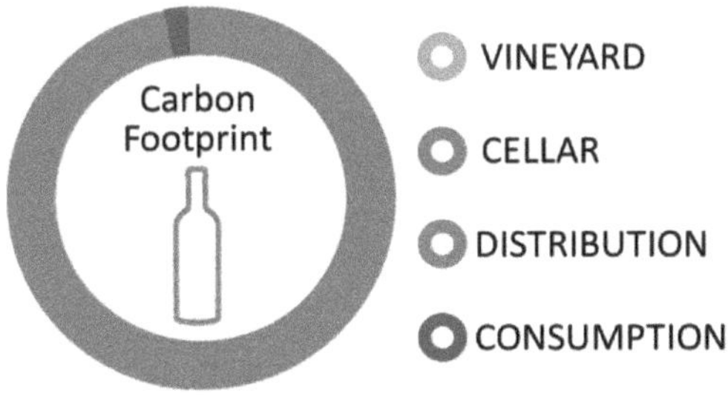

Figure 7.13 Details of carbon footprint results for Red #7.

sequestration naturally takes place, could be considered as a company mitigation strategy, i.e., resulting in a reduction of GHG emission for the company even though there is no relief for the environment.

Details of the results for two wines are reported in Figs. 7.12 and 7.13. The former is relative to Red #6, characterized by a CF of 0.994 $kgCO_{2eq}$/bottle and the latter is relative to Red #7, characterized by a CF of 1.934 $kgCO_{2eq}$/bottle. Very clear differences can be observed analyzing the relative impact of each phase to the total CF. In the first case, the overall performance is much better because of a generally more efficient production during the vinification phase (cellar activities) and distribution. In the second case, even if almost no impact arises from vineyard activities, a less efficient cellar production and distribution leads to a worse environmental performance. This is a clear example of how an LCA approach is essential to correctly estimate the overall sustainability of a product or a process: even a strong reduction of impacts from a single phase is often not sufficient to guarantee that the product is more sustainable than others because other phases might compensate such benefits.

7.5.3 Case Study #3: ISO 14067 for Red Wines

A second pilot project, sponsored by the Umbria region, was developed with the aim of producing a first survey of the sustainability level of the local wine industry (Asdrubali et al. 2014, 2015). The CF of red wines produced by different companies was assessed using the same methodology to compare the results obtained from different vineyard management and cellar procedures. An original approach was developed based on the just-released, at the time, ISO 14067. This two-year project started in 2013 and focused on four red wines produced by fresh local grapes harvested in 2012 and commercialized after 2 year of aging. Together with the CF evaluation, other outcomes of the project were the study and collection of current best practices in wine production and wine making, and the development of a *sustainability plan* for the improvement of the environmental performance of the sector in compliance with national and local rules.

The methodology was setup considering a rigorous cradle-to-grave approach, that, de facto, corresponds to the more modern concept referred to as cradle-to-cradle approach, meaning that recycling and reuse of byproducts and waste were considered within the product lifecycle.

After a deep survey of the company activities, a life cycle model was developed using the PRé Consultants SimaPro 8.0 software (SimaPro LCA Software) supported by the ecoInvent 3.0 database (ecoinvent Version 3). Input data for the model were produced combining site-specific data obtained from the companies and allocations according to mass and volume flows along the product life cycle.

Allocation was performed on a cropped surface basis for impacts arising from tillage in consideration of the fact that treatments are applied using this criterion. For cellar activities including internal transportations, allocation was performed on a mass basis, and for aging, bottling, and distribution it was performed on a volume basis.

In particular, the LCA was divided into the three-module layout as indicated in the relative PCR: upstream, core, and downstream modules. The relevant subprocesses included in each module are the following:

- Upstream module
  - Energywares;
  - Grapes;
  - Packaging;
  - Other materials;
  - Use of fertilizers;

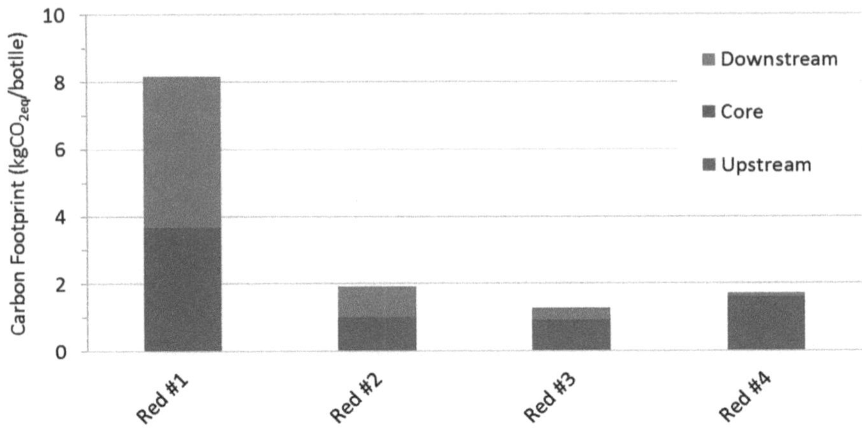

Figure 7.14 Carbon footprint of four red wines computed using the ISO 14067 methodology.

- Core module
    - Materials;
    - Transportation;
- Downstream module
    - Distribution;
    - End of life.

According to this scheme, all of the impacts associated to the production and use of energywares are associated to the upstream phase, including diesel burning for field activities, together with the production of primary, secondary, and third tier packaging materials.

The same system boundaries were used for the four products to make results comparable and to highlight possible differences as discussed in the previous section Fig. 7.5.

As shown in Fig. 7.14, results are very different for Red #1 with respect to the other three wines. For the first product, ~55% of the total CF is because of the distribution phase, and in particular from the fact that it is almost entirely shipped internationally via airplanes. Also, the vineyard contribution is up to 30 times that of the other three wines because of an extraordinary use of treatments to counteract anomalous vine infections that required an excessive consumption of diesel fuel. Neglecting this anomalous behavior, the average CF ranges from 1.27 to 1.93 $kgCO_{2eq}$/bottle.

Details of the results for red wines showing a regular behavior are reported in Fig. 7.15. When they are compared with results from Red #1 (Fig. 7.16), it is clear that the excessive increase of CF in the latter case is

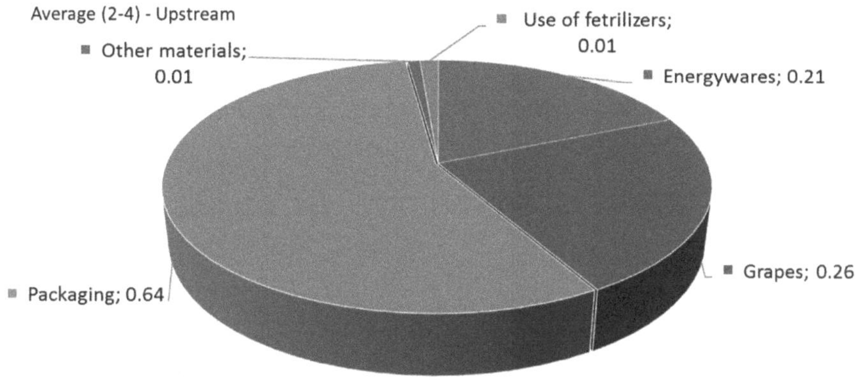

Figure 7.15 Detail of the upstream phase for other wines (average).

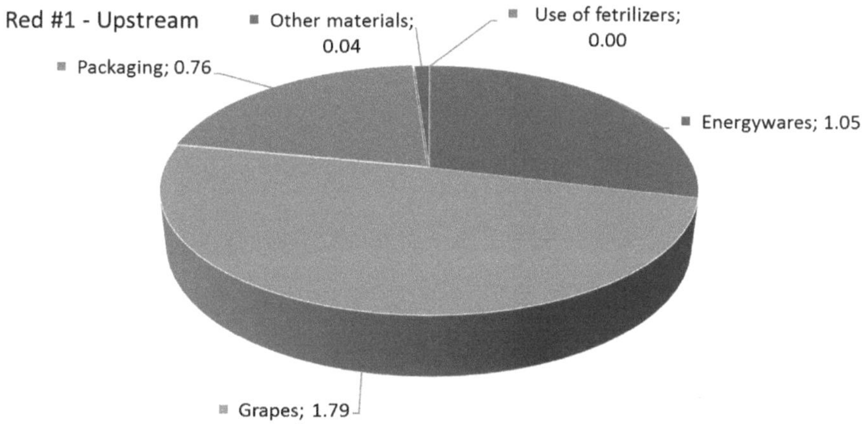

Figure 7.16 Detail of the upstream phase of Red #1.

because of the increase use of diesel that contribute both to the energywares subprocess (diesel production) and the grapes subprocess (diesel burning).

A full result report is shown in Table 7.3 (total CF and contribution from different modules), Table 7.4 (details of the upstream module), Table 7.5 (details of the core module), and Table 7.6 (details of the

Table 7.3 Carbon footprint results for different modules

Wine	Total ($kgCO_{2eq}$)	Upstream ($kgCO_{2eq}$)	Core ($kgCO_{2eq}$)	Downstream ($kgCO_{2eq}$)
Red #1	8.152	3.637	0.035	4.480
Red #2	1.925	0.973	0.025	0.928
Red #3	1.275	0.896	0.023	0.356
Red #4	1.702	1.546	0.054	0.102
Average 2–4	1.634	1.139	0.034	0.462

Table 7.4 Carbon footprint results for the upstream module

Wine	Upstream ($kgCO_{2eq}$)	Energy ($kgCO_{2eq}$)	Grapes ($kgCO_{2eq}$)	Packaging ($kgCO_{2eq}$)	Material ($kgCO_{2eq}$)	Fertilizer ($kgCO_{2eq}$)
Red #1	3.637	1.049	1.788	0.760	0.040	0.000
Red #2	0.973	0.315	0.089	0.560	0.009	0.000
Red #3	0.896	0.123	0.119	0.627	0.015	0.013
Red #4	1.546	0.203	0.582	0.739	0.006	0.017
Avg. 2–4	1.139	0.214	0.263	0.642	0.010	0.015

Table 7.5 Carbon footprint results for the core module

Wine	Core ($kgCO_{2eq}$)	Materials ($kgCO_{2eq}$)	Transportation ($kgCO_{2eq}$)
Red #1	0.035	0.000	0.035
Red #2	0.025	0.005	0.020
Red #3	0.023	0.000	0.023
Red #4	0.054	0.010	0.043
Average 2–4	0.034	0.005	0.028

Table 7.6 Carbon footprint results for the downstream module

Wine	Downstream ($kgCO_{2eq}$)	Distribution ($kgCO_{2eq}$)	End-of-life ($kgCO_{2eq}$)
Red #1	4.480	4.564	−0.084
Red #2	0.928	0.998	−0.070
Red #3	0.356	0.428	−0.071
Red #4	0.102	0.187	−0.085
Average 2–4	0.462	0.538	−0.076

downstream module). The average of products showing a regular behavior (Red #2, Red #3, and Red #4) is also presented.

7.5.4 Case Study #4: Beyond ISO 14067

A final example of CF assessment of the wine industry is given in this section. In this case, we focus on a recent work based on an innovative approach (Bonamente et al., 2016) for the joint evaluation of both CF and WF. The methodology was developed, once again, using a cradle-to-cradle perspective. In this study (Rinaldi et al., 2016), the well-established methodology for the evaluation of the CF of a product is presented along with a preliminary version of a WF approach. The definitive structure of the life cycle of wine production, elaborated during the past years via numerous studies and projects, is used as a starting point to build a solid model for the indicator WF.

The system boundaries are, once again, defined as in Fig. 7.5. A graphic representation of the life cycle is also shown in Fig. 7.17, where red bars show the relative impacts of each subprocess contributing to the total CF, and blue bars show the relative impacts to WF.

The upstream-, core-, and downstream-module structure is preserved; however, an additional subprocess was added to the upstream and core modules, to separately manage the impact on the water resource, which was not needed from a CF only perspective. The recursive approach of an LCA is particularly clear in this choice, as a well-established methodology is reconsidered and improved to include further developments and findings.

According to its general definition, the WF is defined as the sum of green, blue, and gray volumes of freshwater consumed during the product life cycle:

$$WF = WF_{green} + WF_{blue} + WF_{gray}$$

The LCA analysis is carried out following the requirements of international standards ISO 14067 and ISO 14046. A complete review of the WF methodology is available at Rinaldi et al. (2016), as long as guidelines for all the phases of the evaluation procedure, including acquisition and validation of input data, allocation, application of analytic models, and interpretation of the results.

The strength of this approach is the implementation of a side-by-side CF versus WF assessment, based on the same system boundaries, functional unit, and input data, that allows a reliable comparison between the two indicators.

The methodology was applied to a red and a white wine, vintage year 2012, produced in the same company. The FU is a 0.75-L bottle.

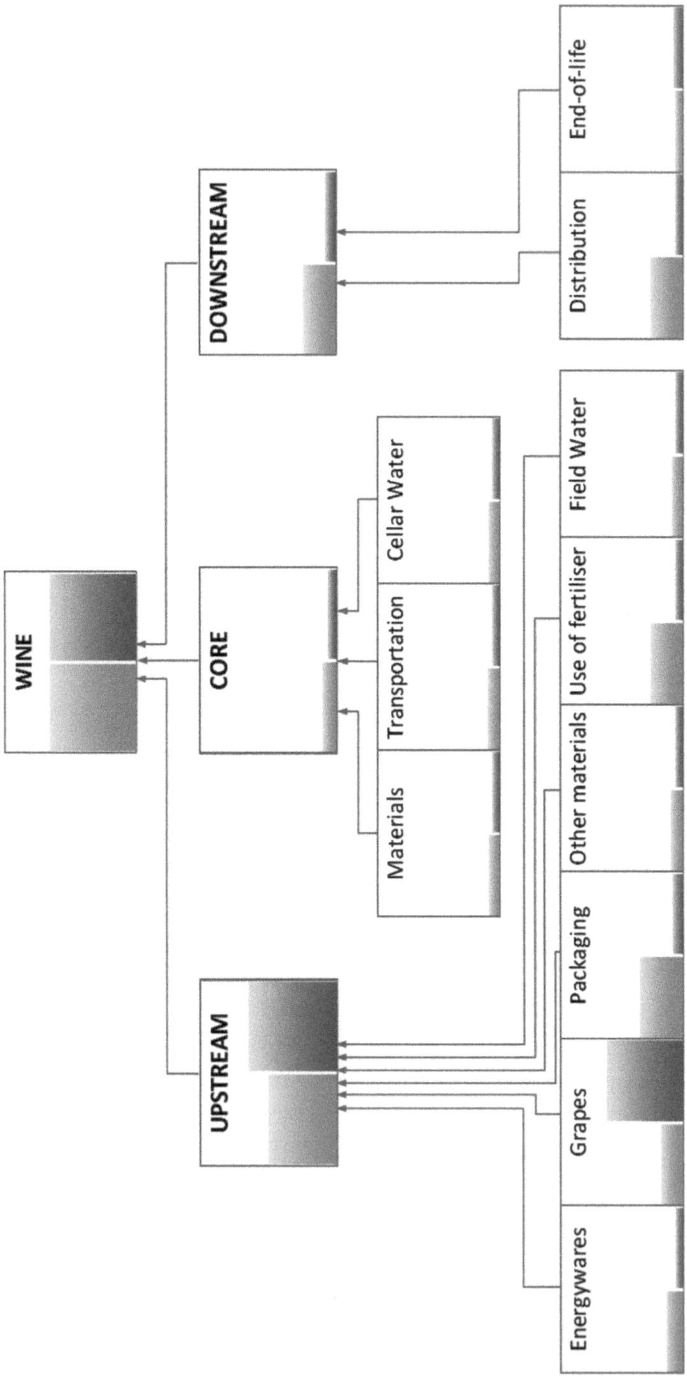

Figure 7.17 Life cycle of wine: carbon footprint versus water footprint.

Table 7.7 Data available for total production

Input	Unit	Amount
Electric energy	kWh	44,561
Diesel	L	5,800
Lubricating oil	kg	56
Glycol	kg	0.75
Potassium metabisulfite	kg	50
Enzyme	kg	0.4
Yeast	kg	10
Carbon dioxide	kg	840
Acetic acid	kg	3.33
Diammonium phosphate	kg	100
Soap	kg	385
Packaging plastic	kg	20.42
Packaging paper	kg	0.4
Refrigerant (R404a)	kg	0.21
Transportation (<3.5 t)	tkm	2700
Transportation (3.5–7.5 t)	tkm	94.08
Transportation (>16 t)	tkm	612.8

The life cycle model was developed using the PRé Consultants SimaPro 8.1 software (SimaPro LCA Software) supported by the ecoInvent 3.1 database (Ecoinvent Version 3).

A comparison between the two products is presented for each LCA phase along with literature results for similar wines. The product life cycle was divided in a total of 11 phases (e.g., subprocesses). An original methodology was setup to correctly take into account all the direct and indirect contributions, and a particular effort was dedicated to the evaluation of the gray component.

A refined allocation procedure was setup to process site-specific data, available at the level of entire production within the company (i.e., including other wines, Table 7.7), at the level of the studied product (Table 7.8), and at the level of the FU (i.e., the bottle, Table 7.9).

The end-of-life process was also modeled considering site-specific data for the percentage of recycled, incinerated, and sent to landfill materials (Table 7.10).

The results, reported in Figs. 7.18–7.20, show a total CF equal to 1.07 ± 0.09 $kgCO_{2eq}$/bottle and a total WF equal to 580 ± 30 L/bottle. Packaging and distribution are the most impacting processes in CF (56.1% and 41.1%, respectively). The WF is dominated by green water that is evapotranspirated by the crop (77.94% of the total).

Table 7.8 Data available for studied product (red wine)

Input	Unit	Amount
Nitrogen fertilizer[a]	kg	15
Phosphate fertilizer[a]	kg	6.252
Potassium fertilizer[a]	kg	18.76
Organic fertilizers[a]	kg	12.5
Packaging plastic	kg	0.5001
Other generic pesticides[a]	kg	1.753
Triazine[a]	kg	0.1313
Fosetyl aluminum[a]	kg	3.359
Sulfur[a]	kg	27.58
Acetamide[a]	kg	0.1309
Copper[a]	kg	2.151
Dichlorophenol[a]	kg	0.3334
Dimethylamine[a]	kg	0.02223
Paper	kg	1.613
Plastic	kg	0.5863
Water	m^3	18.36
Transportation (<3.5 t)	tkm	2.165

[a]Active ingredient.

Table 7.9 Data available for functional unit (red wine)

Input	Unit	Amount
Bottle	kg	0.45
Cork	kg	0.004
Capsule	kg	0.001
Label	kg	0.001
Box	kg	0.048
Packaging film	kg	5.58E-04
Corrugated box	kg	3.00E-04
Plastic	kg	8.61E-03
Transportation (car)	km	3.63E-05
Transportation (<3.5 t)	tkm	0.1539
Transportation (3.5–7.5 t)	tkm	0.2974

Table 7.10 End-of-life scenario for distribution (red wine)

Waste	Material	Recycling (%)	Landfill (%)	Incineration (%)
Box	Cardboard	77.50	22.50	–
Bottle	Glass	70.70	29.30	–
Cork	Cork	–	100	–
Label	Paper	–	29.30	70.70
Capsule	Plastic	34.70	65.30	–

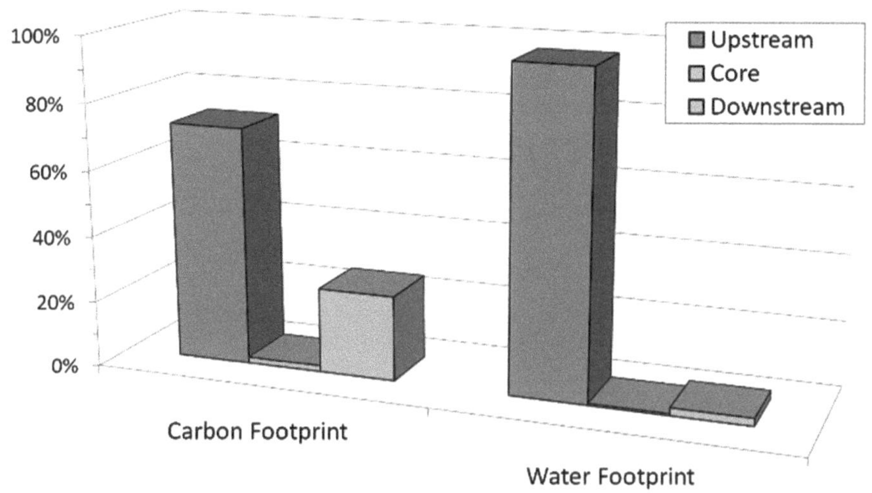

Figure 7.18 Upstream, core, and downstream contributions to the total carbon footprint and water footprint (red wine).

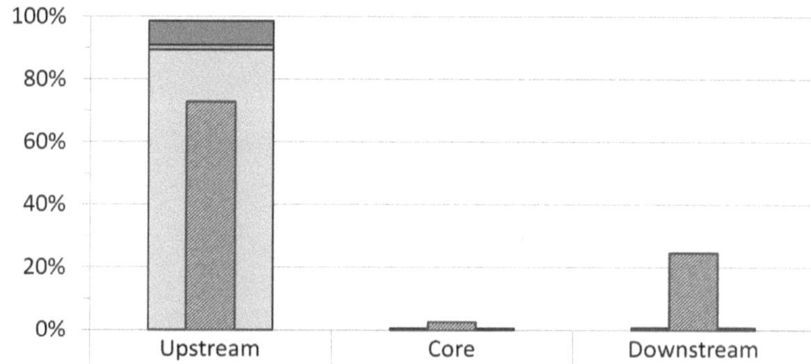

	Upstream	Core	Downstream
Gray WF	7.73%	0.08%	0.82%
Blue WF	1.44%	0.50%	0.05%
Green WF	89.38%	0.00%	0.00%
Carbon Footprint	72.77%	2.53%	24.70%

Figure 7.19 Carbon footprint versus water footprint (WF) for a red wine.

The results are consistent with those of previous analyses of the same product and to other studies relative to red wines produced worldwide. A correlation study between carbon and WF was also performed, and the results indicate a direct proportionality between the total CF and the sum of $WF_{gray,\ indirect}$ and WF_{blue} corresponding to 81.24 L/kgCO_{2eq} (Figs. 7.21 and 7.22).

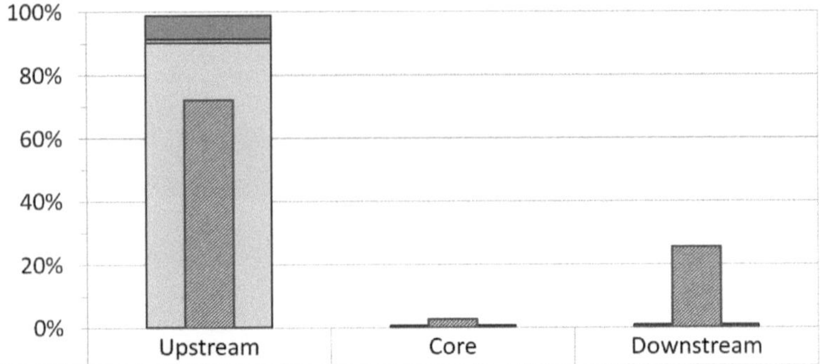

	Upstream	Core	Downstream
Gray WF	7.32%	0.07%	0.71%
Blue WF	1.27%	0.46%	0.05%
Green WF	90.13%	0.00%	0.00%
Carbon Footprint	72.09%	2.55%	25.36%

Figure 7.20 Carbon footprint versus water footprint (WF) for a white wine.

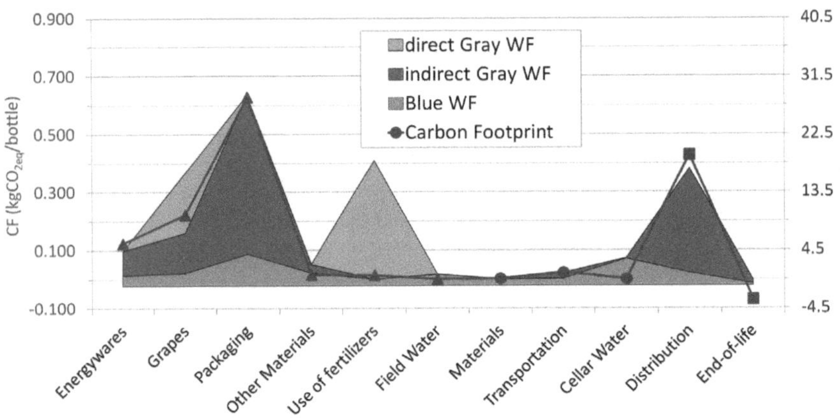

Figure 7.21 Contribution of the 11 phases to the total carbon footprint (CF) (left axis) and the gray + blue water footprint (WF) (right axis) of the red wine.

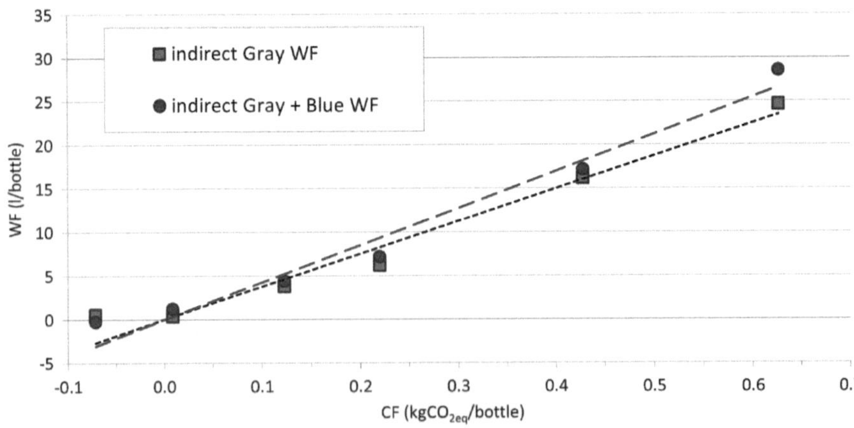

Figure 7.22 Correlation between carbon footprint (CF) and water footprint (WF) for the red wine.

7.6 CONCLUSIONS

The food sector is recognized to be one of the major impacting ones and, within all the food products, wine emerges as one of the most analyzed because of its great relevance both in the economic production and in the world distribution market. The wine industry, in the last years, has increasingly developed an environmental awareness, and it is becoming more and more committed to sustainability.

In this chapter, results regarding CF evaluation for typical Italian wineries through different methodological approach are reported, together with a review of the literature and different evaluation experiences in the wine sector, with the aim of giving a comprehensive view of the environmental pressures related to the life cycle of a wine bottle. In this way, the environmental hot spots of the production process can be clearly identified as a starting point for strategic decisions aimed at the improvement and optimization of the production process itself.

Case study #1 refers to one of the first application of the ISO 14064 standard (i.e., company-oriented evaluation). The CF of wine considering Scope 1 and Scope 2 emissions only and neglecting emission removals is between 0.17 and 0.60 $kgCO_{2eq}$/bottle, which are very similar to the results obtained for analogous phases of the life cycle in the other case studies.

Case study #2 refers to one of the first effort for the evaluation of the CF of the wine sector from a product perspective (VIVA Sustainable Wine project and Lamastra et al. (2014)). The accounting methodology was originally designed from an organization perspective (i.e., based on ISO 14064), and subsequently it was modified and adapted to focus on the product as long as on the corporate performance. The results, related to nine different red wines, were in the range 0.9–2.0 $kgCO_{2eq}$/bottle.

Case study #3 refers to another pilot project aimed at producing a first survey of the sustainability level of the Umbria region wine industry. The calculation methodology was specifically developed based on the just released, at that time, ISO 14067. Neglecting anomalous situation of the specific vintage year, the average CF ranged from 1.27 to 1.93 $kgCO_{2eq}$/bottle.

Case study #4 was a joint evaluation of both carbon and WF carried out according to latest international standards (ISO 14067 and 14046). Results showed a total CF equal to 1.07 ± 0.09 with packaging and distribution being the most impacting processes (56.1% and 41.1%, respectively). Moreover, a direct proportionality between the total CF and the sum of $WF_{gray,\ indirect}$ and WF_{blue} was observed.

Results of the different case studies are aligned, and such evidence can be interpreted as a consequence of the consistency between the different methodologies developed and applied for calculation. Moreover, as a further confirmation of the validity of the adopted calculation procedures, it is possible to note a comparability of the results with those of the available literature.

The major challenge of the presented studies was related to the lack of a dedicated methodology for the quantification of the CF of a product before the release of the ISO 14067. It was necessary to develop and successively improve an original methodology for the transition from an organization point of view to a product-oriented approach. Various efforts were provided, taking care of the consistency of the results at each step. In general, even if the procedure is now standardized, data acquisition still represents a potential issue because data recorded by wineries need to be carefully validated and deeply preprocessed before they can be used to build the life cycle inventory.

Finally, although CF represents probably the most widespread and well-known indicator, it refers to a single impact category. A more comprehensive analysis of the wine production process requires the proper identification of other relevant indicators. First attempts to evaluate such indicators are presented and they include the impacts associated to the water resources, the territory, and the vineyard.

REFERENCES

Agronomic Study SATA – Italian Wine Carbon Calculator. http://www.agronomisata.it/itaca-approffondimento/.

Amienyo, D., Camilleri, C., Azapagic, A., 2014. Environmental impacts of consumption of Australian red wine in the UK. Journal of Cleaner Production 72, 110–119. http://dx.doi.org/10.1016/j.jclepro.2014.02.044.

Aranda, A., Scarpellini, S., Zabalza, I., 2005. Economic and environmental analysis of the wine bottle production in Spain by means of life cycle assessment. International Journal of Agricultural Resources, Governance and Ecology 4 (2), 178–191. http://dx.doi.org/10.1504/IJARGE.2005.007199.

Asdrubali, F., Bonamente, E., Merico, M.C., Scrucca, F., Lunghi, F., 2015. Carbon Footprint nel settore vitivinicolo umbro: implementazione di una metodologia di calcolo e applicazione ad alcuni casi studio. In: 15th CIRIAF National Congress, Perugia, Italy. April 9–11.

Asdrubali, F., Strambaci, A., Bonamente, E., Merico, M.C., Scrucca, F., Lunghi, F., 2014. Implementazione di una metodologia per il calcolo della Carbon Footprint nel settore vitivinicolo umbro. In: 14th CIRIAF National Congress, Perugia, Italy. April 4–5.

Australian Wine Research Institute, 2009. Australian Wine Carbon Calculator. http://www.awri.com.au/industry_support/entwine/carbon-calculator/.

Bonamente, E., Scrucca, F., Rinaldi, S., Merico, M.C., Asdrubali, F., Lamastra, L., 2016. Environmental impact of an Italian wine bottle: carbon and water footprint assessment. The Science of the Total Environment 560–561, 274–283. http://dx.doi.org/10.1016/j.scitotenv.2016.04.026.

Bosco, S., Di Bene, C., Galli, M., Remorini, D., Massai, R., Bonari, E., 2011. Greenhouse gas emissions in the agricultural phase of wine production in the Maremma rural district in Tuscany, Italy. Italian Journal of Agronomy 6, 93–100. http://dx.doi.org/10.4081/ija.2011.e15.

BSI, 2011. PAS 2050:2011 Specification for the Assessment of the Life Cycle Greenhouse Gas Emissions of Goods and Services. http://shop.bsigroup.com/upload/shop/download/pas/pas2050.pdf.

Carballo-Penela, A., García-Negro, M.C., Doménech-Quesada, J.L., 2009. A methodological proposal for corporate carbon footprint and its application to a wine-producing company in Galicia, Spain. Sustainability 1, 302–318. http://dx.doi.org/10.3390/su1020302.

Cholette, S., Kumar, V., 2009. The energy and carbon intensity of wine distribution: a study of logistical options for delivering wine to consumers. Journal of Cleaner Production 17, 1401–1413. http://dx.doi.org/10.1016/j.jclepro.2009.05.011.

Cleary, J., 2013. Life cycle assessments of wine and spirit packaging at the product and the municipal scale: a Toronto, Canada case study. Journal of Cleaner Production 44, 143–151. http://dx.doi.org/10.1016/j.jclepro.2013.01.009.

Climate Change, 2013. IPCC Fifth Assessment Report. The Physical Science Basis.

Comandaru, I.M., Bârjoveanu, G., Peiu, N., Ene, S.-A., Teodosiu, C., 2012. Life cycle assessment of wine: focus on water use impact assessment. Environmental Engineering and Management Journal 11 (3), 533–543.

Corbo, C., Lamastra, L., Capri, E., 2014. From environmental to sustainability programs: a review of sustainability initiatives in the Italian wine sector. Sustainability 6, 2133–2159. http://dx.doi.org/10.3390/su6042133.

ecoinvent Version 3. Available online: http://www.ecoinvent.org/database/ecoinvent-version-3/.

Environdec, 2010. Product Category Rules (PCR) for the Assessment of the Environmental Performance of UN CPC 24212: Wine of Fresh Grapes, Except Sparkling Wine; Grape Must. PCR 2010:02 Version 1.03 2013-07-23. The International EPD ® System, Stockholm, Sweden.

Environdec, 2014. Product Category Category Rules (PCR) for the Assessment of the Environmental Performance of UN CPC 24211: Sparkling Wine of Fresh Grapes. PCR 2014:14 Version 1.00 2017-09-23. The International EPD ® System, Stockholm, Sweden.

European Commission, Directorate-General for Agriculture and Rural Development, 2015. Wine Market Situation, Evolution and Background Information. http://ec.europa.eu/agriculture/sites/agriculture/files/wine/statistics/market-situation-2014-07_en.pdf.

FAO, 2015. Statistical Pocketbook – World Food and Agriculture. http://www.fao.org/3/a-i4691e.pdf.

FAO, 1996. Introduction to Agricultural Water Pollution. Food and Agriculture Organization of United Nations (FAO), Rome. http://www.fao.org/docrep/w2598e/w2598e04.htm.

FIVS, 2016. International Wine Greenhouse Gas Protocol – Version 2.0. http://www.wineinstitute.org/files/FIVS_Intl_Wine_GHG_v2-2016-02-16.pdf.

FIVS, 2008. International Wine Carbon Calculator Protocol – Version 1.2. https://fivs.org/resources/virtualLibrary.htm?a=report&start=0&documentType=&query=IWCC.

Foley, J.A., Ramankutty, N., Brauman, K.A., Cassidy, E.S., Gerber, J.S., Johnston, M., et al., 2011. Solutions for a cultivated planet. Nature 478 (7369), 337–342. http://dx.doi.org/10.1038/nature10452.

Fusi, A., Guidetti, R., Benedetto, G., 2014. Delving into the environmental aspect of a Sardinian white wine: from partial to total life cycle assessment. The Science of the Total Environment 472, 989–1000. http://dx.doi.org/10.1016/j.scitotenv.2013.11.148.

Gazulla, C., Raugei, M., Fullana, P., 2010. Taking a life cycle look at crianza wine production in Spain: where are the bottlenecks? The International Journal of Life Cycle Assessment 15, 330–337. http://dx.doi.org/10.1007/s11367-010-0173-6.

Greenhaigh, S., Mithraratne, N., Sinclair, R., Smith, S., McConachy, E., Barber, A., March 2011. GHG Product Accounting Guidelines for the Wine Industry. MAF Technical Paper No: 2011/16. Ministry of Agriculture and Forestry, New Zealand.

Harris, I., et al., 2016. The impact of alternative routeing and packaging scenarios on carbon and sulphate emissions in international wine distribution. Transportation Research Part D Transport and Environment. http://dx.doi.org/10.1016/j.trd.2016.08.036.

Iannone, R., Miranda, S., Riemma, S., De Marco, I., 2016. Improving environmental performances in wine production by a life cycle assessment analysis. Journal of Cleaner Production 111, 172–180. http://dx.doi.org/10.1016/j.jclepro.2015.04.006.

ISO (International Standard Organization), 2013. ISO/TS 14067:2013 Greenhouse Gases - Carbon Footprint of Products - Requirements and Guidelines for Quantification and Communication. International Organization for Standardization, Geneva, Switzerland.

ISO 14020:2000, Environmental Labels and Declarations – General Principles.

ISO 14024:1999, Environmental Labels and Declarations – Type I Environmental Labelling – Principles and Procedures.

ISO 14025:2006, Environmental Labels and Declarations – Type III Environmental Declarations – Principles and Procedures.

ISO 14040:2006, Environmental Management – Life Cycle Assessment – Principles and Framework.

ISO 14044:2006, Environmental Management – Life Cycle Assessment – Requirements and Guidelines.

ISO 14064–1:2012, Specification with Guidance at the Organization Level for Quantification and Reporting of Greenhouse Gas Emissions and Removals.

Lamastra, L., Suciu, N.A., Novelli, E., Trevisan, M., 2014. A new approach to assessing the water footprint of wine: an Italian case study. Science of the Total Environment 490, 748–756.

Marras, S., Masia, S., Duce, P., Spano, D., Sirca, C., 2015. Carbon footprint assessment on a mature vineyard. Agricultural and Forest Meteorology 214–215, 350–356. http://dx.doi.org/10.1016/j.agrformet.2015.08.270.

Meneses, M., Torres, C.M., Castells, F., 2016. Sensitivity analysis in a life cycle assessment of an aged red wine production from Catalonia, Spain. Science of the Total Environment 562, 571–579. http://dx.doi.org/10.1016/j.scitotenv.2016.04.083.

New Zealand Wine Company press area. http://www.nzwineco.co.nz/press.aspx?pid=10058.

OIV Databases and Statistics. http://www.oiv.int/en/databases-and-statistics.

OIV Resolution: OIV-CST 503AB-2015, 2015. Greenhouse Gases Accounting in the Vine and Wine Sector – Recognized Gases and Inventory of Emissions and Sequestrations, ISBN 979-10-91799-58-4. http://www.oiv.int/public/medias/4521/publication-bilan-ges-en.pdf.

OIV Resolution OIV-CST 431-2011, 2011. General Principles of the OIV Greenhouse Gas Accounting Protocol for the Vine and Wine Sector. http://www.oiv.int/public/medias/2107/oiv-cst-431-2011-en.pdf.

OIV Statistical Report on World Vitiviniculture, 2016. http://www.oiv.int/public/medias/5029/world-vitiviniculture-situation-2016.pdf.

Point, E., Tyedmers, P., Naugler, C., 2012. Life cycle environmental impacts of wine production and consumption in Nova Scotia, Canada. Journal of Cleaner Production 27, 11–20. http://dx.doi.org/10.1016/j.jclepro.2011.12.035.

Rinaldi, S., Bonamente, E., Scrucca, F., Merico, M.C., Asdrubali, F., Cotana, F., 2016. Water and carbon footprint of wine: methodology review and application to a case study. Sustainability 8, 621.

Rinaldi, S., 2010. Sperimentazione di modelli di calcolo per la valutazione delle emissioni di GHG nelle filiere agricole (Ph.D. thesis). University of Perugia, Italy.

Rugani, B., Vázquez-Rowe, I., Benedetto, G., Benetto, E., 2013. A comprehensive review of carbon footprint analysis as an extended environmental indicator in the wine sector. Journal of Cleaner Production 54, 61–77. http://dx.doi.org/10.1016/j.jclepro.2013.04.036.

SimaPro LCA Software. Available online: http://www.pre-sustainability.com/simapro.

Soosay, C., Fearne, A., Dent, B., 2012. Sustainable value chain e a case study of Oxford Landing. Supply Chain Manage 17 (1), 68–77.

Torquati, B., Rinaldi, S., Cotana, F., Fasola, L., Taglioni, C., Mariani, G., 2010. 360 Green Revolution: 0 greenhouse emissions through environmentally responsible agriculture production. In: 119th EAAE Seminar. Capri, Italy. June, 30–July, 2, 2010.

UN – Sustainable Development Solutions Network, 2013. Solutions for Sustainable Agriculture and Food Systems. Technical Report for the Post-2015 Development Agenda. http://unsdsn.org/wp-content/uploads/2014/02/130919-TG07-Agriculture-Report-WEB.pdf.

VIVA Sustainable Wine. Available online: http://www.viticolturasostenibile.org.

Vázquez-Rowe, I., Villanueva-Rey, P., Moreira, M.T., Feijoo, G., 2012. vironmental analysis of Ribeiro wine from a timeline perspective: harvest year matters when reporting environmental impacts. Journal of Environmental Management 98, 73–83. http://dx.doi.org/10.1016/j.jenvman.2011.12.009.

Vermeulen, S.J., Campbell, B.M., Ingram, J.S.I., 2012. Climate change and food systems. Annual Review of Environment and Resources 37, 195–222. http://dx.doi.org/10.1146/annurevenviron020411-130608.

Villanueva-Rey, P., Vázquez-Rowe, I., Moreira, M.T., Feijoo, G., 2014. Comparative life cycle assessment in the wine sector: biodynamic vs. conventional viticulture activities in NW Spain. Journal of Cleaner Production 65, 330–341. http://dx.doi.org/10.1016/j.jclepro.2013.08.026.

World Resources Institute and World Business Council for Sustainable Development, 2004. The Greenhouse Gas Protocol. A Corporate Accounting and Reporting, ISBN 1-56973-568-9.
World Resources Institute, 2014. GHG Protocol Agriculture Guidance. http://www.ghgprotocol.org/standards/agriculture-guidance.
WRAP-Waste Resources Action Programme, 2007. The Life Cycle Emissions of Wine Imported to the UK. Manufacturing/Final Report. Banbury, Oxon, UK, p. 28.

CHAPTER 8

Carbon Footprint of Aluminum Production: Emissions and Mitigation

Meenu Gautam, Bhanu Pandey, Madhoolika Agrawal
Department of Botany, Institute of Science, Banaras Hindu University, Varanasi, India

8.1 INTRODUCTION

Climate change is greatly challenging the sustainability of human society, and there are several efforts that have been made to evaluate the amount of human-induced carbon emissions. The Intergovernmental Panel on Climate Change (IPCC) in its fifth assessment report (AR5) strongly recommended restricting the global temperature rise to well below 2°C as compared with preindustrial climate, to avoid serious ecological as well as economic threats (Thomas et al., 2016; IPCC, 2014). The main factor causing global rise in temperature is increase in global greenhouse gas (GHG) emissions. The link between GHG emissions and its consequent impacts on living beings has drawn the attention of scientists, politicians, and general public through well-known terms "global warming" and "GHG effects" (Fig. 8.1). The successive rise in global temperature has been

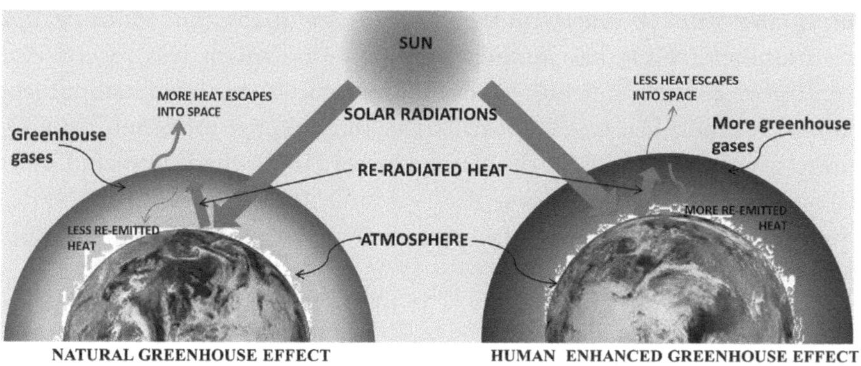

Figure 8.1 Greenhouse effects and global warming. The incoming solar and reemitted infrared radiations.

Environmental Carbon Footprints
ISBN 978-0-12-812849-7
http://dx.doi.org/10.1016/B978-0-12-812849-7.00008-8

accompanied by the changes in weather and climate. Many places have seen changes in rainfall pattern resulting in more floods, drought, or intense rainfall, as well as more frequent and severe heat waves. Oceans and glaciers have also experienced some big changes i.e., oceans are warming and becoming more acidic, ice caps are melting, and sea levels are rising. Based on several studies, future projection of GHG emissions from different entities may likely to cause several other changes in the coming decades (IPCC, 2014; Pandey and Agrawal, 2014; Kerr, 2006).

At global scale, the key GHGs emitted by anthropogenic activities are carbon dioxide (CO_2), methane (CH_4), nitrous oxide (N_2O), and fluorinated gases such as hydrofluorocarbons (HFCs), perfluorocarbons (PFCs), and sulfur hexafluoride (SF_6). Besides, black carbon is not only a solid particle or aerosol (not a gas) but also contributes to warming of the atmosphere (IPCC, 2014). Based on global GHG emissions in 2010, CO_2 emissions from fossil fuel combustion and industrial processes contributed nearly 65%, whereas from forestry and other land-use changes, it accounted ~11% of GHG emissions. Other gases, viz. CH_4, N_2O, and fluorinated gases contributed ~16%, 6.0%, and 2.0% of GHG emissions, respectively (IPCC, 2014). Thus, CO_2 is a major contributory gas among GHGs, which has subsidized the most to climate change between 1750 and 2005 (Trenberth et al., 2007; Forster et al., 2007). The atmospheric CO_2 concentration recorded at Mauna Loa (Fig. 8.2) showed a continuous rise in CO_2 level in the atmosphere with seasonal and monthly variations from 1960 to 2015 (National Oceanic and Atmospheric Administration (NOAA), 2015). The rate of increase has not been steady over time because the annual growth rate of atmospheric CO_2 has roughly tripled from 0.6 ppm/year in the early 1960s to 2.1 ppm during the past 10 years (NOAA, 2015). In 2015, the concentration of atmospheric CO_2 has jumped by 3.05 ppm, which was recorded as the largest annual increase in the observatory's record (State of the Climate: Carbon Dioxide Report, 2015). In 2011, the top identified CO_2 emitters were China (28%), the United States (16%), the European Union (10%), the Russian Federation (6.0%), India (6.0%), and Japan (4.0%), whereas other countries contributed ~30% of global CO_2 emissions (Boden et al., 2015).

The major contributing sectors worldwide in carbon emissions are agricultural, transport, industrial, construction, land-use changes, and wastes (Blanco et al., 2014). The concentration of GHGs in the atmosphere has

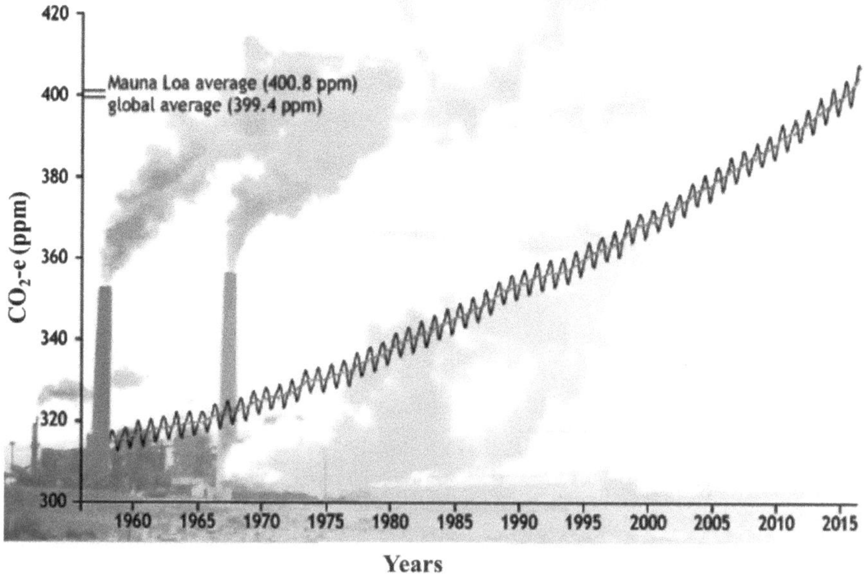

Figure 8.2 The annual growth rate of atmospheric carbon dioxide from 1960 to 2015. Monthly average carbon dioxide concentrations recorded at Mauna Loa [blue (dark gray in print versions) line] with seasonal variations [pink (light gray in print versions) line]. *(Based on State of the Climate: Carbon Dioxide Report, 2015. https://www.climate.gov/news-features/featured-images/2015-state-climate-carbon-dioxide.)*

been rising since the beginning of the Industrial Revolution in 1750 (IPCC, 2014; Etheridge et al., 1996). Energy consumption in industrial sectors accounted for more than 70% of the total energy production; hence energy-related GHG emissions from industrial sectors are likely to cause significant impacts on global climate change (Sims et al., 2007). There are several studies about GHG emissions from mining and metallurgical industries, and aluminum industries are one of the most energy-intensive industries after steel. About 21% of global GHG emissions are contributed by metal industries, of which aluminum industries contribute nearly 1.0% (U.S. Greenhouse Gas Inventory Report, 2016).

In view of the above, this chapter aims to describe the pollutants particularly emission scenario of greenhouse gases from varying processes involved in alumina refineries and aluminum production. The chapter also includes the consequent impacts on ecological system and mitigation steps taken by industries and policy makers in carbon emission reduction.

8.2 ALUMINUM PRODUCTION

8.2.1 Current Scenario of Aluminum Production

Aluminum is the second highest metal in demand after steel for modern societies, and recent decades have witnessed its utilization in packaging, construction, transportation, building, and electrical engineering because of its versatile properties. Global annual production of aluminum is 4,17,712 thousand metric tons, and China contributes maximum proportion followed by Gulf of Cooperation Council and North America (World Aluminium, 2015) (Fig. 8.3). A successive increase in demand for aluminum caused a significant increase in its production in past 10 years (Fig. 8.4). Global annual demand for aluminum is expected to increase twofold to threefold by 2050 (World Aluminium, 2015). The bulk growth in consumption of aluminum is likely to take place in China, India, the Middle East, and other developing countries, where consumption is expected to nearly quadruple by 2025 (Menzie et al., 2010). To meet the increasing demand for aluminum, there is a continuous rise in its production, which is projected to increase from ~51 Mt in 2014 to 89–122 Mt in 2050 (International Energy Agency (IEA), 2012). This increase in aluminum consumption and production will drive significant growth in the industry's absolute energy use and carbon emission. Life cycle assessment (LCA) of aluminum production thus is of paramount

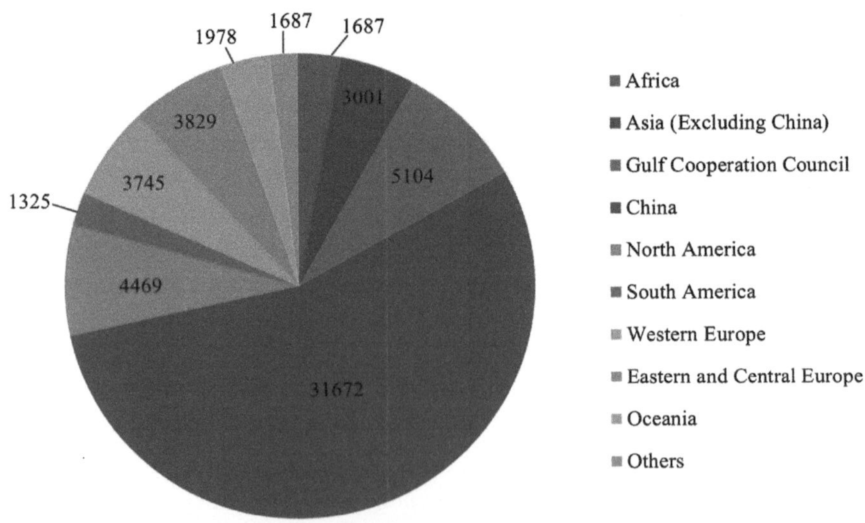

Figure 8.3 Countries with the largest aluminum production (1000 metric tons per annum) in 2015. *(Reproduced from http://www.world-aluminium.org/statistics/#data.)*

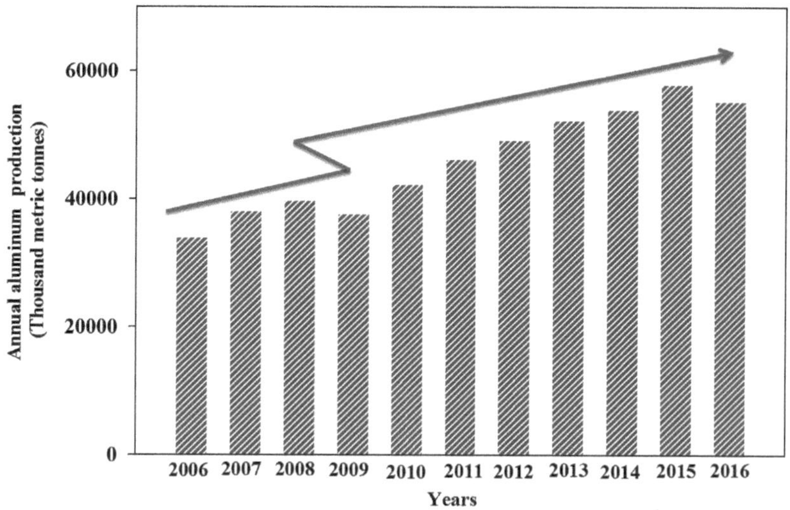

Figure 8.4 Global annual production of aluminum in last 11 years. *(Reproduced from http://www.world-aluminium.org/statistics/#data.)*

importance in identifying the strengths and weaknesses of the methodologies adopted to quantify environmental profiles and pathways toward sustainability goals.

8.2.2 Life Cycle Assessment of Aluminum Production

The LCA is an important tool in environmental management by identifying energy and materials consumed and wastes released into the environment (Russell et al., 2005). It uses a system approach to understand the potential environmental consequences of the product, process, and/or activity from initial extraction of raw materials from the earth until the point when its end products are returned to the earth (US Environmental Protection Agency (U.S. EPA), 2016). Norgate and Haque (2010) performed LCA of aluminum production and investigated the results of energy (coal, natural gas, and hydroelectricity) used in various processing routes of aluminum production and carbon emissions. Aluminum can be produced via two distinctive routes: from natural resources, "bauxite" and from man-made resources and "aluminum scrap." The former process is referred as primary aluminum production, whereas later one includes secondary aluminum production process. Theoretically, aluminum obtained from these two different resources share the same properties and perform the same functions.

8.2.2.1 Primary Aluminum Production

Primary aluminum ingot production is a complex system, which includes bauxite mining, alumina refining, electrolysis (including anode production and smelting), and primary ingot casting (Fig. 8.5). The production of primary aluminum starts with the production of alumina from bauxite ore during Bayer's process. Bauxite is basically aluminum ore, which constitutes aluminum oxide (50%), water (10—20%), and other impurities (Tan and Khoo, 2005). It is the most energy-intensive process, which begins with crushing of large size bauxite lumps (primary raw material) to small size bauxite pieces followed by digestion of bauxite with caustic soda and sodium carbonate in a digester at elevated temperature and pressure (Xue et al., 2016). In digestion, iron and silicon impurities form insoluble oxides called "red mud." The red mud settles out, and a rich concentration of sodium aluminate is filtered and seeded to form hydrated alumina crystals in precipitators. These crystals are then heated in a calciner at 1200°C to drive off any water leaving alumina deposits behind. Anode production is a step prior to electrolysis of alumina to produce aluminum. In the process of making anode, petroleum coke is calcined, pulverized, and blended with coal pitch to form a paste that is subsequently extruded into blocks or briquettes and allowed to cool. Briquettes are then sent directly to the pots for consumption, whereas the blocks are sent to a separate baking furnace to be baked and then sent to Hall—Héroult electrolytic process (Xue et al., 2016).

Electrolysis of alumina takes place at a temperature of ~960°C in a large steel container called pots or cells containing carbon cathode, anode, and electrolyte (alumina, cryolite, and aluminum fluoride) to produce primary aluminum (Gao et al., 2009). When electric current pass through the electrolyte, liquid aluminum is produce at the cathode, whereas at the anode oxygen combines with carbon to form CO_2. The net electrolytic reduction reactions are

$$2Al_2O_3 \rightarrow 3Al^{+3} + 6O^{-2}$$

$$4Al^{+3} + 12e^- \rightarrow 4Al \quad \text{(at cathode)}$$

$$6O^{-2} \rightarrow 3O_2 + 12e^- \quad \text{(at anode)}$$

$$3O_2 + 3C \rightarrow 3CO_2$$

$$2Al_2O_3 + 3C \rightarrow 4Al + 3CO_2$$

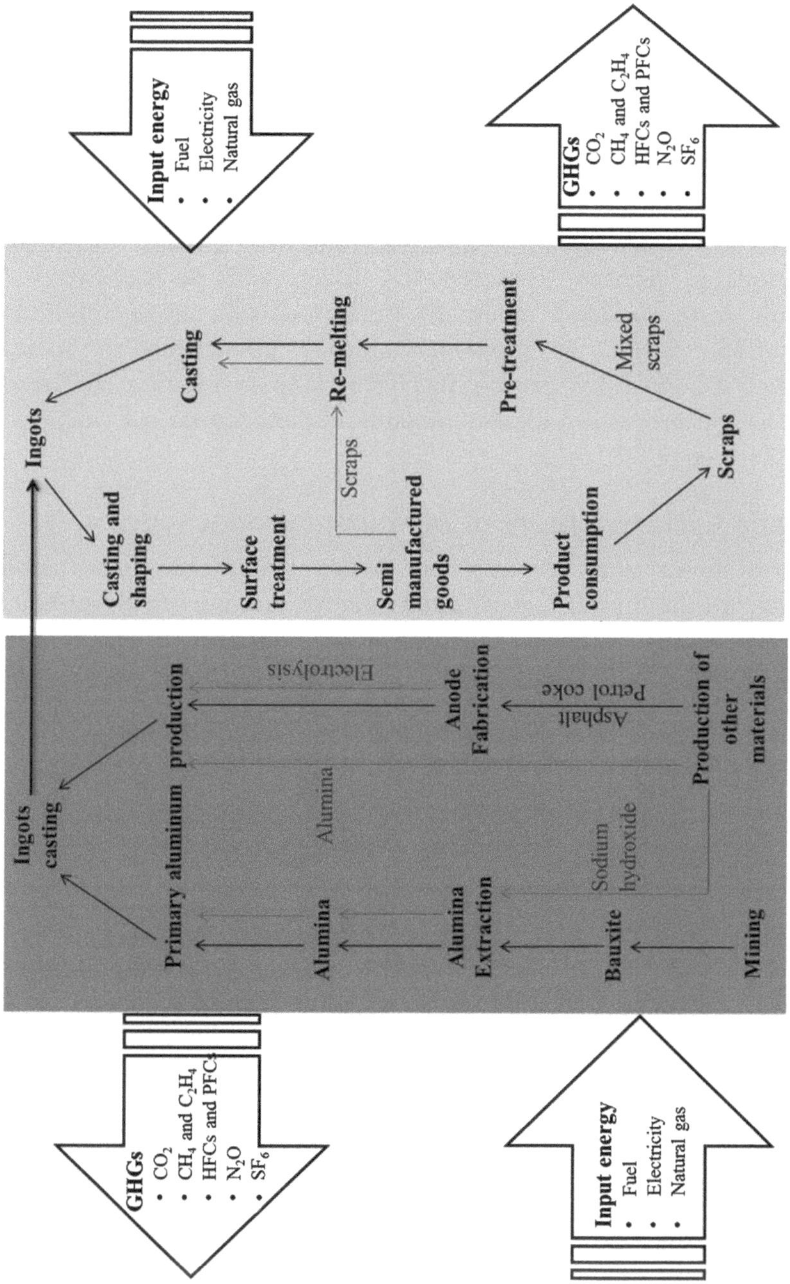

Figure 8.5 System boundary of primary [red (gray in print versions) boundary] and secondary (green (white in print versions) boundary] aluminum production.

After the electrolysis, the metal is refined to remove impurities such as sodium, calcium oxide particles, and hydrogen. Produced molten aluminum is collected from the pots and transported for casting. Casting utilizes water-cooled metal molds to cast slabs, thin sheets, wire rod, T-bars, or billets.

8.2.2.2 Secondary Aluminum Production

Aluminum-bearing scraps or materials (excluding ores) are collected and processed into aluminum alloys for industrial castings and ingots production. In furnace (reverberatory, rotary and electric furnaces), aluminum materials (scrap or ingots) are washed to remove impurities, melted at 450—500°C, and molten aluminum is then refined to produce aluminum followed by casting (Fig. 8.5). Inputs required for scrap collection and processing include gasoline, diesel, natural gas, and electricity (Zare et al., 2016).

8.2.3 Life Cycle Inventory of Aluminum Production

Because of the size and complexities of the supply chain, it is important to emphasize on the inputs (materials and energy) and outputs (by-products and/or emissions) of the aluminum production process (Table 8.1).

Table 8.1 Inputs and outputs (by-products and/or emissions) from varying processes involve in aluminum production

Process	Inputs	Outputs (by-products and/or emissions)
Primary aluminum production		
Bauxite mining	Bauxite ore, water, electricity, and heavy fuel oil.	Bauxite, overburden, hazardous and nonhazardous wastes, wastewater, and water vapor.
Alumina refining	Bauxite, sodium hydroxide, sodium carbonate, water, coal, electricity, heavy fuel oil, steam, and natural gas.	Aluminum oxide (alumina), red mud, hazardous and nonhazardous wastes, particulate matter, SO_2, NOx, mercury (II), water vapor, oil and grease.
Anode production	Petrol coke, heavy fuel oil, electricity, natural gas, hard coal pitch, recycled anode	Anode, refractories, steel, carbon wastes, particulate matter, SO_2, NOx, fluorides, benzo{a}pyrene, polycyclic

Table 8.1 Inputs and outputs (by-products and/or emissions) from varying processes involve in aluminum production—cont'd

Process	Inputs	Outputs (by-products and/or emissions)
	butts, refractories, steel, and cooling water.	aromatic hydrocarbons, wastewater, oil and grease, scrubber sludge, and other hazardous and nonhazardous wastes.
Aluminum smelting	Alumina, electricity, coal, natural gas, aluminum fluoride, anode, cathode, refractories, steel, and cooling water.	Aluminum (molten), alumina waste, refractories, used anode, steel scraps, carbon wastes, particulate matter, fluorides, benzo{a}pyrene, polycyclic aromatic hydrocarbons, CO_2, SO_2, CH_4, NOx, CF_4, HFCs, PFCs, carbonyl sulphide, wastewater, oil, and scrubber sludge.
Ingot casting	Molten aluminum, alloy components, chlorine, NOx, cooling water, light and heavy fuel oil, electricity, natural gas, and propane.	Aluminum cast ingots, dross, filter dust, solid wastes, refractories, particulates, hydrochloric acid (HCl), SO_2, NOx, wastewater, oil, and suspended solids.
Secondary aluminum production		
Aluminum scrap processing	Aluminum scrap, filter media, oil, quicklime, refractories, sodium silicate, water, electricity, and natural gas.	Processed aluminum scrap, shredded fluffs, metal scraps, oil, filter dust, refractories, solid wastes, lime, and wastewater.
Ingot casting	Electricity, natural gas, processed aluminum scrap, argon, chlorine, cryolite, filter media, oil, quicklime, refractories, potassium chloride (KCl), sodium chloride (NaCl), sodium hydroxide (NaOH) NOx, O_2, sulfuric acid (H_2SO_4), plastics, steel, wood, and water.	Aluminum ingots, metal scraps, dross, salt cake, oil and grease, filter dust, refractories, solid wastes, water, chloride, heavy metals (Pb, Hg, Al, Cu, and Zn), chlorides, fluorides, CO_2, CH_4, NOx, SO_2, and volatile organic carbon.

HFCs, hydrofluorocarbons; *PFCs*, perfluorocarbons; *SF_6*, sulfur hexafluoride.

Alumina refining and aluminum production are one of the most energy exhaustive processes, where primary aluminum production is ~10 times more energy intensive than the secondary process (Liu et al., 2016). On a unit basis, a ton of aluminum production requires ~126 GJ of energy (Liu et al., 2016). The sources of energy associated with the entire process of aluminum production are coal, hydroelectricity, and natural gas (International Aluminium Institute (IAI), 2011). IAI (2016) reported that per ton of aluminum production on an average consumes 13.56–15.75 MWh of electricity. As an output, the major pollutants that are likely to be emitted into atmosphere during aluminum production are particulate matter, sulfur dioxide (SO_2), nitrogen oxide, mercury (II), water vapor, fluorides, benzo{a}pyrene, CO_2, carbon monoxide (CO), PFCs, HFCs, and polycyclic aromatic hydrocarbons (Table 8.1) (Liu et al., 2016). The solid wastes generated from the industry are sludge from effluent treatment plant and red mud (as high as 2 tons of mud per ton of aluminum produced), which contains oxides of alumina, silicon, iron, titanium, sodium, calcium, and other organic and inorganic constituents (Xue et al., 2016).

8.3 CARBON FOOTPRINTS

To meet the commitment of global endeavor for reducing GHG emissions, many companies are integrating the environmental issues into their management systems, with potential effects in their entire production chains. Several tools and metrics have been developed to measure the environmental impact of the product in the life cycle perspective of the whole product chain. A metric that has gained prominence in recent times is carbon footprint, which quantifies the climate change impact of GHG emissions in a life cycle perspective (Laurent et al., 2010).

Carbon footprint, also termed as "carbon profile" is the measurement of overall amount of CO_2 and other GHG emissions (e.g., CH_4, ethane (C_2H_6), N_2O, HFCs, PFCs, and SF_6) associated with the product along with its supply chain, utilization, end-of-life recovery, and disposal. As per Kyoto Protocol, carbon footprint measures the direct and indirect emissions of GHGs from defined population, system, and activity (U.S. EPA 2016; United Nation Framework Convention on Climate Change (UNFCCC, 2008)). Mining, alumina refining, smelting and casting of primary aluminum have been reported to release about 0.4 billion tons of

CO_2 equivalent (CO_2e) per year (IEA, 2010). Indeed, 5.0% of this result from PFCs released during the anode production process (IAI, 2010). GHG emissions from primary and secondary processes of aluminum productions result mainly from following sources:

- Direct energy related emissions: CO_2 released from the combustion of fossil fuels used in mining, alumina refining, anode production, smelting, casting, transportation, semifabrication, and recycling contributes about 25% of total GHG emissions (Tyabji and Nelson, 2012).
- Indirect energy related emissions: CO_2 emissions occurred from electricity inputs required for the electrolysis process, accounting for about 55% of total GHG emissions (Tyabji and Nelson, 2012).
- CO_2 and PFCs emissions are a result of "anode effect" in electrolysis process, which contributes around 20% of total emissions (Tyabji and Nelson, 2012).

8.3.1 Estimation of Carbon Footprint

Carbon footprint, a quantitative expression of GHG emissions plays a significant role in emission management and evaluation of mitigation measures (Carbon Trust, 2007). For the evaluation of carbon footprint of a product, the opening step is to assess the amount of GHGs emitted or embodied in product's life cycle, technically known as "GHG accounting." Life cycle includes all the stages involved in the manufacturing of the product, thus gives a complete picture of inputs and outputs during the entire process with respect to energy consumption, GHG emissions, generation of air, water and soil pollutants, and cost-benefit initiatives (Table 8.1). GHG accounting for estimation of carbon footprint can be assessed following standard methods and guidelines:

- The GHG protocol of World Resource Institute and World Business Council on Sustainable Development forms a basis for GHG accounting guidelines including International Organization for Standardization (ISO) 14064. It provides a sector specific calculation tools and deals with quantification of reduced GHG because of adoption of mitigation methods, given in its Project protocol (WRI/WBCSD, 2004, 2005).
- ISO 14064 is useful in determining system boundaries and quantification of GHG emissions. Also, it provides underlying standards in designing GHG mitigation projects (ISO, 2006a,b)

- Publicly Available Specifications-2050 (PAS, 2050) of British Standard Institution (BSI) was introduced in 2008, with the aim of providing a consistent internationally applicable method for assessing the life cycle of GHG emissions of goods and services (BSI, 2011).
- The 2006 IPCC guidelines for National GHG inventories is divided into five volumes, which include general guidance and reporting of GHG emissions due to energy, industrial processes and product use, agriculture, forestry and other land use, and wastes. The 2013 IPCC guidelines for National GHG inventories of wetlands are a supplement to 2006 IPCC guidelines. It provides internationally agreed methodologies to the countries to estimate GHG emissions from anthropogenic sources and their removals by sinks. This facilitates different countries to provide their reports to the United Nations Framework Convention on Climate Change (IPCC, 2006).
- According to ISO 14025, Product Category Rules provides a set of specific rules, requirement and guidelines for environmental declarations with respect to life cycle analysis of one or more product categories. It includes instructions for gathering data about consumption of resources (energy, water, and renewable resources) and emissions to air, water, and soil (ISO, 2006a).
- ISO 14067 provides information regarding requirements and guidelines for quantification and communication with respect to carbon footprint of products and services. It has been developed to increase transparency in quantifying and reporting CO_2 emissions over the entire life cycle of products and services (ISO, 2013).

Carbon footprint is a stepwise process, which requires a selection of GHGs, the system boundary setting and collection of emitted/removed GHGs data. Greenhouse gas emissions are largely dependent on the type of process involved in the manufacture of the product. For an instance, CO_2 released during aluminum production is a predominant GHG, and emissions of other gases into atmosphere are relatively less. Therefore, only CO_2 emission measurement is feasible to estimate GHG emission from aluminum production. However, several studies conducted on carbon footprint of aluminum production included the six Kyoto gases, viz. CO_2, CH_4, C_2H_6, N_2O, HFCs, and PFCs (Liu et al., 2016; Zhang et al., 2015). In the second step, for carbon footprint study of large scale industries such as aluminum industries, the "Greenhouse Gas Protocol-A Corporate Accounting and Reporting Standard," is generally used for the development of system boundaries (Ranganathan et al., 2004). The step is entirely based on the objectives and characteristics of the entity for which footprinting is to be done.

To facilitate convenient estimation of carbon footprint of aluminum production, tiers or scopes have been suggested (BSI, 2011; Carbon Trust, 2007; WRI/WBCSD, 2004) (Fig. 8.6). On the basis of GHG emissions, carbon footprint can be primary/basic (calculated from either direct emissions (Tier I) or emissions embodied in energy purchase (Tier II)) and full carbon footprint (calculated from both direct and indirect emissions from the process) (Carbon Trust, 2007; Lynas, 2007). Small to large scale industries and companies preferably report their GHG emissions from the production of one or more product categories up to Tier II, which is characterized by well-defined boundary and accurate measurement of carbon footprint of product or product categories (Carbon Disclosure Project, 2008).

The final step of carbon footprint measurement includes collection of GHG data from defined boundaries and calculation. Collection of GHGs data is merely based on the objectives, credibility, feasibility, cost incurred, and capacity considerations. To achieve the same, data can either be

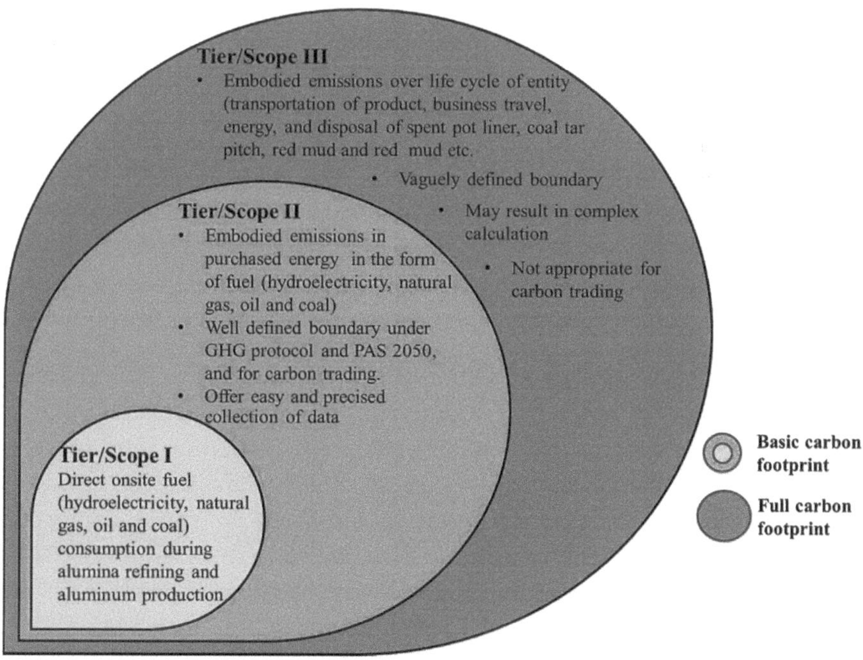

Figure 8.6 Boundaries for calculation of carbon footprint from aluminum industries.

collected directly through on-site real time measurement or through emission factors and models (Pandey et al., 2011).

- Direct on-site real time measurement: GHGs of interest are measured through several techniques, viz. photo acoustic infrared sensors, IR spectroscopy (for CO_2 measurement only), and gas chromatography for measuring all the six Kyoto gases (Berg et al., 2006; U.S. Climate Change Technology Program, 2005). Eddy covariance or flux towers are used to measure the flux from the entire landscape affected by GHG emissions (Velasco et al., 2005), whereas cavity ring-down spectrometers are commonly used for aerial measurements (Kelly et al., 2009). Secondary data on GHG emissions for carbon footprint estimation can also be obtained from various sources such as Global Trade Analysis Project (Dimaranan, 2006), IEA, United Nation Development Programme (Brown et al., 2009), "Vulcan" (joint project of NASA and US Department of Energy) (Gurney et al., 2009; Kelly et al., 2009), national GHG inventories, and other government offices.
- Emission factors and models are the most suggested techniques for the calculation of carbon footprint using data on emissions from consumption of fuels, energy, and other inputs. Emission factors for a wide range of industrial processes in GHG protocol, PAS-2050 (IPCC, 2006, 2014) as well as country wise emission factors may be obtained from UNFCCC, US EPA, and UK Department of Environment, Food and Rural Affairs (IPCC, 2006, 2014; WRI/WBCSD, 2004).

After data collection, calculation of carbon footprint in terms of CO_2e using conversion factors provided by IPCC (BSI, 2011; WRI/WBCSD, 2004) may be estimated. However, some organizations reported carbon footprint as carbon equivalent (Wiedmann and Minx, 2007), but CO_2e is widely used and commonly accepted unit for carbon footprint.

8.3.2 Carbon Footprint of Aluminum Production

Bauxite mining and mineral processing are known to produce less energy related GHG emissions than alumina refining and aluminum production (Ranganathan et al., 2004). Table 8.2 summarizes a range of GHG emissions from varying processes involved in the manufacture of aluminum. Moreover, concentrations of emitted GHGs, viz. CO_2, CH_4, CF_4, and C_2F_6 from varying steps of aluminum production are illustrated in Table 8.3 (Gao et al., 2009). As per the report by IAI (2009), about 547 Mt of CO_2e is released per ton of aluminum production, of

Table 8.2 Greenhouse gas emissions from varying processes of aluminum production

Process	GHG emission ($kg\ CO_2e\ kg^{-1}$ aluminum) Minimum	Maximum
Primary aluminum production	5.92[a]	41.10[b]
Secondary aluminum production	0.32[c]	0.74[d]
Rolling	0.20[e]	1.35[c]
Extrusion	0.28[e]	0.74[f]
Shape casting	0.48[e]	0.62[f]

GHG, greenhouse gas.
[a]Schmidt and Thrane (2009).
[b]Steen-Olsen (2009).
[c]EAA (2008).
[d]Hong et al. (2012).
[e]Green (2007).
[f]IAI (2000).

which smelting process (direct and indirect) during electrolysis contributes ~71.3% and alumina refining 10.8% of total CO_2 emission. Other processes such as ingot casting, transport of materials and final products, semifabrication and recycling processes cumulatively contribute about 14.8% of total emissions (Fig. 8.7). During the process, large quantities of potent GHGs, viz. CO_2, CH_4, C_2H_6, N_2O, HFCs, and PFCs (CF_4 and C_2F_6) are generated in anode reduction cell as a result of "anode effects" (Fig. 8.5). Furthermore, not all GHGs have equal capacity to cause global warming, but their strengths merely depend on radioactive forcing and residence time in the atmosphere. The combined effects of these two together decide the global warming potential (GWP) of a gas

Table 8.3 Emissions of carbon dioxide, methane, tetrafluoromethane, tetrafluoroethylene, and global warming potential from varying processes of aluminum production

Process	CO_2 (kg/t)	CH_4 (kg/t)	CF_4 (kg/t)	C_2F_4 (kg/t)	GWP (kg CO_2e/t)
Mining	22.9	0.028	0	0	23.5
Alumina refining	4682.4	4.3	0	0	4772.7
Anode production	506.6	0.26	0	0	512.1
Smelting	12,150.8	35.7	0.35	0.035	15,497.5
Ingot casting	821.5	0.92	0	0	755.8

CO_2, carbondioxide; *CF_4*, carbon tetrafluoride; *CH_4*, methane; *C_2F_4*, tetrafluoroethylene; *GWP*, global warming potential.
Based on Gao, F., Nie, Z., Wang, Z., Li, H., Gong, X., Zuo, T., 2009. Greenhouse gas emissions and reduction potential of primary aluminium production in China. Science in China Series E: Technological Sciences 52 (8), 2161–2166.

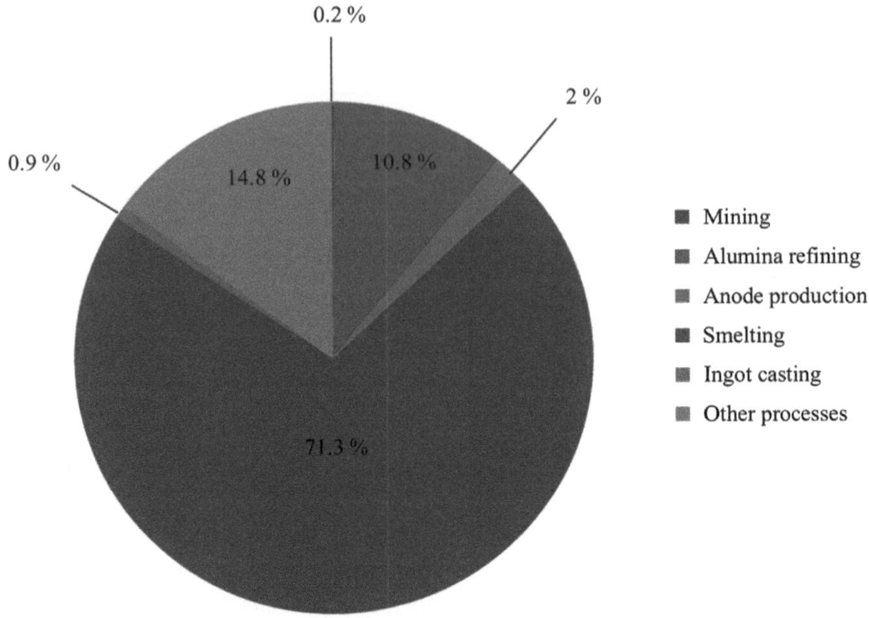

Figure 8.7 Percentage contribution in greenhouse gas emissions from major processes of aluminum production. *(Modified from the International Aluminium Institute (IAI), 2009. Aluminium for Future Generations: Contribution to UNFCCC COP-15 Copenhagen Climate Change Negotiations, p. 2.)*

molecule. World average level of GWP in 2009 was 23.96 tons CO_2e per ton of aluminum production (Steen-Olsen, 2009). Aluminum smelting contributes 72% of overall GWP followed by alumina refining and ingot casting (Sanchez and Stern, 2016). Among gases (Table 8.4), CF_4 and C_2F_6 pose very high GWP (7390 and 12,200 times equivalent of CO_2,

Table 8.4 Global warming potentials of selected greenhouse gases emitted from aluminum industries

GHGs	GWP_{100} (ton CO_2e per ton of aluminum)
CO_2	1
CH_4	25
N_2O	298
SF_6	22,800
HFCs	124 to 14,800
PFCs	7390 to 12,200

CO_2, carbondioxide; *GHGs*, greenhouse gases; *GWP*, global warming potential; *HFCs*, hydrofluorocarbons; *CH_4*, methane; *N_2O*, nitrous oxide; *PFCs*, perfluorocarbons; *SF_6*, sulfur hexafluoride.

respectively) with long lifetimes in the atmosphere (50,000 and 1,00,000 years, respectively) (IAI, 2016; Sanchez and Stern, 2016).

8.4 SOCIOECONOMIC AND ECOLOGICAL THREATS

The portion presents the potential impacts associated with carbon emissions from alumina refining and aluminum production, which may be characterized into socioeconomic and environmental impacts. The former is concerned with the range of impacts on human sociocultural system and other associated economic activities particularly in China and India. Environmental impacts are concerned mainly with air, water, soil and noise pollution, changes in land-use pattern, plant community structure and animal diversity, and other environmental nuisance.

8.4.1 Socioeconomic Problems

There are limited numbers of studies available on the socioeconomic impacts of carbon emissions from aluminum industries. Most of the studies have dealt with the impacts of emitted pollutants on environment and living beings. Main social and economic impacts associated with carbon emissions from the aluminum production include:

- Degradation of vegetation and soil erosion, which ultimately affect agrosystems in nearby areas, thereby affecting the crop production and economic development (Murthy and Patra, 2006).
- Increase in energy demand and requirement for energy storage for continuous supply to the industry (Liu et al., 2016).
- Internal disparity within and among countries (Hoque et al., 2010)
- Health hazard and risk during manufacturing phase (Krewski et al., 2007).
- Economically detrimental subsidies, viz. uncontrolled and miscalculated fee-in-tariff mechanisms (Murthy and Patra, 2006).

8.4.2 Environmental Problems

The main environmental concern of the industry is related to primary aluminum production, where GHGs including CO_2, fluorinated compounds, SO_2, etc., are generated as a result of the anode effects during the process of aluminum production.

- **CO_2**: A major GHG in the environment, which may lead to various health impacts within the context of global warming (Norgate et al., 2007).

- **SO_2**: Conifers, lichens, and mosses are among the plants, which are most sensitive to SO_2, a principal pollutant contributing to acid rain. Thus, SO_2 emission in atmosphere is the one of the major causes of atmospheric acidification, thereby affecting living beings on the earth (Murthy and Patra, 2006).
- **Fluorinated compounds**: 99% of these emissions are produced by electrolytic cells. Fluorine is a highly reactive and toxic halogen readily absorbed by plants (particularly conifers). Animals, including livestock that feed on contaminated plants can develop osteoporosis or even osteosclerosis (Murthy and Patra, 2006). Also, it may cause dental fluorosis during the critical tooth development period in children less than 6 years of age. The permissible limit for fluorine in forage is usually in the order of 40 ppm. Continuous exposure to fluorine compounds for 8 h/day may lead to eye and skin irritations (>900 $\mu g/m^3$), of the respiratory tract irritations (>2100 $\mu g/m^3$), dental fluorosis, and osteoporosis (>1780 $\mu g/m^3$). Aluminum plant workers are the most exposed, because ambient concentrations in pot lines can sometimes reach 1500 $\mu g/m^3$. Chlorofluorocarbon compounds emission into atmosphere may cause ozone layer depletion (Murthy and Patra, 2006).
- **Polycyclic aromatic hydrocarbons**: Several polycyclic aromatic hydrocarbons (PAHs) are considered potential carcinogens. Workers are most at risk, and bladder cancer has been associated with PAH exposure. Benzo{a}pyrene is a typical PAH, and its ambient air concentration should not exceed 0.9 ng/m^3 (Yu, 2002).
- **Spent pot lining**: The refractory materials in the electrolytic cells are commonly referred as spent pot lining. Because of mainly carbon, cyanide, and fluoride contents, spent pot lining is considered hazardous waste and must be stored safely, when exposed to water, acids, or alkaline substances, pot lining can emit toxic, flammable, and potentially explosive gases (fluorides, ammonia, hydrogen cyanide, sulfur oxides, hydrogen, and acetylene) (Murthy and Patra, 2006).
- **Coal tar pitch**: Volatiles from coal tar pitch in the workroom atmosphere of aluminum smelters contain PAHs. Atherosclerosis is a characteristic lesion in ischemic heart disease and peripheral arteriosclerosis (Murthy and Patra, 2006).

Other human health problems associated with toxic emissions from aluminum production are cancer, senile psychoses, hypertensive disease, Ischemic heart disease, cerebrovascular disease, peripheral arteriosclerosis, pneumonia, chronic bronchitis, emphysema, asthma, peptic ulcer, liver cirrhosis, chronic nephritis, and in extreme cases may cause sudden death (Wesdock and Arnold, 2014; Krewski et al., 2007).

8.5 MITIGATION STRATEGIES IN CARBON EMISSIONS REDUCTION

Carbon emissions are generated in almost all the processes of aluminum industry, i.e., right from the extraction of aluminum from the ore to its end-of-life management. Because of increasing demand for aluminum, there is a continuous increase in carbon emissions, but necessary improvements in industrial processes to mitigate emissions are still lagging behind. Many countries have established a long-term carbon emission reduction targets such as the United States and China have committed to lower the carbon emissions by 17 and 45% of 2005 levels by 2020, respectively; whereas the United Kingdom is committed to abate the emissions by 80% of 1990 levels by 2050 (Huisingh et al., 2015). India has aimed of reducing its emissions by 20–25%, whereas Brazil is committed to reduce carbon emissions by 38–42% of 2005 levels by 2020 (Huisingh et al., 2015). Thus to achieve their goals, government and policy makers have set a series of energy conservation policies and regulations (Fig. 8.8). Fortunately, the process technology of industry is continuing to evolve with some new carbon emission reduction technologies, if effectively applied sector wide, these will promisingly help the societies to make the progress in alleviating the growing climate change crisis (Slowak and Taticchi, 2015). At present, the principal approaches to decrease industrial carbon emissions include adjusting energy and industrial structures, improving technology and energy efficiency, and implementing a strict management system. Liu et al. (2011) indicated that reformation in industrial technology could contribute in 12–14% reduction in carbon emissions from aluminum production. Apart from technical innovation, the design and timing of policy interventions are crucial for reducing innovation barriers and in improving the energy efficiency (Slowak and Taticchi, 2015). Keeping in view the following technological innovations and mitigations strategies, significant improvements are expected in energy use efficiency and in reducing carbon emissions from aluminum production.

8.5.1 Emerging Electrode Technologies

In the conventional process (Hall–Héroult process) of aluminum production, energy efficiency improvements are limited by the extremely

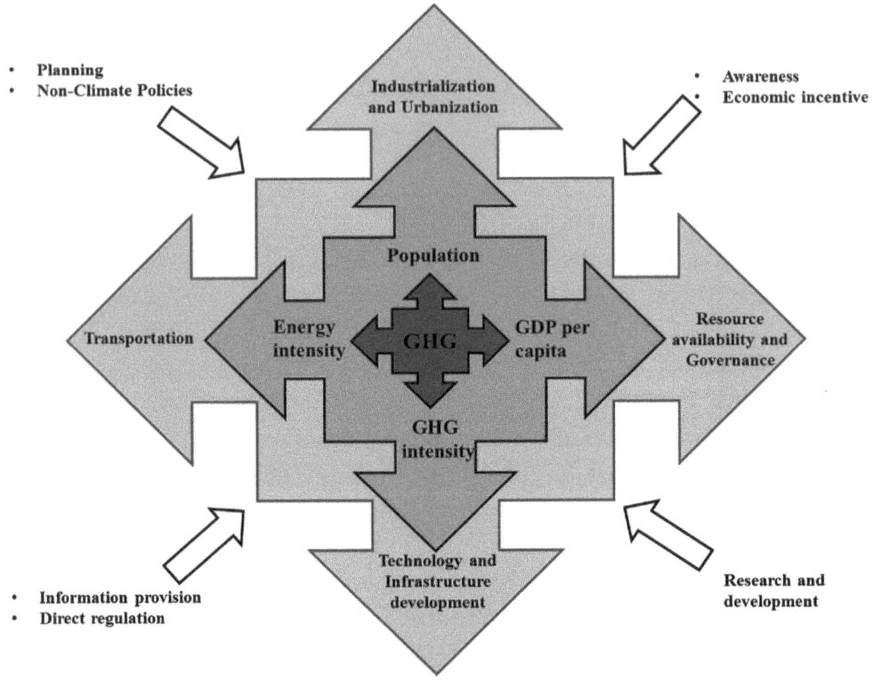

Figure 8.8 Interconnections among greenhouse gas emissions, immediate and underlying drivers, and preventive measures. *GHG*, greenhouse gas.

corrosive and aggressive electrolytic cell environment and by the lack of materials to withstand the cryolite electrolyte at 950–1000°C (operating temperature). Thus, modification in production of electrode and design could be the main pathway for making the Hall–Héroult process more energy efficient. Replacement of conventional electrodes (anode and cathode) with advanced (inert/sloped and perforated/slotted anodes, and wetted cathode) ones have been reported to efficiently improve the efficiency of Hall–Héroult process coupled with reduced GHG emissions from the process (Kvande and Drabløs, 2014).

Carbon anodes, which are traditionally used in Hall–Héroult process, require a replacement at regular time interval because of its high reactivity and shorter lifetime. Therefore, it is high energy and cost intensive and lead to greater emission of GHGs from electrolytic processes (anode production and smelting) (Table 8.5). According to Kvande and Drabløs (2014), replacement of carbon anodes with inert one may reduce the energy requirement and is cost effective. Inert anode mainly constitutes metals,

Table 8.5 Differences between conventional and advanced methods of aluminum production

Hall–Héroult's process	Conventional methods	Advanced methods
Anodes	• Carbon and anodes are highly reactive, and require a regular replacement. • Shorter lifetime. • High energy requirement. • High capital cost in production and replacement. • Higher greenhouse gas emissions. • Oxygen generated reacts with carbon anode to produce carbon dioxide. • Risk to occupational health due to anode effects. • Less efficient.	**(a) Inert anode** • Chemically nonreactive. • Have relatively longer lifetime. • Saves 3–4% more energy than conventional process. • Capital costs for inert anodes are 10–30% less than conventional carbon anodes. • Inert anodes produce oxygen. • Reduce greenhouse gas emissions. • Improving occupational health. • More efficient. **(b) Sloped anodes and perforated or slotted anodes** • Allow proper and quick circulation of electrolyte. • Vertical cells reduce the anode current density and cell voltage. • Savings of 2–2.5 kWh of power consumption saved per kilogram of aluminum production. • Can be easily combined with inert anode materials and/or wetted cathodes. • Prevent anode's harmful effects. • More efficient process.

Continued

Table 8.5 Differences between conventional and advanced methods of aluminum production—cont'd

Hall–Héroult's process	Conventional methods	Advanced methods
Cathode	• Solid carbon material with less durability. • Unusual loss of energy with impeded circulation within electrolyte. • High energy demand. • May cause contamination of electrolytic bath. • Easily corroded and not withstand high temperature. • Lower cell life. • High amount of spent pot liner generation.	• Wetted cathode (preferably of titanium diboride) with more durability. • Reduce about 20% of energy consumption by lowering the anode cathode distance. • Withstand the corrosion and high temperature. • Extended cell life by preventing contamination with bath chemicals and lowering the formation of undissolved alumina sludge. • Less amount of spent pot liner generation.
Electrolytic cell	• Composed of 28–48 anodes, a horizontal cathode and electrolyte. • High temperature requirement (1200°C) and low current density. • High capital-intensive. • Low aluminum production. • Chance of contaminated end product, i.e., aluminum. • High emission of harmful gases.	• Multiple electrodes in a single reaction area. • Works with inert anodes. • Require less temperature ($\approx$ 700°C) and generate high current densities. • Energy savings of around 40%. • High aluminum production. • An electrode operating life of nearly 3 years. • Minimizes evolution of harmful gases. • Highly efficient and durable.

ceramics, and cermet; therefore it is chemically inert and is not consumed by electrolytic process. Besides, inert anode is relatively more durable and efficiently works with same lifetime as of electrolytic cell (Kvande and Drabløs, 2014). Utilization of inert anode also enhances the quality of aluminum production without a risk of aluminum contamination. Unlike conventional anode, inert anode produces O_2 instead of CO_2, which may mitigate the anode effects, thereby improving plant operating system and occupational health (Kvande and Drabløs, 2014). Similarly, the design and composition of anode plays a significant role in qualitative and quantitative production of aluminum during smelting process. The design of anodes in aluminum production cells can be altered to reduce anode—cathode distance, which may lower the energy requirements. Several anode designs have been recently commercialized such as sloped and perforated anode or slotted anodes, which allow gaseous by-product to be bubbled out safely from the molten cryolite bath thereby minimizing anode effects (Zhou et al., 2007). These designs also allow the cryolite bath to circulate more quickly, making the electrolysis of alumina more efficient. Vertical electrode cells featured with vertically oriented anode improve the electrical conductivity by allowing gaseous bubbles to escape more easily, thus reducing anode effects (Hryn et al., 2014). Vertical cells reduce the anode current density and cell voltage, saving energy, and reducing heat loss.

Cathode is negatively charged electrode of Hall—Héroult cell on, which aluminum gets deposited. Replacement of carbon cathode with wetted cathodes has been found to decrease the distance between electrodes, improve the electrical contact between molten aluminum and cathode as well as risk of anode effects (Green, 2007). It is inert to the cell bath and requires less energy to carry out the process (Table 8.5). It is durable and can withstand corrosive and high temperature condition within a cell. Although there are certain limitations of utilizing wetted cathode as a part of Hall—Héroult cell, viz. wetted cathode is generally made from titanium bromide and related composites, which are very expensive materials (Welch, 1999). But above all, its utilization lowers the energy requirement and generation of toxic waste (spent pot liner) and carbon emissions to a great extent (Green, 2007). Also, replacement of conventional cathode with wetted carbon has been found to improve the cell performance by $\approx$15%.

8.5.2 Multipolar Electrolytic Cell

Hall–Héroult cell basically has horizontal bottom surface with 24–48 anodes and a cathode immersed in electrolytic cryolite bath. This type of cell poses a single pole arrangement, which is highly capital intensive and energy consuming. Modification of unipolar cell to multipolar cell could greatly increase productivity within a cell by placing multiple electrodes in a single reaction area (Table 8.5). These modifications in electrolytic process have been reported as an efficient reformation in aluminum production at low temperature with less evolution of toxic gases (U.S. DOE, 2007).

8.5.3 Alternate Methods to Reduce Carbon Emissions From Aluminum Production

Emerging new technologies in aluminum production pathways rather than improving the existing processes are as follows:

8.5.3.1 *Carbothermic Reduction*

Unlike the Hall–Héroult process, carbothermic reduction is a nonelectrolytic reaction, where alumina reacts with carbon to form aluminum and CO (Fig. 8.9). The process requires high temperature (2000°C), which is achieved through multichambered furnace (White et al., 2012). The high temperature requirement for carbothermic reduction can allow

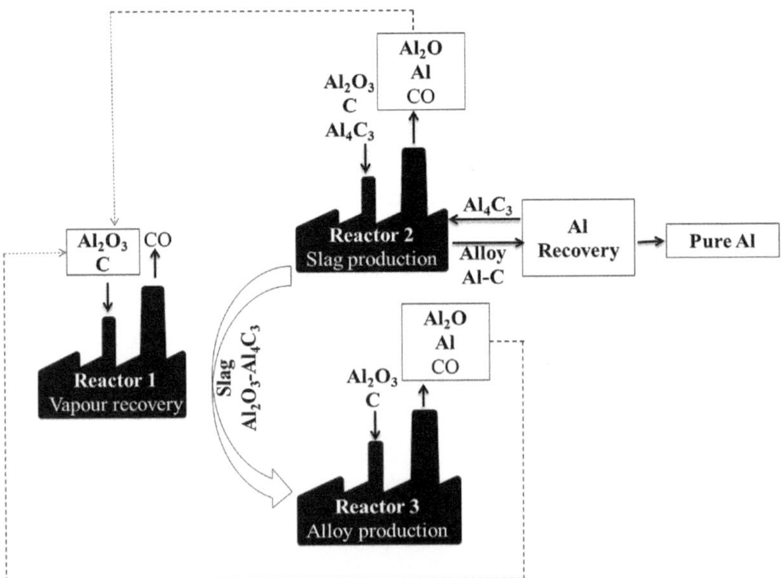

Figure 8.9 Carbothermic reaction process.

integration of a greater range of alumina qualities and even scrap metal (White et al., 2012). This concept is highly achievable, feasible for small to medium scale industry and may save 20–30% of energy consumption. Many companies such as Aluminum Company of America and Energy Exergy Aluminum are experimenting with carbothermic reduction using advanced reactor process and improved electric arc furnace, respectively (Balomenos and Panias, 2011). Besides, these companies are now focusing on concentrated solar energy utilization to power the furnace, which have successively reduced fossil fuel requirement for energy, capital cost incurred in aluminum production and CO_2 emissions (Vishnevetsky et al., 2014).

8.5.3.2 Kaolinite Reduction

Kaolinite reduction offers an alternative way to both Bayer's and Hall–Héroult process (Fig. 8.10). The process involves multipolar cell and multiple stacked bipolar graphite electrode in a chloride bath. In spite of bauxite, kaolin (a clay mineral formed from silicone dioxide and aluminum oxide) is used in carbochlorination process in the presence of clay oxide and coke to form aluminum chloride. Thereafter, aluminum chloride is electrolyzed to aluminum followed by casting of aluminum ingots. The process offers faster and more efficient conversion of kaolin into aluminum chloride and subsequently to pure aluminum in presence of low temperature, thereby reducing the loss of energy efficiently. The overall

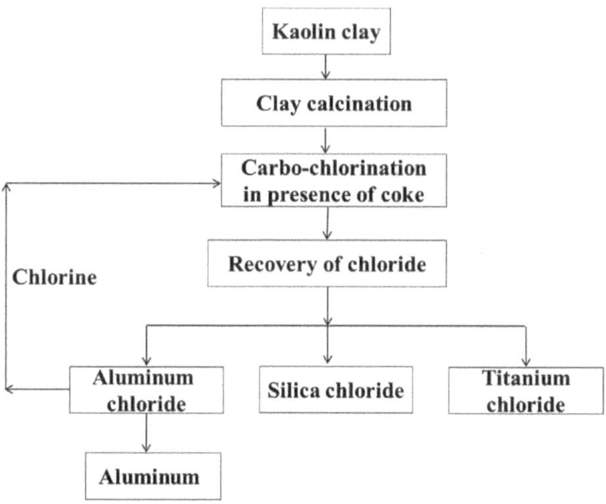

Figure 8.10 Kaolinite reduction process.

energy saving during the process is around 12–46% of conventional process (U.S. DOE, 2007).

8.5.3.3 Low-Temperature Reduction Using Ionic Liquid

Commonly used electrolyte in aluminum smelting process is molten aluminum fluoride and sodium bath. Therefore, a very high temperature (900–1000°C) is required to keep the bath in molten phase, which is a high energy–consuming process. Ionic liquids (nonconventional organic solvents, electrolytes, and molten salts) exhibit low melting points and could dramatically reduce the necessary smelting temperature for aluminum production (Zhang et al., 2003). There are several classes of ionic liquids, viz. chloroaluminate molten salts, fluoro-methane salts, and imidazolium sulfate compounds, which are commonly used in low temperature–reduction process (Poulimenou et al., 2015). The process has been found to efficiently reduce the emission of polluting gases (CO_2, CO, HFCs, and PFCs) and generation of solid wastes from spent linings in current Hall–Héroult process (Zhang et al., 2003) and increase the efficiency of bauxite to alumina conversion (Poulimenou et al., 2015). Also, low temperature–reduction process has been found to efficiently reduce 30–85% of energy consumption as compared with Hall–Héroult smelting process.

8.5.4 Recycling

Recycling of aluminum could save up to 95% of energy and emissions as compared with primary aluminum production. Therefore, the need now is to reemphasize recycling of not only used beverage cans but also all other unused aluminum forms. For the future, the most important area for recycling will be automotive aluminum because the volume of recycled aluminum from automotive components has surpassed the aluminum recycling from unused beverage cans (Das and Green, 2010). A key issue in recycling of aluminum scraps is the control on alloy addition during secondary aluminum production process to retain the value of secondary metal.

8.5.5 Carbon Trading

Although carbon trading scheme associated with aluminum industries is not yet developed and far from clear, but development of such system would definitely benefit the growth of countries in near future. Another expected outcome of carbon trading system is reduction in more than 90% of energy consumption and capital cost incurred in aluminum production

(Das and Green, 2010). Thus, carbon trading might favor more remelters construction in consumption rich countries such as India and China, where aluminum scrap is more likely to be available.

8.6 CONCLUSIONS

Aluminum is the second highest metal in demand after steel and its production involves high energy—intensive processes. LCA of aluminum production indicated coal, hydroelectricity, oil, and natural gases as main sources of energy related greenhouse gas emissions from primary as well as secondary aluminum production. Main emitted greenhouse gases during the processes are CO_2, CH_4, C_2H_6, HFCs, PFCs, and SF_6. For estimation of carbon footprint of aluminum production, greenhouse gas accounting is appropriate using system boundary method up to Tier II. Carbon footprint of aluminum production showed that emission during the process accounts for ~1.0% of global greenhouse gas emissions, and primary aluminum production augments a significant contribution to it. There are certain mitigation strategies such as reformation of conventional technologies in aluminum production, recycling of waste aluminum products, and carbon trading to alleviate carbon emissions, which may be adopted by aluminum industries to cut down the greenhouse gas emissions into the atmosphere. It also requires awareness to motivate collaborative approaches with associated industries to deal with the emerging issues related to greenhouse gas emissions. Furthermore, the aluminum industry may take a global leadership position by actively developing internationally accepted and approved carbon footprint credit protocols.

ACKNOWLEDGMENTS

Authors are thankful to DST-FIST and CAS programme running in the Department of Botany for necessary facilities and HINDALCO Industries, Renukoot, India, for process-related information on aluminum production. Meenu Gautam and Bhanu Pandey are thankful to the Council of Scientific and Industrial Research and University Grants Commission, New Delhi, India, respectively, for financial assistance.

REFERENCES

Balomenos, E., Panias, D., 2011. Carbo-thermic reduction of alumina: a review of developed processes and novel concepts. In: Proceedings — European Metallurgical Conference (EMC), p. 3.

Berg, W., Brunsch, R., Hellebrand, H.J., Kern, J., June 2006. Methodology for measuring gaseous emissions from agricultural buildings, manure, and soil surfaces. In: Workshop on Agricultural Air Quality, Washington, DC, pp. 5—8.

Blanco, G., Gerlagh, R., Suh, S., Barrett, H.C., et al., 2014. Drivers, trends and mitigation. In: Edenhofer, O., Pichs-Madruga, R., Sokona, Y., Farahani, E., et al. (Eds.), Climate Change 2014: Mitigation of Climate Change, Contribution of Working Group III to the Fifth Assessment Report of the Intergovernmental Panel on Climate Change, Cambridge, United Kingdom and New York. Cambridge University Press, NY, USA, pp. 351–411.

Boden, T., Marland, G., Andres, R., 2015. Global, Regional, and National Fossil-Fuel CO_2 Emissions in Trends. Carbon Dioxide Information Analysis Centre (CDIAC), UK. http://cdiac.ornl.gov/trends/emis/meth_reg.html.

Brown, M.A., Southworth, F., Sarzynski, A., 2009. The geography of metropolitan carbon footprints. Policy and Society 27, 285–304.

British Standards Institute (BSI), 2011. Publicly Available Specification 2050. Specification for the Assessment of the Life Cycle Greenhouse Gas Emissions of Goods and Services.

Carbon Trust, 2007. Carbon Footprint Measurement Methodology. Version 1.1. The Carbon Trust, London, UK. http://www.carbontrust.co.uk.

Carbon Disclosure Project (CDP), 2008. Carbon Disclosure Project Report-2008: Global 500, on Behalf of 385 Investors with Assets of $57 Trillion. http://www.cdproject.net/reports.asp.

Das, S.K., Green, J.A., 2010. Aluminum industry and climate change—assessment and responses. Journal of the Minerals, Metals and Materials Society 62 (2), 27–31.

Dimaranan, B.V., 2006. Global Trade, Assistance, and Production: The GTAP6 Database. Centre for Global Trade Analysis, Purdue University, West Lafayette, USA.

Etheridge, D.M., Steele, L.P., Langenfelds, R.L., Francey, R.J., et al., 1996. Natural and anthropogenic changes in atmospheric CO_2 over the last 1000 years from air in Antarctic ice and firm. Journal of Geophysics Research-atmosphere 101 (D2), 4115–4128.

European Aluminium Association (EAA), 2008. Environmental Profile Report for the European Aluminium Industry: Life Cycle Inventory Data for Aluminium Production and Transformation Processes in Europe. European Aluminium Association, Brussels.

Forster, P., Ramaswamy, V., Artaxo, P., Berntsen, T., et al., 2007. Changes in atmospheric constituents and in radiative forcing. In: Solomon, S., Qin, D., Manning, M., et al. (Eds.), Climate Change 2007: The Physical Science Basis, Contribution of Working Group I to the Fourth Assessment Report of the Intergovernmental Panel on Climate Change, Cambridge, United Kingdom and New York. Cambridge University Press, NY, USA, pp. 129–234.

Gao, F., Nie, Z., Wang, Z., Li, H., Gong, X., Zuo, T., 2009. Greenhouse gas emissions and reduction potential of primary aluminium production in China. Science in China Series E: Technological Sciences 52 (8), 2161–2166.

Green, J.A.S., 2007. Green Aluminum Recycling and Processing for Energy Conservation and Sustainability. ASM International, Ohio, pp. 191–196.

Gurney, K.R., Mendoza, D.L., Zhou, Y., Fischcer, M.L., et al., 2009. High resolution fossil fuel combustion CO_2 emission fluxes for the United States. Journal of Environmental Science and Technology 43 (14), 5535–5541.

Hong, J., Zhou, J., Hong, J., Xu, X., 2012. Environmental and economic life cycle assessment of aluminum-silicon alloys production: a case study in China. Journal of Cleaner Production 24, 11–19.

Hoque, S., Dwyer, L., Forsyth, P., Spurr, R., et al., 2010. Economic Impacts of Greenhouse Gas Reduction Policies on the Australian Tourism Industry: A Dynamic CGE Analysis. CRC for sustainable tourism, Technical Report, Gold Coast, Australia.

Hryn, J.N., Tkacheva, O.Y., Spangenberger, J.S., 2014. Ultra-High-Efficiency Aluminum Production Cell. Argonne National Laboratory (ANL-14/09). http://www.osti.gov/scitech/biblio/1156937.

Huisingh, D., Zhang, Z., Moore, J.C., Qiao, Q., et al., 2015. Recent advances in carbon emissions reduction: policies, technologies, monitoring, assessment and modeling. Journal of Cleaner Production 103, 1–12.

Intergovernmental Panel on Climate Change (IPCC), 2006. National Greenhouse Gas Inventories: Land Use, Land Use Change and Forestry. Institute of Global Environmental Strategies, Hayama, Japan.

Intergovernmental Panel on Climate Change (IPCC), 2014. In: Pachauri, R.K., Meyer, L.A. (Eds.), Climate Change 2014: Synthesis Report Contribution of Working Groups I, II and III to the Fifth Assessment Report of the IPCC. Intergovernmental Panel on Climate Change, Geneva, Switzerland, p. 151.

International Aluminium Institute (IAI), 2000. Life Cycle Inventory of the Worldwide Aluminium Industry with Regard to Energy Consumption and Emissions of Greenhouse Gases. International Aluminium Institute, London.

International Aluminium Institute (IAI), 2009. Aluminium for Future Generations: Contribution to UNFCCC COP-15 Copenhagen Climate Change Negotiations, p. 2.

International Aluminium Institute (IAI), 2010. Results of the 2009 Anode Effect Survey: Report on the Aluminium Industry's Global PFC Gases Emissions Reduction Programme. International Aluminium Institute.

International Aluminium Institute (IAI), 2011. Primary Aluminium Smelting Power Consumption, 2010 Data. International Aluminium Institute. http://www.world-aluminium.org/statistics/primary-aluminium-smelting-power-consumption/.

International Aluminium Institute (IAI), 2016. Results of the 2015 Anode Effect Survey: Report on the Aluminium Industry's Global Perfluorocarbon Gases Emissions, 2016 Data. International Aluminium Institute. http://www.world-aluminium.org/media/filer_ public/2016/08/08/2015_anode_effect_survey_result_2016.pdf.

International Energy Agency (IEA), 2010. Energy Technology Perspectives. International Energy Agency, Paris, p. 194.

International Energy Agency (IEA), 2012. Energy Technology Transitions for Industry: Strategies for the Next Industrial Revolution. The International Energy Agency, Paris.

International Organization for Standardization (ISO) 14064-1:2006, 2006a. Greenhouse Gases Part 1: Specification with Guidance at the Organization Level for Quantification and Reporting of Greenhouse Gas Emissions and Removals. http://www.iso.org/iso/catalogue_detail? csnumber=38381.

International Organization for Standardization (ISO) 14064-2:2006, 2006b. Greenhouse Gases-Part 2: Specification with Guidance at the Project Level for Quantification, Monitoring and Reporting of Greenhouse Gas Emission Reductions or Removal Enhancements. http://www.iso.org/iso/iso_catalogue/catalogue_tc/catalogue_ detail.htm?csnumber=38382.

International Organization for Standardization (ISO) 14067, 2013. Greenhouse Gases — Carbon Footprint of Products: Requirements and Guidelines for Quantification and Communication. ISO 14067, Geneva. https://www.iso.org/standard/59521.html.

Kelly, L.M., Shepson, P.B., Strim, B.P., Karion, A., et al., 2009. Aircraft-based measurements of the carbon footprint of Indianapolis. Environmental Science & Technology 43, 7816–7823.

Kerr, A.R., 2006. A worrying trend of less ice higher seas. Science 311, 1698–1701.

Krewski, D., Yokel, R.A., Nieboer, E., Borchelt, D., et al., 2007. Human health risk assessment for aluminium, aluminium oxide, and aluminium hydroxide. Journal of Toxicology and Environmental Health, Part B 10 (S1), 1–269.

Kvande, H., Drabløs, P.A., 2014. The aluminum smelting process and innovative alternative technologies. Journal of Occupational and Environmental Medicine 56 (Suppl. 5), S23–S32.

Laurent, A., Olsen, S.I., Hauschild, M.Z., 2010. Carbon footprint as environmental performance indicator for the manufacturing industry. CIRP Annals-Manufacturing Technology 59 (1), 37–40.
Liu, Q., Jiang, P., Zhao, J., Zhang, B., et al., 2011. Life cycle assessment of an industrial symbiosis based on energy recovery from dried sludge and used oil. Journal of Cleaner Production 19, 1700–1708.
Liu, Z., Geng, Y., Adams, M., Dong, L., et al., 2016. Uncovering driving forces on greenhouse gas emissions in China's aluminium industry from the perspective of life cycle analysis. Applied Energy 166, 253–263.
Lynas, M., 2007. Carbon Counter. Harper Collins Publishers, Glasgow.
Menzie, W.D., Barry, J.J., Bleiwas, D.I., Bray, E.L., et al., 2010. The Global Flow of Aluminum from 2006 Through 2025 (No. 2010–1256). U.S. Geological Survey.
Murthy, A., Patra, H.S., 2006. Ecological, Socio-Economic and Health Impact Assessment Due to Aluminum Smelters: A Case Study of Hindalco in Orissa (Bhubaneswar: Vasundhara).
National Oceanic, Atmospheric Administration (NOAA), 2015. Department of Commerce, Earth System Research Laboratory, Global Monitoring Division, U.S. https://www.esrl.noaa.gov/gmd/obop/mlo/.
Norgate, T., Haque, N., 2010. Energy and greenhouse gas impacts of mining and mineral processing operations. Journal of Cleaner Production 18 (3), 266–274.
Norgate, T.E., Jahanshahi, S., Rankin, W.J., 2007. Assessing the environmental impact of metal production processes. Journal of Cleaner Production 15 (8), 838–848.
Pandey, D., Agrawal, M., 2014. Carbon footprint estimation in the agriculture sector. In: Assessment of Carbon Footprint in Different Industrial Sectors, vol. 1. Springer, Singapore, pp. 25–47.
Pandey, D., Agrawal, M., Pandey, J.S., 2011. Carbon footprint: current methods of estimation. Environmental Monitoring and Assessment 178 (1–4), 135–160.
Poulimenou, N., Giannopoulou, I., Panias, D., 2015. Use of ionic liquids as innovative solvents in primary aluminum production. Materials and Manufacturing Processes 30 (12), 1403–1407.
Ranganathan, J., Corbier, L., Bhatia, P., Schmitz, S., et al., 2004. The Greenhouse Gas Protocol: A Corporate Accounting and Reporting Standard (Revised Edition). World Resources Institute and World Business Council for Sustainable Development, Washington, DC.
Russell, A., Ekvall, T., Baumann, H., 2005. Life cycle assessment–introduction and overview. Journal of Cleaner Production 13 (13), 1207–1210.
Sanchez, L.F., Stern, D.I., 2016. Drivers of industrial and non-industrial greenhouse gas emissions. Ecological Economics 124, 17–24.
Schmidt, J.H., Thrane, M., 2009. Life Cycle Assessment of Aluminium Production in New Alcoa Smelter in Greenland. http://uk.nanoq.gl/Emner/News/News_from_Government/2009/09/w/media/274FEC1E8D684F21808A5667896D755A.ashx.
Sims, R.E.H., Schock, R.N., Adegbululgbe, A., Fenhann, J., et al., 2007. Energy supply. In: Metz, B., Davidson, O.R., Bosch, P.R., Dave, R., et al. (Eds.), Climate Change 2007: Mitigation. Contribution of Working Group III to the Fourth Assessment Report of the IPCC, Cambridge, Cambridge University Press, United Kingdom and New York, NY, USA.
Slowak, A.P., Taticchi, P., 2015. Technology, policy and management for carbon reduction: a critical and global review with insights on the role played by the Chinese Academy. Journal of Cleaner Production 103, 601–619.
State of the Climate: Carbon Dioxide Report, 2015. https://www.climate.gov/news-features/featured-images/2015-state-climate-carbon-dioxide.

Steen-Olsen, K., 2009. Environmental Assessment of Aluminium Production in Europe: Current Situation and Future Scenarios. Department of Energy and Process Engineering, Norwegian University of Science and Technology, Trondheim.

Tan, R.B., Khoo, H.H., 2005. An LCA study of a primary aluminium supply chain. Journal of Cleaner Production 13 (6), 607–618.

Thomas, R., Graven, H., Hoskins, B., Prentice, I.C., 2016. What is meant by "balancing sources and sinks of greenhouse gases" to limit global temperature rise? Briefing Note (3), 1–5.

Trenberth, K.E., Jones, P.D., Ambenje, P., Bojariu, R., et al., 2007. Observations: surface and atmospheric climate change. In: Solomon, S., et al. (Eds.), Climate Change 2007. The Physical Science Basis. Cambridge University Press, New York, pp. 235–336.

Tyabji, N., Nelson, W., 2012. Mitigating Emissions from Aluminum: The Gncs Factsheets. Columbia Climate Centre, Earth Institute, Columbia University. http://climate.columbia.edu/files/2012/04/GNCS-Aluminum-Factsheet.pdf.

U.S. Climate Change Technology Program, 2005. Technology Options for Near and Long Ter U.S. Climate Change Technology Programm Future. http://www.climatetechnology.gov/library/2005/tech-options/index.htm.

U.S. Department of Energy (U.S. DOE), 2007. U.S. Energy Requirements for Aluminum Production: Historical Perspective, Theoretical Limits and Current Practice.

U.S. Environmental Protection Agency (U.S. EPA), 2016. Inventory of U.S. Greenhouse Gas Emissions and Sinks 1990–2014. U.S. EPA 430-R-16–002, Pennsylvania, Washington, U.S.A. https://www.epa.gov/sites/production/files/2016-04/documents/us-ghg-inventory -2016-main-text.pdf.

U.S. Greenhouse Gas Inventory Report: 1990–2014, 2016. https://www.epa.gov/ghgemissions/sources-greenhouse-gas-emissions.

United Nation Framework Convention on Climate Change (UNFCCC), 2008. Offsetting the Carbon Footprint for Poznan. http://unfccc.int/press/news_room/newsletter/items /4603.php.

Velasco, E., Pressley, S., Allwine, E., Westberg, H., et al., 2005. Measurements of CO_2 fluxes from the Mexico City urban landscape. Atmospheric Environment 39 (38), 7433–7446.

Vishnevetsky, I., Epstein, M., Rubin, R., 2014. Solar carboreduction of alumina under vacuum. Energy Procedia 49, 2059–2069.

Welch, B.J., 1999. Aluminum production paths in the new millennium. Journal of the Minerals, Metals and Materials Society 51 (5), 24.

Wesdock, J.C., Arnold, I.M., 2014. Occupational and environmental health in the aluminum industry: key points for health practitioners. Journal of Occupational and Environmental Medicine 56 (Suppl. 5), S5.

White, C.V., Mikkelsen, Ø., Roha, D., 2012. Status of the Alcoa carbothermic aluminum project. In: International Smelting Technology Symposium: Incorporating the 6th Advances in Sulfide Smelting Symposium. John Wiley & Sons, Inc., pp. 81–88

Wiedmann, T., Minx, J., 2007. A Definition of Carbon Footprint. ISAUK Research Report 07–01. ISAUK Research & Consulting, Durham.

World Aluminium, 2015. http://www.world-aluminium.org/statistics/#data.

World Resource Institute/World Business Council for Sustainable Development (WRI/WBCSD), 2004. The Greenhouse Gas Protocol: A Corporate Accounting and Reporting Standard Revised Edition. World Business Council for Sustainable Development and World Resource Institute, Geneva.

World Resource Institute/World Business Council for Sustainable Development (WRI/WBCSD), 2005. The Greenhouse Gas Protocol: Project Accounting. World Business Council for Sustainable Development and World Resource Institute, Geneva.

Xue, S., Zhu, F., Kong, X., Wu, C., et al., 2016. A review of the characterization and revegetation of bauxite residues (red mud). Environmental Science and Pollution Research 23 (2), 1120–1132.

Yu, H., 2002. Environmental carcinogenic polycyclic aromatic hydrocarbons: photochemistry and phototoxicity. Journal of Environmental Science and Health, Part C 20 (2), 149–183.

Zare, R., Nouri, J., Abdoli, M.A., Atabi, F., 2016. Life cycle assessment of secondary extruded aluminum production process in industrial city of Arak. Applied Ecology and Environmental Research 14 (2), 125–135.

Zhang, M., Kamavarum, V., Reddy, R.G., 2003. New electrolytes for aluminum production: ionic liquids. Journal of the Minerals, Metals and Materials Society 55 (11), 54–57.

Zhang, W., Li, H., Chen, B., Li, Q., Hou, X., Zhang, H., 2015. CO_2 emission and mitigation potential estimations of China's primary aluminium industry. Journal of Cleaner Production 103, 863–872.

Zhou, N., Xue, Y., Chen, J.J., Taylor, M.P., 2007. Numerical simulation of electrolyte two-phase flow induced by anode bubbles in an aluminum reduction cell. Chemical Product and Process Modeling 2 (2), 11.

CHAPTER 9

Carbon Footprint of Utility Consumption and Cleaning Tasks in Buildings

Alejandro Martínez-Rocamora[1], Jaime Solís-Guzmán[2], Madelyn Marrero[2]
[1]University of Bío-Bío, Concepción, Chile; [2]University of Seville, Seville, Spain

9.1 INTRODUCTION

The recent recession in the activity of the construction sector in Spain has forced companies to innovate in their business model, leading to the majority of the current business being focused on the maintenance and renovation of existing buildings (Kohler and Hassler, 2002). In Spain there has been a period of extremely intense new construction, during which construction companies failed to take the time to consider and carefully study the maintenance costs of their buildings. As a result of this thoughtless design, there are numerous buildings in Spain that imply an exorbitant maintenance cost, such as the City of Sciences in Valencia, the Bankia Tower in Madrid, and the Cajasol Tower in Seville.

With the new field of business focused on building maintenance, the role of the facility manager (FM) has become essential and is responsible for the control and coordination of numerous activities in mainly tertiary buildings, such as offices, hotels, and hospitals (Heng et al., 2005). The maintenance of the installations for their correct functioning, the repair of damages, the cleaning of the building, and the management of the costs of the utility consumption are among those tasks, and to manage them, the FM usually works with his/her own estimates based on experience. Databases of operation and maintenance costs also exist, but these are mostly accessible only from the business sector and require payment of an expensive license fee. These databases include Building Life Cycle Cost Program (National Institute of Standards and Technology, 2013), CostLab (CBRE, 2015), ECONPACK (U.S. Army Corps of Engineers, 2014),

Environmental Carbon Footprints
ISBN 978-0-12-812849-7
http://dx.doi.org/10.1016/B978-0-12-812849-7.00009-X

Building Cost Information Service Online (Royal Institute of Chartered Surveyors, 2015), and, at national level, CYPE's decennial maintenance module (CYPE Ingenieros, 2015). Without an open and free-access database, the estimation of maintenance costs from the project phase becomes extremely complicated. Needless to say, this information would be invaluable to both the owner and the FM and would provide a considerable advance in the evaluation of the costs of buildings during their service life, also known as life-cycle costs (LCC) (ISO 15686-5, 2008).

In recent years, numerous studies have been carried out on the application of environmental indicators to the building sector. A number of these studies use the ecological footprint (Bastianoni et al., 2007; Solís-Guzmán et al., 2013; González-Vallejo et al., 2015), emergy analysis (Marchi et al., 2015; Pulselli et al., 2014), or carbon footprint (CF) (Solís-Guzmán et al., 2014, 2015) as the indicator, usually focused only on the construction phase of buildings. These indicators have gained importance because of their ease of implementation and communication of results to nonspecialized audiences. Although this environmental analysis has sometimes been extended to other phases of the life cycle of buildings (Freire and Marrero, 2014; Alba Rodríguez et al., 2015), the use-and-maintenance phase has yet to be fully addressed following the guidelines set by UNE-EN 15978 (UNE-EN 15978, 2012). Certain studies focus on the maintenance of specific constructive elements, such as different flooring types (Minne and Crittenden, 2014), but none takes on the assessment of an entire building.

All the aforementioned models carry out an estimation of the environmental impact from the bill of quantities of the project, in which the various processes are divided into tasks, which are then broken down into three basic categories: manpower, materials, and machinery (Marrero and Ramirez-De-Arellano, 2010). Once the quantities of each element have been obtained, it is possible to apply environmental indicators to translate the consumption of resources into environmental impact. However, the phase has no open-access cost database available as does the construction phase, and hence it remains impossible to quantify the resources necessary for the maintenance or cleaning tasks in a disaggregated way, encompassed within this phase.

According to Fuller (2010), studies that are focused on the use-and-maintenance phase of buildings classify the sources of impact of this stage into three categories: utility consumption (electricity, water, and fuel), periodic cleaning, and maintenance and renovation of finishes and installations. Nevertheless, most studies focus solely on utility consumption, whereas a small number contemplate the renovation of finishing materials and assign these materials a specific service life (Adalberth, 1997; Pulselli et al., 2007), while the manpower and machinery necessary to carry out these tasks are often totally disregarded.

This chapter addresses the quantification of the environmental impact produced by the utility consumption and cleaning tasks through the CF indicator. First, the system boundaries are defined, followed by the study of the differentiating points of these tasks with respect to those studied in the construction phase. The methodology and formulae necessary for the application of the CF indicator to this phase are then explained. Finally, a case study of a university Hall of Residence in Seville is presented to illustrate the results obtained.

9.2 SYSTEM BOUNDARIES

As stated above, the activities carried out during the use-and-maintenance phase of the building fall into three categories: utility consumption, cleaning, and maintenance. Maintenance is divided into four branches: the renovation of finishes and installations at the end their service life; predictive maintenance, which includes inspections and periodic monitoring to anticipate possible future failures for their prevention; preventative maintenance, encompassing adjustments, calibrations, and other conservation tasks to maintain the functioning of the elements; and corrective maintenance, which consists of repairs once a failure has occurred.

Leaving aside the utility consumption because this is obtained either from bills or simulations, the maintenance and cleaning tasks are classified into the five aforementioned categories. However, it is necessary to establish boundaries for the sources of impact that should be considered in the environmental assessment of this phase. The UNE-EN 15978 standard (UNE-EN 15978, 2012) establishes that the use-and-maintenance phase of buildings begins with the completion of the construction work and comes to an end when the building is to be demolished or deconstructed. At this point, it is essential to specify that the end of this phase is fixed as the time when the building ceases to be habitable, and it is considered that when a major rehabilitation takes place, significant changes also occur in its use profile, installations, materials, and energy demand, and the building can no longer be considered the same as it was before the intervention (Spain ME, 2003). In this way, a rehabilitated building is considered as starting a new service life, and therefore a new use-and-maintenance phase, with which the system's longitudinal limits are strictly defined.

Regarding the transversal boundaries, those aspects and elements that interact with the building during this phase must be studied, and those that should be omitted must be specified. Three possible sectors have been identified to which these elements can be assigned: industry, building, and occupants (Fig. 9.1). The impacts produced during the manufacture of furniture, electrical appliances, decoration, and other objects belonging to the occupants of the building are assigned to the industry sector, and hence their impacts must be computed when calculating the CF of the corresponding industry. On the other hand, it is considered that the impacts caused through the consumption of food, mobility, and municipal solid waste (MSW) generated by the occupants should be part of their CF because they are not a direct consequence of the existence of the building. In short, utility consumption and the manpower, materials, and machinery necessary to carry out the maintenance and cleaning tasks should also be attributed to the building.

When studying the impact of manpower, the consumption of food by workers and the generation of MSW at work are quantified. Similarly, the

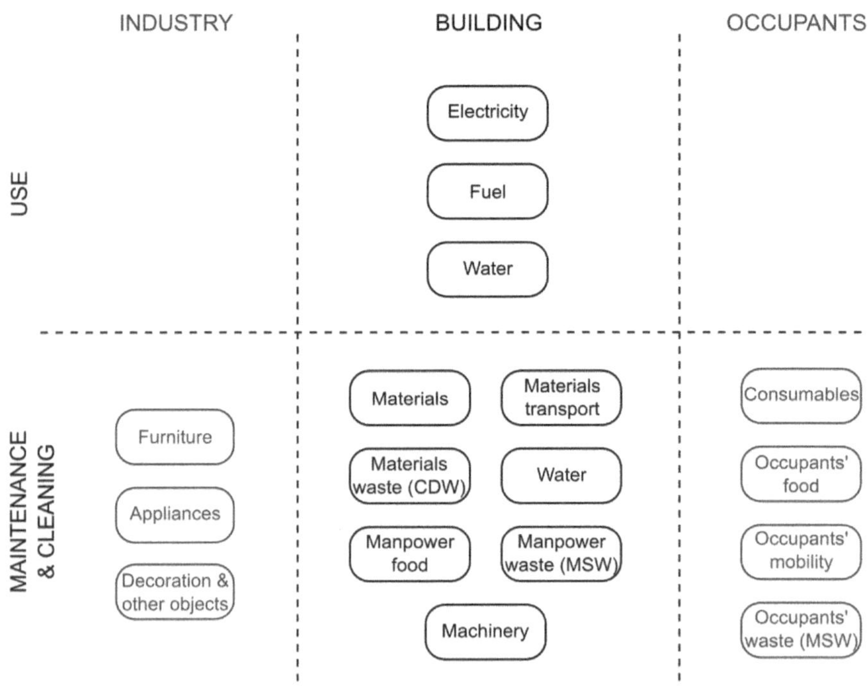

Figure 9.1 System boundaries. *CDW*, construction and demolition wastes; *MSW*, municipal solid waste.

impact of manufacturing and transporting materials and the construction and demolition wastes (CDW) generated by losses or demolition of constructive elements are also considered as impact sources. In the following section, a method to study cleaning tasks in detail, as well as the peculiarities that differentiate them from tasks in the construction phase and from maintenance tasks, is proposed. The CF calculation methodology applied to the utility consumption and cleaning tasks in buildings is then defined.

9.3 METHODOLOGY

9.3.1 Peculiarities of Cleaning Tasks

Cleaning tasks exhibit a different nature from all those related to construction, and, as such, they present numerous singularities that require separate treatment for their adaptation to the CF indicator. First, it is necessary to establish a method for the quantification of the resources required for each task. To this end, this study uses the structure of the Andalusian Construction Cost Database (ACCD) (Andalusia Government, 2014), which classifies resources into the aforementioned three basic categories, as already defined in previous developments of this study (Martínez-Rocamora et al., 2015).

Numerous products and tools are used in cleaning tasks, depending on the element or surface to be cleaned. In addition, various combinations of these products and tools can be used to clean the same element. However, when these tasks are analyzed from the point of view of costs and environmental impact studies, it makes no sense to provide a breakdown of the cost for each of these combinations because their number would be excessive and contrary to the ease of adaptation to the model proposed here.

As a simplified solution to this problem, the study of toolkits is proposed, in this case the cleaning trolley used by professional staff (Table 9.1). First, frequency of replacement, cost, weight, and base material for each product or cleaning tool is specified. From this data and the CF of each material from Ecoinvent, which has been identified as the best life cycle assessment (LCA) database for the study of complete buildings (Martínez-Rocamora et al., 2016a), it is possible to obtain the CF for the cleaning trolley as a whole. By taking 1515 as the number of hours worked per year by a cleaning worker (Andalusia Government, 2013), the CF per hour is calculated as 0.286 $kgCO_2eq/h$ and 0.057 kgMSW/h (bottom line in Table 9.1).

Table 9.1 Detailed economic and environmental data on the contents of a cleaning trolley

Tool/product	Repl. (years)	Cost (€)	Hourly cost (€/h)	Main material	Emission factor ($kgCO_2eq/kg$)	Weight (kg)	Hourly emissions ($kgCO_2eq/h$)	Annual weight (kg/year)	Hourly MSW weight (kg MSW/h)
Cleaning trolley									
Cleaning trolley	5.000	137.09	0.02	Polypropylene	3.250	14.000	0.006	2.800	0.002
Wide mop	0.043	4.83	0.07	Cotton	12.100	0.515	0.096	11.977	0.008
Dust cloths (3)	0.167	3.70	0.01	Polyester	13.130	0.117	0.006	0.701	0.000
Buckets (2)	1.000	15.92	0.01	Polypropylene	3.250	0.350	0.001	0.350	0.000
Mop or broom sticks (4)	0.500	14.52	0.02	Aluminum	8.914	0.984	0.012	1.968	0.001
Mops (2)	0.083	2.42	0.02	Cotton	12.100	0.440	0.042	5.301	0.003
Broom	0.250	1.20	0.00	PET	8.340	0.225	0.005	0.900	0.001
Dustpan	1.000	1.44	0.00	Polypropylene	3.250	0.135	0.000	0.135	0.000
Large rubbish bag	0.004	0.16	0.03	Polyethylene (LDPE)	2.530	0.010	0.004	2.500	0.002
Rubbish bags	0.009	0.91	0.07	Polyethylene (LDPE)	2.530	0.150	0.028	16.667	0.011
Wiper blade	2.000	10.32	0.00	Chromium steel	7.910	0.211	0.001	0.106	0.000
Wiper blade rubber	0.250	1.72	0.00	Synthetic rubber	2.630	0.070	0.000	0.280	0.000
Wiper tool	2.000	11.35	0.00	Polypropylene	3.250	0.120	0.000	0.060	0.000
Wiper tool fleece	1.000	6.46	0.00	Cotton	12.100	0.050	0.000	0.050	0.000
Disposable gloves	0.004	0.01	0.00	Polyethylene (LDPE)	2.530	0.005	0.002	1.250	0.001
Paper hand towel roll 120 m	0.065	2.71	0.03	Paper	−0.600	1.512	−0.009	23.262	0.015
Ostrich-feather duster	0.250	5.45	0.01	Polypropylene	3.250	0.055	0.000	0.220	0.000
Antistatic duster	0.250	1.33	0.00	Polypropylene	3.250	0.115	0.001	0.460	0.000
Bleach (5 L)	0.043	6.62	0.10	Surfactants (15%)	0.420	5.000	0.032	116.279	0.004

Neutral cleaner (1.5 L)	0.043	2.24	0.03	Surfactants (15%)	0.420	1.500	0.010	34.884	0.001
Bath descaler (750 mL)	0.250	2.77	0.01	Surfactants (15%)	0.420	0.750	0.001	3.000	0.000
Floor cleaner (1 L)	0.022	1.50	0.04	Surfactants (15%)	0.420	1.000	0.013	45.455	0.002
Furniture cleaner spray (450 mL)	0.167	4.68	0.02	Surfactants (15%)	0.420	0.450	0.001	2.695	0.000
Mop cleaner spray (750 mL)	0.250	2.15	0.01	Surfactants (15%)	0.420	0.750	0.001	3.000	0.000
Soapy cleaner (1 L)	0.167	3.52	0.01	Surfactants (15%)	0.420	1.000	0.002	5.988	0.000
Glass cleaner (750 mL)	0.083	0.90	0.01	Surfactants (15%)	0.420	0.750	0.003	9.036	0.000
Ink remover (750 mL)	0.043	3.65	0.06	Surfactants (15%)	0.420	0.750	0.005	17.442	0.001
Liquid soap refill (5 L)	0.065	5.12	0.05	Surfactants (15%)	0.420	5.000	0.021	76.923	0.003
Air freshener (1 L)	0.083	4.36	0.03	Surfactants (15%)	0.420	1.000	0.003	12.048	0.000
Total			0.69				0.286	395.734	0.057

LDPE, low-density polyethylene; *PET*, polyethylene terephthalate.

Secondly, the performance for cleaning a surface depends on its roughness, position, presence of obstacles, and degree of disorder. In fact, the number of obstacles depends directly on the type of room. For example, an assembly hall has more obstacles than a living room, and a kitchen more than a hallway. In addition, not all the elements are cleaned every time a room is cleaned. Instead these are cleaned with different periodicities. Therefore, if, in the construction phase, the tasks are classified by constructive element, then for cleaning it is necessary to quantify the spaces by type and to quantify the cleaning of these spaces in accordance with the various periodicities in which specific elements contained within are cleaned. The building is then divided into functional spaces or cells, as described in the MaClar method (Revuelta Marchena et al., 2015) and in international standards, such as Omniclass (2012) and ISO 12006-2 (2015).

Thus, the structure for the classification of cleaning tasks (Fig. 9.2) has a higher level where the space and periodicity are defined (functional costs (FC)). The second level specifies the elements to be cleaned (simple costs (SC)), and the third level defines the basic elements (materials, manpower, and machinery) needed to carry out those tasks (basic costs (BC)).

The coding of these costs is based on the ACCD classification system, with 10 alphanumeric characters: two figures representing the stage of the construction process, three letters to specify the category of the constructive element, and five more digits for uniquely numbering the item or entry. With this system, a one-to-one correspondence between codes and entries is established (Barón Cano et al., 2014).

In this classification system, all the stages of the construction process are covered in separate chapters, ranging from demolition and previous

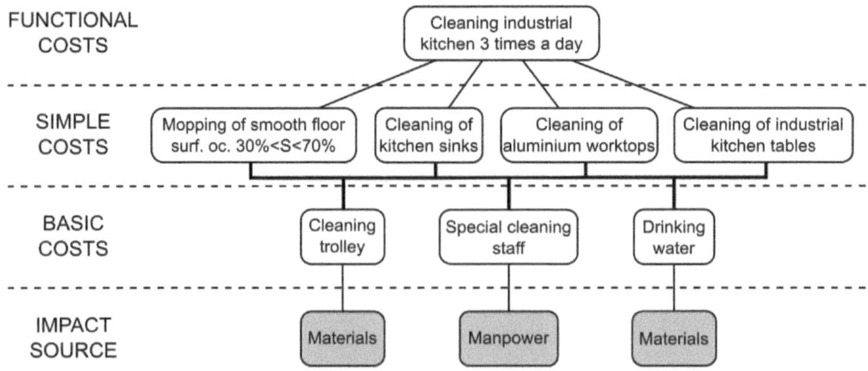

Figure 9.2 Cost structure for cleaning tasks (Martínez-Rocamora et al., 2016b).

work to painting and installation of equipment, as well as chapters focused on urbanization, waste management, and health and safety measures, the latter being Chapter 19. Because all existing chapters are related to the construction phase, it seems appropriate that use and maintenance, as another phase of the building's life cycle, should be assigned to Chapter 20. This chapter must refer not only to cleaning but also to maintenance tasks.

The assignation of the following three characters poses the first problem in the structure because the use of all three characters to define the category of the constructive element causes the groups to remain small. To avoid this situation, it is proposed that, in this chapter, constructive elements should be defined by the first two letters, thus forming more general sets (Table 9.2), whereas the third letter specifies the type of task to be performed (Table 9.3).

In addition, the last five digits of the code can be given an extra mission: the first digit can be employed to identify the frequency with which the tasks are carried out (Table 9.4). The second digit can be used to express the years of the repair dates (between 1 and 9, 10 and 19, etc.). This makes the various entries easier to find.

9.3.2 Carbon Footprint

The CF indicator measures the total amount of greenhouse gas emissions (GHG) caused directly and indirectly by an individual, organization, or product and is expressed in equivalent units of mass of CO_2 (Weidema et al., 2008). To convert the remaining GHG into equivalent CO_2, the ratio of their global warming potential to that of CO_2 is used. The CF is usually calculated according to the methodology defined in the GHG Protocol and in PAS 2050 (Pérez Leal, 2012). The integration of this indicator with the Kyoto Protocol has provided it with significant support for its implementation in companies and public institutions. In addition, as with the ecological footprint indicator, the ease with which the nonspecialized public understands this concept enhances its communicative ability.

The methodology used in this study (Fig. 9.3) divides the resource consumption of this phase of the building's life cycle into two main branches: utility consumption, where data is obtained either from bills in the case of existing buildings or through simulations in the case of projected buildings and maintenance and cleaning, where data is given by the project's bill of quantities.

Table 9.2 Coding for subchapters and sections of Chapter 20 (Martínez-Rocamora et al., 2015)

Subcapítulos/*subchapters*		**Apartados/*sections***	
20A	Aislamientos/*Insulation*	20AV	Aislamientos vistos/*Visible insulation*
		20AW	Varios/*Various*
20C	Cimentaciones/*Foundations*	20CL	Losas/*Concrete slabs*
		20CM	Muros pantalla/*Retaining walls*
		20CW	Varios/*Various*
20D	Divisiones interiores/*Interior partitions*	20DB	Barandillas/*Railings*
		20DD	Divisiones/*Partitions*
		20DM	Mamparas/*Screens*
		20DP	Puertas/*Doors*
		20DW	Varios/*Various*
20E	Estructuras/*Structures*	20EA	Acero/*Steel*
		20EF	Forjados/*Floor slabs*
		20EH	Hormigón/*Concrete*
		20EM	Madera/*Wood*
		20EW	Varios/*Various*
20F	Fachadas/*Façades*	20FC	Carpintería y protección/*Windows, doors, and weather protection*
		20FP	Paredes/*External walls*
		20FW	Varios/*Various*
20I	Instalaciones/*Installations*	20IA	Aire acondicionado/*Air conditioning*
		20IC	Calefacción/*Heating*
		20IE	Electricidad/*Electricity*
		20IF	Fontanería/*Plumbing*
		20II	Iluminación/*Illumination*
		20IK	Comunicación/*Communications*
		20IL	Gases y licuados/*Gases and liquefied fuels*
		20IM	Electro-mecánicas/*Electromechanical devices*

		20IN	Energía solar/*Solar energy*
		20IP	Protecciones/*Weather Protection*
		20IR	Evacuación de residuos/*Waste disposal*
		20IS	Saneamiento/*Sewers*
		20IV	Ventilación/*Ventilation*
		20IW	Varios/*Various*
20M	Mobiliario/*Fixtures and fittings*	20MA	Aseos y baños/*Toilets and bathrooms*
		20MC	Cocinas/*Kitchens*
		20MD	Dormitorios/*Bedrooms*
		20ME	Estanterías y armarios/*Shelves and wardrobes*
		20MK	Decoración/*Decoration*
		20ML	Lámparas/*Lamps*
		20MM	Mesas/*Tables*
		20MS	Sillas, sofás y sillones/*Chairs, sofas, and armchairs*
		20MW	Varios/*Various*
20P	Espacios interiores/*Interior spaces*	20PA	Aseos y baños/*Toilets and bathrooms*
		20PB	Bibliotecas y aulas/*Libraries and classrooms*
		20PC	Cocinas/*Kitchens*
		20PD	Dormitorios/*Bedrooms*
		20PE	Escaleras y rampas/*Stairs and ramps*
		20PK	Comedores y cafeterías/*Dining rooms and cafeterias*
		20PL	Laboratorios/*Laboratories*
		20PM	Almacenes/*Store rooms*
		20PO	Oficinas y despachos/*Offices*
		20PP	Pasillos y distribuidores/*Corridors and hallways*
		20PS	Salones y salas/*Living, meeting, and conference rooms*
		20PV	Vestíbulos/*Foyers*
		20PW	Varios/*Various*

Continued

Table 9.2 Coding for subchapters and sections of Chapter 20 (Martínez-Rocamora et al., 2015)—cont'd

Subcapítulos/*subchapters*		**Apartados/*sections***	
20Q	Cubiertas/*Roofs*	20QH	Horizontales/*Horizontal roofs*
		20QI	Inclinadas/*Sloped roofs*
		20QW	Varios/*Various*
20R	Revestimientos/*Finishings*	20RA	Aplacados/*Cladding*
		20RC	Continuous/*Continuous finishings*
		20RL	Ligeros/*Lightweight finishings*
		20RP	Pinturas/*Paint*
		20RS	Suelos/*Floors*
		20RT	Techos/*Ceilings*
		20RW	Varios/*Various*
20U	Urbanizaciones/*Urbanizations*	20UA	Alcantarillado/*Sewer system*
		20UD	Instalaciones deportivas y de ocio/*Sports facilities*
		20UE	Electricidad/*Electrial system*
		20UJ	Jardinería/*Gardening*
		20UM	Mobiliario/*Furniture*
		20UP	Pavimentos/*Pavements*
		20UR	Sistemas de riego/*Irrigation systems*
		20US	Suministro de agua/*Water supply*
		20UW	Varios/*Various*
20W	Varios/*Various*	20WW	Varios/*Various*

Table 9.3 Coding for groups (type of task)

Code	Task
20**C	Conservaciones/*Conservation*
20**L	Limpiezas/*Cleaning*
20**N	Renovaciones/*Renovations*
20**R	Reparaciones/*Repairs*
20**V	Revisiones/*Revisions*
20**W	Varios/*Various*

Table 9.4 Coding proposal for task periodicity

Code	Periodicity
0****	Hourly
1****	Daily
2****	Weekly
3****	Monthly
4****	Annual
5****	Decennial
6****	Each 20 years
7****	Each 30 years
8****	>40 years
9****	Undetermined

As can be observed in the diagram, electricity, fuel, and water consumption are included in the use branch, whereas in the maintenance and cleaning branch, the manpower, materials, and machinery necessary to carry out the tasks are quantified The methodology herein described transforms these resource consumptions into quantities of CF.

9.3.2.1 Electricity

The CF of electricity consumption depends exclusively on the energy mix of the country or region where the building under study is located. Energy generation obtained from renewable sources decreases emissions, thereby also decreasing the CF, and hence the future trend of energy production points toward a reduction in CF. In Eq. (9.1), the electricity consumption in kWh is multiplied by the emission factor of the energy mix. Consumption data, as mentioned above, is either obtained from actual bills in the case of existing buildings or from simulations in the case of projected buildings. The more precise the data on the energy mix, the more accurate the CF calculation.

$$CF_{el} = C_{el} \cdot E_{el} \tag{9.1}$$

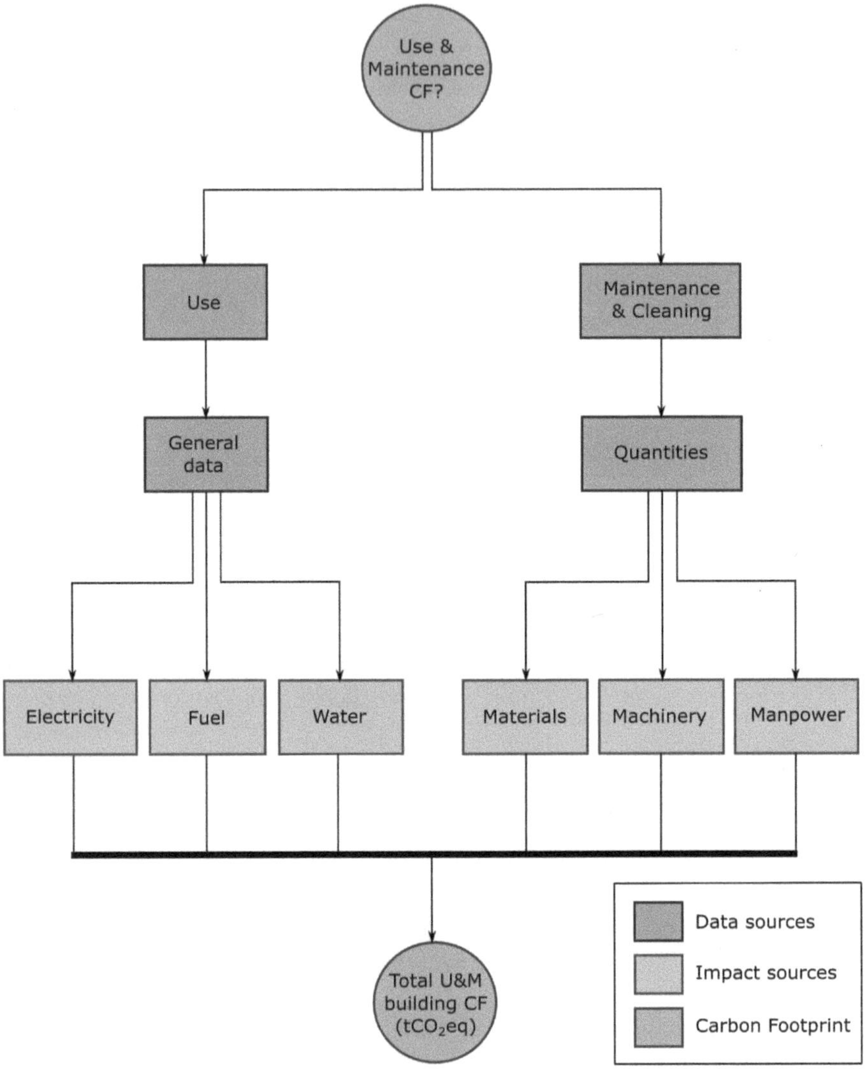

Figure 9.3 Methodology flowchart for the application of the carbon footprint (CF) indicator to the use-and-maintenance phase of buildings.

where CF_{el} is the annual carbon footprint of electricity consumption (tCO_2eq/year); C_{el} is the annual electricity consumption (kWh/year); and E_{el} is the weighted emission factor of the energy mix (0.000248 tCO_2/kWh) (REE, 2014).

9.3.2.2 Fuel

During the use phase of the building, fuel is employed for heating and/or to produce domestic hot water (DHW). The fuels consumed can be either

Table 9.5 Emission factors of fuel (IDAE, 2011b)

Fuel	Unit	Emission factor (tCO_2/u)
Hard coal	t	2104
Black lignite	t	1325
Coal coke	t	3034
Agricultural biomass	t	Neutral
Forest industrial biomass	t	Neutral
Petroleum coke	t	3194
Coke oven gas	t	1676
Diesel C	L	0.0028
Fuel oil	L	0.0028
Natural gas	Nm^3	0.0026
Liquefied petroleum gas	L	0.0015
Butane	L	0.0016
Propane	L	0.0015
Refinery gas	t	2706

gases or liquefied fuels. Their emission factor (Table 9.5) is much higher than that of electricity, as is their energy efficiency.

The calculation of the CF of fuel consumption is carried out in a similar way to that of electricity by simply replacing the consumption and emission factor with those corresponding to each fuel used. Given that it is possible to use several types of fuel at the same time in the same building, Eq. (9.2) provides a summation of their CF.

$$CF_{fu} = \sum_{i} C_{fu_i} \cdot E_{fu_i} \tag{9.2}$$

where CF_{fu} is the annual carbon footprint due to fuel consumption (tCO_2eq/year); C_{fu_i} is the annual consumption of fuel i (u/year); and E_{fu_i} is the emission factor of fuel i (tCO_2eq/u) (Table 9.4) (IDAE, 2011b), with t, L, or Nm^3 as the possible units for fuel consumption.

9.3.2.3 Water

As indicated in the UNE-EN 15978 standard (UNE-EN 15978, 2012), the model considers water consumption in the building, including the energy to bring it to the point of consumption, as well as the necessary treatment before and after its use. In the calculation of its CF, it is assumed that the energy used for these processes is electric, thus converting the volume of water consumed from bills or estimates in equivalent CO_2 emissions (Eq. 9.3).

$$CF_{wa} = C_{wa} \cdot EI_{wa} \cdot E_{el} \tag{9.3}$$

where CF_{wa} is the annual carbon footprint due to water consumption ($tCO_2eq/year$); C_{wa} is the annual water consumption in the building ($m^3/year$); EI_{wa} is the energy intensity of drinking water ($0.44\ kWh/m^3$) (EMASESA, 2005); and E_{el} is the emission factor of electricity ($0.000248\ tCO_2/kWh$) (REE, 2014).

9.3.2.4 Materials

Once the use branch of the methodology diagram (Fig. 9.3) is completed, corresponding to utility consumption, the second branch focuses on impact sources, which are quantifiable through the project's bill of quantities. The first factor to evaluate is the consumption of materials, which produces environmental impact due to three processes: manufacturing, transportation, and waste. In the study of cleaning tasks as proposed in this study, the materials used are those contained in the cleaning trolley or in the domestic cleaning kit. The consumption of these two sets of materials is accounted for hours of use, whereby it is necessary to convert these hours into equivalent CO_2 emissions by applying the data obtained in Table 9.1 for the emission factor, the weight of materials, and the generation of MSW per hour. Thus, to obtain the CF of the manufacture of materials, Eq. (9.4) is used.

$$CF_{ma} = \sum_i H_{ma_i} \cdot E_{ma_i} \tag{9.4}$$

where CF_{ma} is the annual carbon footprint of materials manufacture ($tCO_2eq/year$); H_{ma_i} represents the hours of use per year of the set of materials i (h/year); and E_{ma_i} is the emission factor for the set of materials i (tCO_2eq/h).

The environmental impact of the transport of cleaning tools and products must be added to that above because LCA databases use a cradle-to-gate analysis, excluding delivery of the finished product. For the calculation of the CF of transport (Eq. 9.5), it is necessary to consider the following hypotheses:

- A single purchase of cleaning products is made each year
- The average return distance from the factory to the building (back and forth) is 30 km
- Transport vehicle:
  - Maximum load: 2.1 t
  - Fuel consumption: 7.9 L/100 km
  - Type of fuel: Diesel

Table 9.6 Emission factors of transport and machinery fuel (IDAE, 2011a)

Fuel	Unit	Emission factor (tCO_2/u)
Petrol	L	0.0022
Diesel A and B	L	0.0026
Natural gas	Nm^3	0.0026
Biodiesel	L	Neutral
Bioethanol	L	Neutral
Liquefied petroleum gas	L	0.0015
Kerosene	L	0.0025

- Fuel emission factor: Table 9.6 (IDAE, 2011a)

$$CF_{tr} = \sum_i \frac{H_{ma_i} \cdot W_{ma_i}}{V_{cap}} \cdot D_{ma} \cdot V_{con} \cdot E_{fu} \quad (9.5)$$

where CF_{tr} is the annual carbon footprint of material transport (tCO_2eq/year); H_{ma_i} represents the annual hours of use of the set of materials i (h/year); W_{ma_i} is the weight of materials consumed per hour of use of the set i (kg/h); V_{cap} is the carrying capacity of the transport vehicle (kg/trip); D_{ma} is the average return distance (back and forth) per trip (km/trip); V_{con} is the fuel consumption of the vehicle (l/100 km); and E_{fu} is the fuel emission factor (tCO_2eq/L).

Unlike the waste produced in the construction phase, waste generated from cleaning products and tools cannot be considered CDW. This is counted as MSW, which must be managed by the municipal company. According to Almasi and Milios (2013), the emission factor of MSW in Spain is 0.244 tCO_2/ton of MSW. Included in this calculation are the benefits of recycling and energy recovery.

The CF of the MSW generated must be calculated through Eq. (9.8), in which the generation of MSW per hour of use of the cleaning trolley is obtained from Table 9.1. It is assumed that the entire weight of tools becomes waste at the end of their service life, whereas for cleaning products only the empty bottles become waste.

$$CF_{MSW\ mat} = \sum_i H_{ma_i} \cdot G_{ma_i} \cdot E_{MSW} \quad (9.6)$$

where $CF_{MSW\ mat}$ is the annual carbon footprint of the MSW generation from materials (tCO_2eq/year); H_{ma_i} represents the annual hours of use of the set of materials i (h/year); G_{ma_i} is the generation of MSW per hour of use of the set of materials i (kg/h); and E_{MSW} is the emission factor for the management of MSW (tCO_2eq/kg).

9.3.2.5 Machinery

The machinery used in cleaning tasks is always electric. The first step to calculate the CF of the machinery is to transform the hours of use into electrical energy consumption through the power of its engine (Freire and Marrero, 2014). Once energy consumption is obtained, the emission factor of the energy mix is applied as shown in Eq. (9.7).

$$CF_{mn} = \sum_i H_{mn_i} \cdot C_{mn_i} \cdot E_{el} \tag{9.7}$$

where CF_{mn} is the annual carbon footprint of machinery (tCO_2eq/year); H_{mn_i} represents the annual hours of use of the machine i (h/year); C_{mn_i} is the power of machine i (kWh/h); and E_{el} is the emission factor of electricity (0.000248 tCO_2eq/kWh) (REE, 2014).

9.3.2.6 Manpower

Manpower is probably one of the most significant impact sources on cleaning tasks due to the high number of hours required for the proper cleaning of a building. This causes environmental impact for two different reasons: food consumption by workers and MSW generation. Although workers' mobility was taken into account in previous models (González-Vallejo et al., 2015; Solís-Guzmán et al., 2014, 2015), the UNE-EN 15978 standard (UNE-EN 15978, 2012) states that, for the construction phase, worker's mobility should not be included in the analysis, and therefore, following this line of reasoning, it makes sense to omit mobility in this study for the cleaning staff.

To obtain the CF of food consumption, previous models considered that all workers consume a significant part of their daily diet in the workplace, and this gives them the energy necessary to carry out their tasks, as applied by Domenech Quesada (2007) for corporations. This hypothesis has to withstand the common argument that people will consume food even if they stay at home without working.

The solution to this problem is to calculate the difference in metabolic rate between that produced at work and that developed at home. Thus, the difference in kilocalories between the two activities, which can be attributable to the actual work, is obtained. In this study, a moderate metabolic rate is considered for cleaning tasks, which in a continuous regime involves certain physical effort. The CF of food consumption is obtained through Eq. (9.8) by transforming this consumption of kilocalories into CO_2 emissions due to food processing. For a better understanding of the equation, the H_d factor has been left visible, despite the fact that it is canceled out.

$$CF_{fo} = \frac{H_w}{H_d} \cdot \frac{M_{dif} \cdot H_d}{C_c} \cdot E_{fo} \quad (9.8)$$

where CF_{fo} is the annual carbon footprint for food consumption (tCO_2eq/year); H_w is the total hours worked per year (h/year); H_d is the number of hours worked per person per day (8 h/person per day); M_{dif} is the increment in the metabolic rate of manpower (kcal/h) (Table 9.7) (NTP 1011, 2014); C_c is the average daily calorie consumption per person (kcal/person per day) (FAO, 2014); and E_{fo} is the emission factor of processed food consumed per person per day (tCO_2eq/person per day).

Regarding the generation of MSW, according to EUROSTAT (EUROSTAT, 2015), in 2013 each person in Spain generated 449 kg of MSW. Taking into account that people sleep 8 h a day, during which no waste is generated, an MSW generation of 0.077 kg/h is obtained. This is combined in Eq. (9.9) with the emission factor for the aforementioned waste management, applied to the hours worked by manpower.

$$CF_{MSW\ mp} = H_w \cdot G_w \cdot E_{MSW} \quad (9.9)$$

Table 9.7 Metabolic rate by type of activity (NTP 1011, 2014)

Activity	**Metabolic rate (kcal/h)**	**Average (kcal/h)**
Sedentary work	85.4–108.7	97.1
Administrative work	108.7–155.3	132.0
Mason	170.8–248.5	209.7
Carpenter	170.8–271.8	221.3
Glazier	139.8–194.1	166.9
Painter	155.3–201.9	178.6
Blacksmith	139.8–310.6	225.2
Welder	116.5–194.1	155.3
Turner	116.5–194.1	155.3
Miller	124.2–217.4	170.8
Gardener	178.6–295.1	236.8
Machinery driver	132.0–170.8	151.4
Crane operator	100.9–225.2	163.1
Low metabolic rate	110.3–201.9	155.3
Moderate metabolic rate	203.4–310.6	256.2
High metabolic rate	312.2–403.8	357.2

where $CF_{MSW\ mp}$ is the annual carbon footprint of MSW generation by manpower (tCO_2eq/year); H_w is the annual number of hours worked (h/year); G_w is the MSW generation per hour (0.000077 t/h) (EUROSTAT, 2015); and E_{MSW} is the emission factor by MSW management (0.245 tCO_2eq/t) (Almasi and Milios, 2013).

9.3.3 Carbon Footprint Assessment in Long-Term Scenarios

When applying the defined methodology to long-term cases, such as the use-and-maintenance phase of buildings, it is necessary to use environmental discount rates to update the emission quantities. The longevity of buildings leads to the need for the prediction of changes in the amount of GHG emissions, advances in power generation technology, and the emergence of new policies to mitigate climate change (Riahi et al., 2007).

For the purposes of this study, the BLUE map scenario by the International Energy Agency (IEA, 2010) is used, which assumes that by 2050, emissions will stabilize at 50% of 2005 levels. From this scenario, discount rates are extracted for the CF indicator using 2016 as baseline year, which enables the impacts caused in other years to be translated into current CF (Eq. 9.10).

$$CF = \sum_i CF_i \cdot \frac{CF_{w,i}}{CF_{w,b}} \tag{9.10}$$

where CF is the total carbon footprint of the building in terms of the baseline year b (tCO_2eq); CF_i is the carbon footprint of the building in year i (tCO_2eq); $CF_{w,i}$ is the global carbon footprint in year i (tCO_2eq); and $CF_{w,b}$ is the global carbon footprint in the baseline year b (tCO_2eq).

In the next section, the case study of a building is presented to illustrate the usefulness of the methodology described above, followed by the discussion of results.

9.4 CASE STUDY

The building studied is located in the southern sector of Seville (Spain). It is a university Hall of Residence built in 1948 on the Square of Conquerors of the Ibero-American exhibition of 1929. In this study, 100 years of service life since the year of construction are assumed for this building. It is a detached building with 7187 m^2 of built-up surface on a 10,501 m^2 parcel with gardens, car park, sports facilities, and a swimming pool (Fig. 9.4).

Figure 9.4 Front and back façades of Hernando Colón university Hall of Residence.

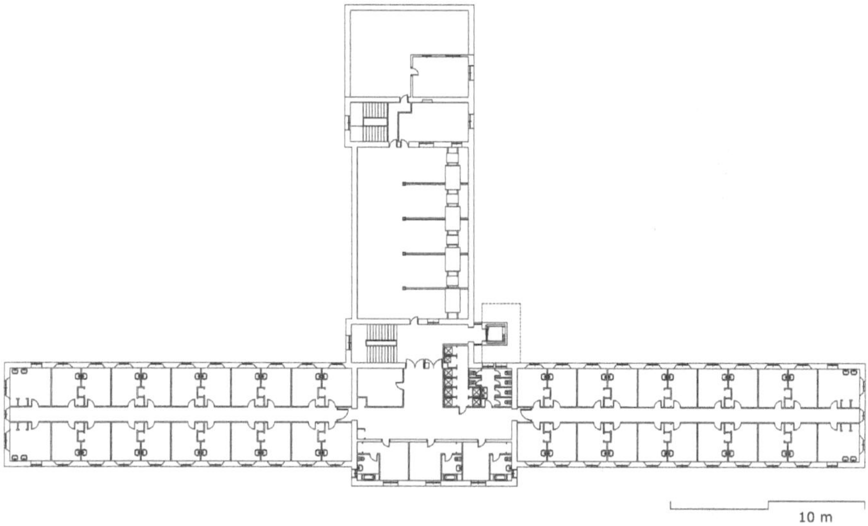

Figure 9.5 First floor of Hernando Colón university Hall of Residence.

The building is T-shaped (Fig. 9.5), with its central core and lateral wings reaching four storeys above ground, whereas the rear wing has only one storey. There is also a semibasement floor occupied by a library, warehouses, a gym, and engine rooms. On the ground floor there are offices, assembly halls, a kitchen, and a common dining room, with the left wing occupied by the director's residence. The upper three floors contain students' bedrooms occupying both wings, with a capacity of up to 139 residents.

Floors are covered mostly by terrazzo, although certain areas have marble or ceramic tiles. Several walls have a marble plinth, and textured or smooth paint is found throughout the entire building. The windows are single-glazed with anodized aluminum frames. Interior doors are mostly made of painted light wood.

The heating system with radiators and DHW is fed by two gas—oil boilers with 160,000 kcal/h of useful power, located in the semibasement. On the roof of the rear wing there are four heat pumps with 33 kW of cooling capacity and 40 kW of heating capacity, which feed a fan coil system that is distributed throughout the rooms. There is also a 1029-L propane tank that feeds the industrial kitchen.

9.5 RESULTS

Given that the case study is based on an existing building, it is impossible to separate the environmental impact caused by the consumption due to auxiliary appliances present in the building from that of the energy demand of the building itself. Because consumption data have been obtained from actual bills, these are not differentiated. Therefore, these impacts cannot be reported separately as required by the UNE-EN 15978:2012 standard (UNE-EN 15978, 2012).

According to the use branch defined in the diagram of Fig. 9.3, electricity, drinking water, diesel C, and propane are consumed in the building (Table 9.8). Following the methodology described in Section 3, all these consumptions cause an annual CF of 228.73 tCO_2eq/year, of which 45% is due to electricity consumption and 54% to consumption of diesel C for heating and DHW. The consumption of propane and drinking water causes no significant impact.

Regarding cleaning tasks, Table 9.9 shows the breakdown in terms of the different impact sources classified into materials, manpower, machinery, and indirect costs. In this case study, the indirect costs of cleaning consist of only a supervisor of the cleaning team, who is assumed to intervene in all tasks simultaneously as an organizer.

Table 9.8 Annual carbon footprint of the "use" branch

Impact driver	Consumption	Unit	Carbon footprint (tCO_2eq/year)
Electricity	416.000	kWh	103.17
Drinking water	7.480	m^3	0.82
Diesel C	44.000	L	123.20
Propane	1.029	L	1.54
Total			228.73

Table 9.9 Annual carbon footprint of the "cleaning tasks" branch

Resource consumption for cleaning (2016)				Carbon footprint of cleaning tasks (2016)
Description	**Concept**	**Total**	**Unit**	**(tCO_2eq/year)**
Materials	Total			5,734
	Manufacture			5,372
	Transport			0.097
	Municipal solid waste (MSW) materials			0.265
Drinking water	Manufacture	3,12,495	m^3	0.034
	Transport			0.000
	MSW			0.000
Cleaning trolley	Manufacture	18,764.066	h	5,255
	Transport			0.096
	MSW			0.261
Domestic cleaning kit	Manufacture	2,93,524	h	0.082
	Transport			0.001
	MSW			0.004
Manpower	Total			0.967
	Food			0.609
	Manpower MSW			0.358
Special cleaning worker	Food	18,764.066	h	0.600
	MSW			0.353
Domestic employee	Food	2,93,215	h	0.009
	MSW			0.006
Machinery	Total			0
Ind. costs	Total			0.077
Supervisor	Food	1515.000	h	0.048
	MSW			0.028
Total				6,778

It is important to remember that the consumption of water for cleaning tasks must be subtracted from the total consumption obtained from bills to avoid double counting. In the case of using electrical machinery for cleaning, its consumption should also be subtracted from that of the billing data.

As can be observed in Fig. 9.6, the source of highest impact is the manufacture of materials contained in the sets of tools and products at 79%. It is to be highlighted that food consumption and MSW generation by

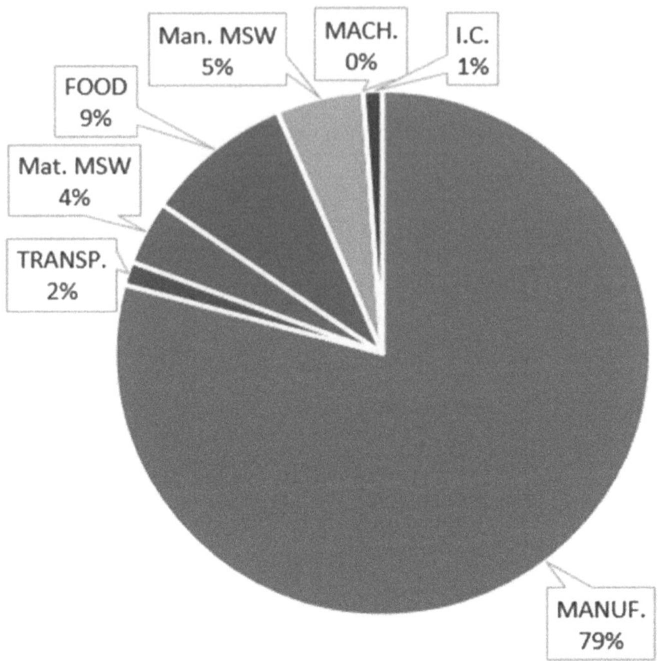

Figure 9.6 Annual carbon footprint (CF) of cleaning in terms of impact driver. *MSW*, municipal solid waste.

workers represent 14% of the CF of cleaning tasks. Taking into account that the modifications of the methodology regarding CF of food consumption reduced its impact by 85%, it can be considered a significant percentage. This is due to the high needs for manpower in cleaning tasks given their manual nature.

The total results (Table 9.10) are summarized with 228.73 tCO_2eq/year caused by utility consumption and 6.78 tCO_2eq/year by cleaning tasks, which represents 2.96% of the CF of the use branch. This is a considerable share because the CF of maintenance tasks remains to be quantified in the

Table 9.10 Annual carbon footprint of use and cleaning

	Annual carbon footprint	
Impact driver	**(tCO_2eq/year)**	**(%)**
Utility consumption	228.73	97.12
Cleaning tasks	6.78	2.88
Total	235.51	100.00

future progress of this study, and together, these two factors will probably exceed 10% of the total impact of the use-and-maintenance phase.

If 139 residents are assumed, the CF per person for this phase is 1.69 tCO_2eq/year. In the case of the director's residence, assuming four family members, the CF per member is ~0.026 tCO_2eq/year. This difference in the order of magnitude is because of the fact that domestic environments are normally cleaned much less frequently than public spaces.

Discount rates have been applied to the CF according to the BLUE map scenario explained in Section 3.3. These factors have been calculated from 1948, year of construction of the building, to 2047, at the end of its assumed service life, taking 2016 as the baseline year (Fig. 9.7). The total CF for these 100 years is 15420 tCO_2eq, with 97.12% due to utility consumption and 2.88% to cleaning tasks.

To show the sensitivity of the model, 35 scenarios have been studied and ordered in terms of their associated CF (Fig. 9.8). Each scenario represents a possible distribution of the available area for bedrooms, en-suite and communal bathrooms, and shower rooms (~1840 m^2) in four types of room: single (11 m^2), single en-suite (13.5 m^2), twin (15 m^2), and twin en-suite (18 m^2). A communal bathroom and a shower room (22.5 m^2 each) are assumed for every 20 and 40 rooms without their own bathrooms, respectively.

The results of this sensitivity analysis show that the CF of cleaning tasks is certainly sensitive to changes in the distribution of spaces. Distributions with only en-suite rooms resulted in lower CF. In contrast, those with a

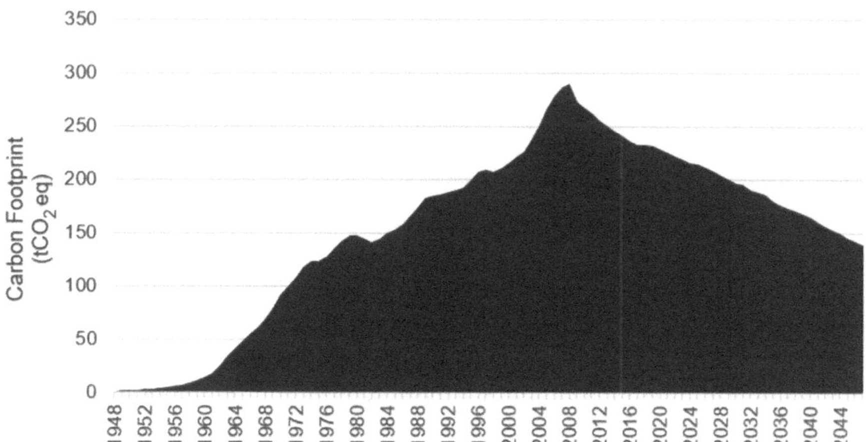

Figure 9.7 BLUE map scenario applied to the case study for 100 years of service life.

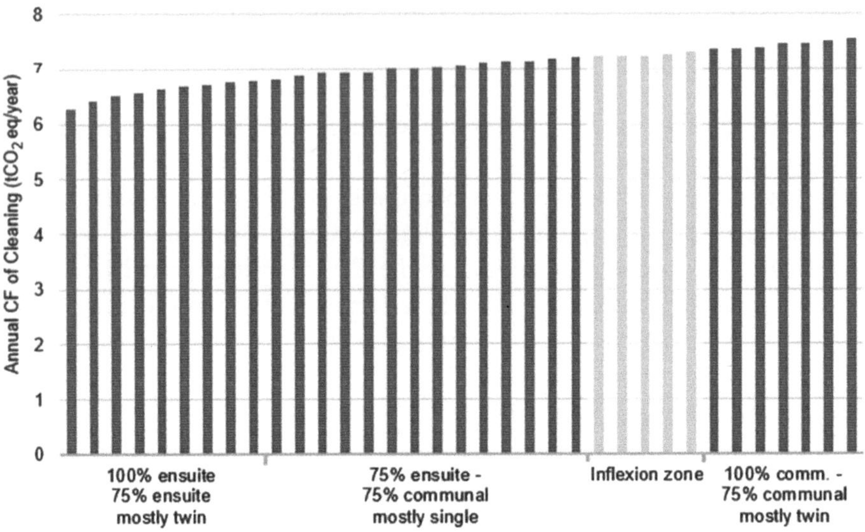

Figure 9.8 Sensitivity analysis of the carbon footprint (CF) of 35 cleaning scenarios.

majority of rooms with a communal bathroom show a larger CF. This is because of the higher cleaning frequency of communal bathrooms and shower rooms compared with private bathrooms. An inflexion zone is produced where the number of communal bathrooms and shower rooms increases at the same time due to the assumptions made above. Those cases remain thus unclassifiable because they are not representative of the set of scenarios taken.

9.6 CONCLUSIONS

In this chapter, the information requirements to calculate the CF of a building during the use-and-maintenance phase are first determined. A robust and structured methodology for the generation of a new chapter of the ACCD that includes the use and maintenance of buildings is presented. Specifically, a suitable method has been designed for the development of cleaning costs, with descriptive codes for the various categories in which they are included, as well as a method for encoding or classifying the periodicities with which tasks are performed. Once the necessary resources have been determined through the classification system herein described, these are then broken down into materials, manpower, and machinery.

This way of classifying these resources simplifies their implementation by professionals in the construction sector. To the best of our knowledge, these estimates, which managers of hotels or other tertiary buildings may well already be implementing internally in a reasonably accurate way, had not, to date, been addressed from the point of view of a project's bill of quantities. The methodology described and the work announced herein certainly constitute a leap forward in LCC assessment.

Second, a methodology is developed to determine the CF of utility consumption and cleaning tasks in buildings. The methodology presented in previous publications is revised, and, given the numerous peculiarities of the activities studied herein, the formulae have had to be adjusted. The CF indicator is proven flexible to be adapted to the use-and-maintenance phase of buildings.

Because this is a first proposal of a CF calculation model for this phase of the life cycle of buildings, it has been applied to a single case study. In future developments of this research, not only will the model be developed to include its application to maintenance tasks and the renovation of constructive elements but also the study will involve a representative set of cases of the building sector in Spain, regarding a wide variety of building typologies. At this point, the case study used enables the results to be analyzed in terms of their order of magnitude.

In the case study presented herein, the impact of cleaning tasks represents 2.96% of that of utility consumption. This is a percentage that, taken on its own, could be considered negligible, but, together with the CF of maintenance tasks and renewal of elements, it must be quantified because it could exceed 10% of the total CF of this phase. As expected, utility consumption caused 97% of the total CF of this phase. Manpower leads to 14% of the CF of cleaning tasks, whereas manufacture materials (cleaning tools and products) cover 79% of this impact. It is important to bear in mind that manpower is a necessary impact, and that its reduction would be difficult because of the manual nature of these tasks. On the other hand, this encourages alternative designs and distributions of buildings to be studied that would allow these areas to be cleaned more efficiently, thereby reducing their economic and environmental impact.

The sensitivity analysis carried out with 35 scenarios of space distribution of bedrooms proves that these variations cause significant changes in the results. In addition, this analysis is applied to determine which configuration allows the annual CF of cleaning tasks to be reduced. It is concluded that distributions dominated by twin en-suite bedrooms cause less CF than the existing distribution in the building.

The methodology presented is considered useful for the analysis of the environmental and economic impact of all types of buildings, especially tertiary buildings (offices, hotels, etc.), because cleaning incurs a greater incidence in these cases. Furthermore, FMs and companies who wish to use the results as a sustainability-based marketing strategy may well find this methodology to be a valuable tool.

REFERENCES

Adalberth, K., 1997. Energy use during the life cycle of single-unit dwellings: a method. Building and Environment 32, 321–329. http://dx.doi.org/10.1016/S0360-1323(96)00069-8.

Alba Rodríguez, M.D., Marrero, M., Solís-Guzmán, J., Llácer Pantión, R., Lucas Ruíz, R., 2015. Ecological footprint of building rehabilitation: economic and environmental assessment. In: 6th Int. Congr. Energy Environ. Eng. Manag., Paris.

Almasi, A.M., Milios, L., 2013. Municipal Waste Management in Spain.

Andalusia Government, 2013. Convenio Colectivo del Sector de Limpieza de Edificios y Locales de Sevilla y Provincia. http://www.dipusevilla.es/export/bop/201310/25.pdf.

Andalusia Government, 2014. Andalusia Construction Cost Database (ACCD). http://www.juntadeandalucia.es/fomentoyvivienda/portal-web/web/areas/vivienda/texto/7a0899c8-0038-11e4-8cc4-27ee69a25823.

Barón Cano, J.L., Conde Oliva, J., Osuna Rodríguez, M., Ramírez-de-Arellano Agudo, A., Solís Burgos, J.A., 2014. Clasificación sistemática de Precios Básicos, Auxiliares y Unitarios (Systematic classification for Basic, Auxiliary and Unitary Costs). https://ws147.juntadeandalucia.es/obraspublicasyvivienda/davwebviv/Banco de precios/BCC.

Bastianoni, S., Galli, A., Pulselli, R.M., Niccolucci, V., 2007. Environmental and economic evaluation of natural capital appropriation through building construction: practical case study in the Italian context. Ambio 36, 559–565. http://dx.doi.org/10.1579/0044-7447(2007)36{[}559:EAEEON]2.0.CO;2.

CBRE, 2015. Whitestone Research, CostLab Pro. http://whitestoneresearch.com/CBRE-Store/Store.html.

CYPE Ingenieros, 2015. Mantenimiento decenal de CYPE Ingenieros. Calendario de mantenimiento y Valoración de mantenimiento decenal del edificio. http://mantenimiento-decenal.cype.es/.

Domenech Quesada, J.L., 2007. Huella ecológica y desarrollo sostenible (Ecological Footprint and Sustainable Development). AENOR, Madrid.

EMASESA, 2005. Sostenibilidad y gestión. Así éramos, así somos. 1975–2005 (Sustainability and Management. How We Were, How We Are. 1975–2005).

EUROSTAT, 2015. Municipal waste generated by country in selected years (kg per capita). http://ec.europa.eu/eurostat/statistics-explained/index.php/File:Municipal_waste_generated_by_country_in_selected_years_%2528kg_per_capita%2529_new.png.

Freire, A., Marrero, M., 2014. Analysis of the ecological footprint produced by machinery in construction. In: World Sustain. Build. 14, Barcelona.

Fuller, S., 2010. Life-Cycle Cost Analysis (LCCA).

González-Vallejo, P., Marrero, M., Solís-Guzmán, J., 2015. The ecological footprint of dwelling construction in Spain. Ecological Indicators 52, 75–84. http://dx.doi.org/10.1016/j.ecolind.2014.11.016.

Heng, H.K.S., McGeorge, W.D., Loosemore, M., 2005. Beyond strategy: exploring the brokerage role of facilities manager in hospitals. Journal of Health Organization and Management 19, 16–31. http://dx.doi.org/10.1108/14777260510592112.

IDAE, 2011a. Factores de emisión de CO_2 (CO_2 Emission Factors). http://www.idae.es/uploads/documentos/documentos_Factores_Conversion_Energia_y_CO2_2011_0a9cb734.pdf.

IDAE, 2011b. Guía práctica de la energía. Consumo eficiente y responsable (Practical Guide of Energy. Efficient and Responsible Consumption). Instituto para la Diversificación y Ahorro de la Energía, Madrid. www.idae.es.

IEA, 2010. Energy Technology Perspectives 2010-Scenarios and Strategies to 2050. Paris.

ISO 12006-2, 2015. Building Construction – Organization of Information about Construction Works – Part 2: Framework for Classification.

ISO 15686-5, 2008. Buildings and Constructed Assets – Service-Life Planning – Part 5: Life-Cycle Costing.

Kohler, N., Hassler, U., 2002. The building stock as a research object. Building Research and Information 30, 226–236. http://dx.doi.org/10.1080/09613210110102238.

Marchi, M., Pulselli, R.M., Marchettini, N., Pulselli, F.M., Bastianoni, S., 2015. Carbon dioxide sequestration model of a vertical greenery system. Ecological Modelling 306, 46–56. http://dx.doi.org/10.1016/j.ecolmodel.2014.08.013.

Marrero, M., Ramirez-De-Arellano, A., 2010. The building cost system in Andalusia: application to construction and demolition waste management. Construction Management and Economics 28, 495–507. http://dx.doi.org/10.1080/01446191003735500.

Martínez-Rocamora, A., Solís-Guzmán, J., Marrero, M., 2015. A structure for the quantity surveillance of costs and environmental impact of cleaning and maintenance in buildings. In: Sustain. Renov. Build. Neighb. Bentham Science Publishers, pp. 103–118.

Martínez-Rocamora, A., Solís-Guzmán, J., Marrero, M., 2016a. LCA databases focused on construction materials: a review. Renewable and Sustainable Energy Reviews 58, 565–573. http://dx.doi.org/10.1016/j.rser.2015.12.243.

Martínez-Rocamora, A., Solís-Guzmán, J., Marrero, M., 2016b. Toward the ecological footprint of the use and maintenance phase of buildings: utility consumption and cleaning tasks. Ecological Indicators 69, 66–77. http://dx.doi.org/10.1016/j.ecolind.2016.04.007.

Minne, E., Crittenden, J.C., 2014. Impact of maintenance on life cycle impact and cost assessment for residential flooring options. The International Journal of Life Cycle Assessment 36–45. http://dx.doi.org/10.1007/s11367-014-0809-z.

National Institute of Standards and Technology, 2013. Building Life Cycle Cost Program (BLCC). http://www.energy.gov/eere/femp/building-life-cycle-cost-programs.

NTP 1011, 2014. Determinación del metabolismo energético mediante tablas (Determination of Energetic Metabolism through Tables).

Omniclass, 2012. Omniclass: A Strategy for Classifying the Built Environment – Table 13: Spaces by Function.

Pérez Leal, M.M., 2012. Huella de carbono. Herramienta de gestión ambiental, empresarial y social (Carbon Footprint. Environmental, Business and Social Management Tool). University of Seville.

Pulselli, R.M., Simoncini, E., Pulselli, F.M., Bastianoni, S., 2007. Emergy analysis of building manufacturing, maintenance and use: em-building indices to evaluate housing sustainability. Energy and Buildings 39, 620–628. http://dx.doi.org/10.1016/j.enbuild.2006.10.004.

Pulselli, R.M., Pulselli, F.M., Mazzali, U., Peron, F., Bastianoni, S., 2014. Emergy based evaluation of environmental performances of living wall and grass wall systems. Energy and Buildings 73, 200–211. http://dx.doi.org/10.1016/j.enbuild.2014.01.034.

REE, 2014. El sistema eléctrico español (The Spanish Electric System).

Revuelta Marchena, M.P., Ramírez-de-Arellano Agudo, A., García Vázquez, P., 2015. Control de costes por anticipación. Editorial Universidad de Sevilla, Sevilla.
Riahi, K., Grübler, A., Nakicenovic, N., 2007. Scenarios of long-term socio-economic and environmental development under climate stabilization. Technological Forecasting and Social Change 74, 887–935. http://dx.doi.org/10.1016/j.techfore.2006.05.026.
Royal Institute of Chartered Surveyors, 2015. BCIS Building Running Costs Online. http://www.rics.org/es/knowledge/bcis/online-products/bcis-building-running-costs-online/.
Solís-Guzmán, J., Marrero, M., Ramírez-de-Arellano, A., 2013. Methodology for determining the ecological footprint of the construction of residential buildings in Andalusia (Spain). Ecological Indicators 25, 239–249. http://dx.doi.org/10.1016/j.ecolind.2012.10.008.
Solís-Guzmán, J., Martínez-Rocamora, A., Marrero, M., 2014. Methodology for determining the carbon footprint of the construction of residential buildings. In: Assess. Carbon Footpr. Differ. Ind. Sect. Springer, pp. 49–83.
Solís-Guzmán, J., González-Vallejo, P., Martínez-Rocamora, A., Marrero, M., 2015. The carbon footprint of dwelling construction in Spain. In: Carbon Footpr. Handb. CRC Press - Taylor & Francis Group, pp. 261–283.
Spain ME, 2003. Orden ECO/805/2003, de 27 de marzo, sobre normas de valoración de bienes inmuebles y de determinados derechos para ciertas finalidades financieras. Madrid, Spain.
Statistic division of the FAO, FAOSTAT, 2014. http://faostat3.fao.org/home/E.
UNE-EN 15978, 2012. Sustainability of Construction Works. Assessment of Environmental Performance of Buildings. Calculation Method.
U.S. Army Corps of Engineers, 2014. Economic Analysis Package (ECONPACK). http://www.wbdg.org/tools/econpack.php.
Weidema, B.P., Thrane, M., Christensen, P., Schmidt, J., Løkke, S., 2008. Carbon footprint: a catalyst for life cycle assessment? Journal of Industrial Ecology 12, 3–6. http://dx.doi.org/10.1111/j.1530-9290.2008.00005.x.

CHAPTER 10

Greenhouse Gas Emissions From Coal Mining Activities and Their Possible Mitigation Strategies

Bhanu Pandey, Meenu Gautam, Madhoolika Agrawal
Department of Botany, Institute of Science, Banaras Hindu University, Varanasi, India

10.1 INTRODUCTION

How consumption of energy is affecting the change in climate is one of the most important topics in energy economics. The demand for energy is directly affected by the changes in weather and climatic conditions. The energy requirement of the world is increasing day by day to attain and maintain development criteria. Gross domestic product (GDP), per capita income, level of industrialization, amount of widespread infrastructure, and general standard of living are the criteria for evaluation of development in a country. Population growth has always been and will remain one of the key drivers of energy demand, along with economic and social developments. Hence, economic development and population growth of the world are closely related to energy consumption because high population and progressive development have customarily been accompanied by rapid increase in energy demand. By 2030, the world is projected to consume two-third of more energy than its present consumption, thereby enabling the developing countries to replace the industrialized world as the largest group of energy consumers (Birol, 2010). To meet the world's demand for energy, much faster rate of energy production and utilization of resources are required.

10.1.1 Increasing Demand for Coal as a Source of Energy

The world's energy resources can be broadly divided into fossil fuel, nuclear fuel, and renewable resources. The contributions of different energy sources to the total supply of energy in the world are of fossil fuels >> renewable resources (other than hydro) > nuclear > hydro (Fig 10.1). The global energy demand is expected to rise from 533

Environmental Carbon Footprints
ISBN 978-0-12-812849-7
http://dx.doi.org/10.1016/B978-0-12-812849-7.00010-6

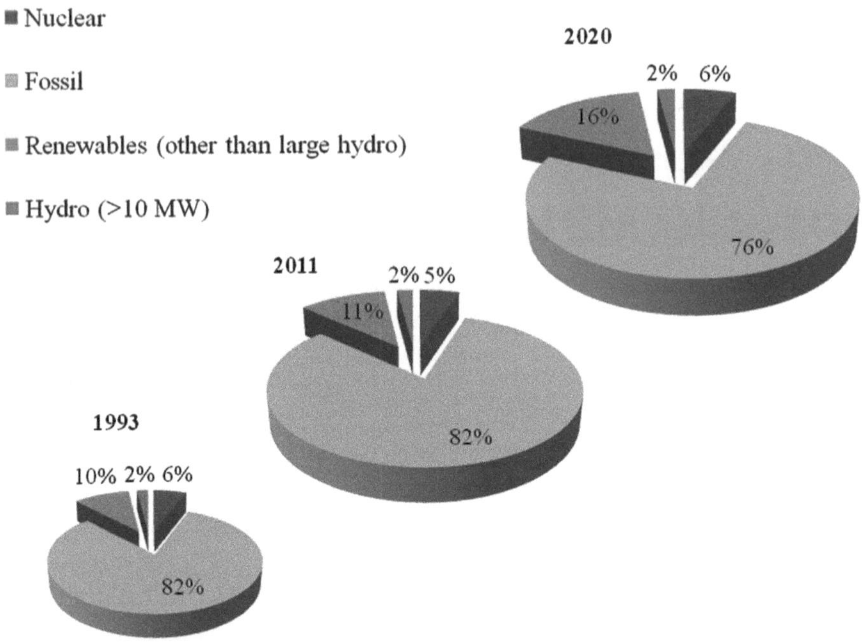

Figure 10.1 Total primary energy supply by various resources in 1993, 2011, and 2013. *(Reproduced from World Energy Council, 2013. World Energy Resources, Survey. Available at: http://www.worldenergy.org.)*

exajoules in 2008 to 653 exajoules in 2020 and 812 exajoules in 2035 (Conti et al., 2011). Among all the nonrenewable energy resources, coal accounted for nearly 28% of primary energy consumption as well as world's largest source of electricity, accounting for around 40% of global electricity production (IEA, 2014). At present coal is the world's second largest source of primary energy (Fig 10.2), and it is widely expected to replace oil as the world's largest source of primary energy within a few years because of the fact that coal is far more abundant, widely distributed across the globe and affordable (World Energy Council, 2013). To meet the energy requirement, the overall coal production and coal mining have tremendously increased around globe (Fig 10.3).

The mining techniques used for extraction of coal depend on the quality, depth of the coal seam, and the geology of the coal deposits. The coal mining procedure can be classified into two categories depending on the way of operation: surface (opencut) and underground mining. Generally, surface mining is practiced when a coal seam is within 70 m from the earth's surface, whereas coal bed beyond 70 m from the earth's surface is

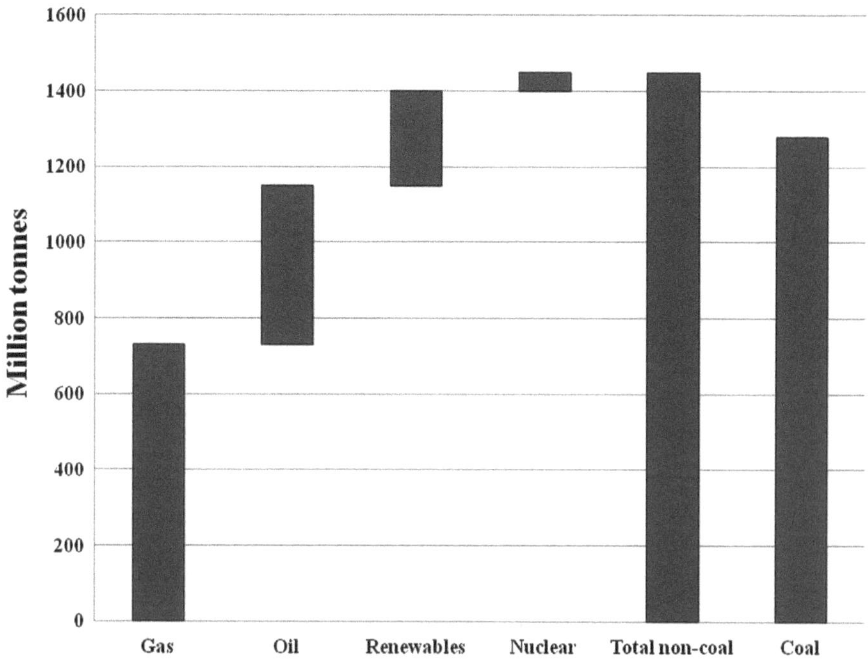

Figure 10.2 Increase in primary energy demand worldwide from renewable and nonrenewable energy resources. *(Reproduced from International Energy Agency (IEA), 2014. Key World Energy Statistics. OECD/International Energy Agency, Paris.)*

mined by adopting underground mining methods. Surface mining also known as strip mining, is the most common type of mining of coal in the world responsible for ~8% of coal production in China, 80% in Australia, and 67% in the United States (IEA, 2014). The bed of soil and rock, covering the coal bed (commonly referred as "overburden") is first removed using draglines, power shovels, bulldozers, and front-end loaders to expose the coal seam for mining. In general, surface mining is the cheapest and the most productive method of coal mining. Surface mining techniques can extract up to 95% of the coal in a particular deposit.

10.1.2 Sources of Air Pollutants Due to Coal Mining

10.1.2.1 Coal-Mining Activities

Coal is considered as the most polluting source of energy, which creates environmental problems at various stages of its procurement from mining, transportation, stock piling, coal preparation, and utilization (Pandey et al., 2014a). From mining to coal cleaning, from transportation to electricity generation and finally to its disposal as end product, it releases numerous

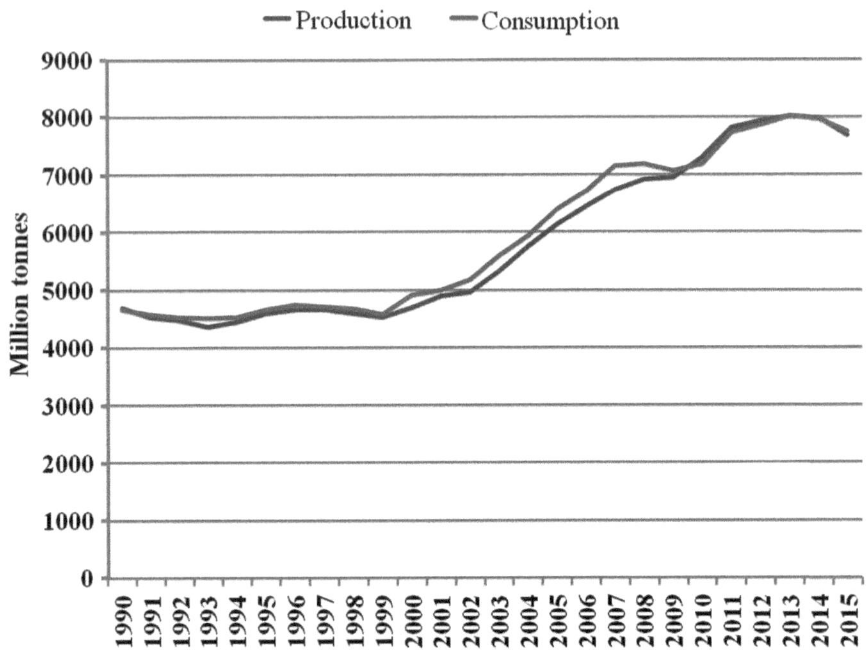

Figure 10.3 Worldwide coal production and consumption from 1990 to 2015. *(Reproduced from International Energy Agency (IEA), 2016. International Energy Outlook 2016. With Projections to 2040. International Energy Agency, Paris, France.)*

toxic pollutants into the environment. Mining operations have the potentiality to demolish flora and fauna and contaminate soil, air, and water in the surrounding areas (Fig 10.4). Deterioration of air quality due to coal mining, land degradation, water pollution and siltation, loss of vegetation and driving out fauna, noise and vibration, reduction in aesthetics, and rehabilitation of people are some of the major issues associated with coal-mining activities (Maiti, 2013).

Coal mining starts with land cleaning for the removal of vegetation and topsoil by using dozers and scrapers, which may create damages to soil quality and plant community as well as releases a large amount of dust, which is a serious concern of coal mining areas (Ghose, 2007). After removal of vegetation and top soil, drilling and blasting actions are performed to reach to the coal seam. Blast holes generally drilled from the surface, are vertical, and vary in diameter from 25 to 100 cm. In some mines, horizontal holes are drilled into the overburden with the drill sitting on the coal surface. The holes are charged with explosives that are based on a mixture of ammonium nitrate and fuel oil in dry mix, slurry, or emulsion forms. It is

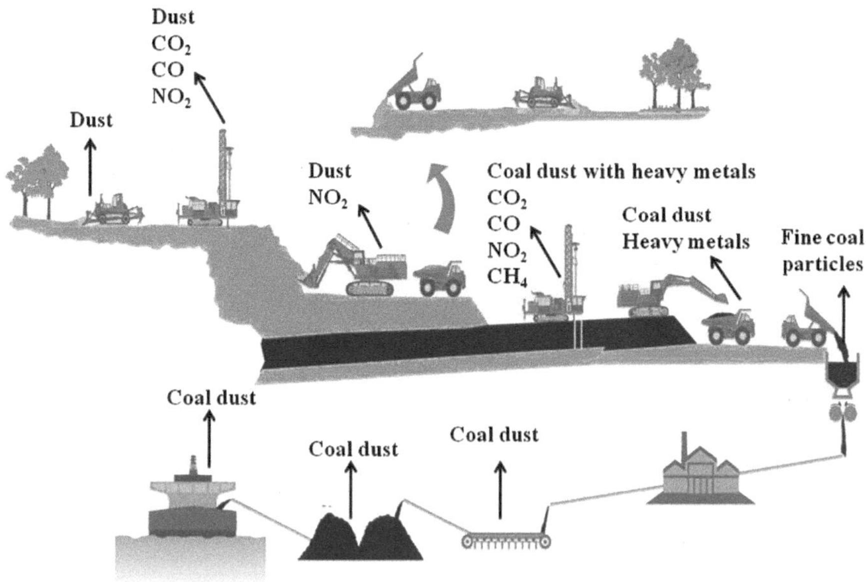

Figure 10.4 Air pollutants released from different processes involved in surface coal mining.

common to have a bulk-explosive truck drive into the area where holes have been drilled to fill holes with custom-designed explosive mixtures, to break the rock into small pieces. The main objective of this operation is to extract the maximum quantity of coal at minimum cost (Kuzu and Ergin, 2005). The drilling process includes crushing and grinding of rocks, which result in formation of a cloud of dust around the drilling operation. The dust concentrations varied between 65 and 107 mg/m (Millgate and Hagan, 2015). Drilling is also the next contributing source of fugitive dust after haul roads (Pandey et al., 2014a). According to Ghose and Majee (2000), 9.4 ton dust per day is generated because of mining operations in an opencast mine of Jharia coalfield, out of which 0.12 ton of dust is released because of drilling and blasting operations. Under ideal conditions, the main gaseous products released from the explosion are carbon dioxide (CO_2), water (H_2O), and nitrogen (N_2) as shown in the following reaction:

$$3NH_4NO_3 + CH_2 \rightarrow 3N_2 + CO_2 + 7H_2O$$

However, even small changes in the stoichiometry (either in the bulk material or caused by localized conditions such as moisture in the blast hole, mineral matter, or other factors) can lead to the formation of substantial amounts of the toxic gases such as carbon monoxide (CO) and nitric oxide (NO):

$$2NH_4NO_3 + CH_2 \rightarrow 2N_2 + CO + 5H_2O$$

$$5NH_4NO_3 + CH_2 \rightarrow 4N_2 + 2NO + CO_2 + 7H_2O$$

In addition, some of the NO formed may oxidize in the presence of oxygen (O_2) to produce NO_2.

$$2NO + O_2 \rightarrow 2NO_2$$

Often in practice, large quantities of NO_2 are released from blasts, which are observed as intense orange plumes (Attalla et al., 2008).

After drilling and blasting, the topsoil is carried through the haul road for about 0.5 km for dumping, which requires excavator, loader, scraper, and dumper, thereby resulting in massive discharge of fine particulates from overburden materials. Vehicular traffic on the haul road of mechanized opencast mines is also identified as the most prolific source of fugitive dust emitted from surface coal mining (Pandey et al., 2014a). The haul road dust is formed of a spectrum of sizes from submicroscopic particles to large size particles, whereas dumper moving at a high speed crushes the large pieces to finer sized fugitive dust. Studies conducted around surface coal mining suggest that about 50% of dust is released during the journey time of a dumper on an unpaved haul road, whereas loading and unloading of dumper contributes 25% (Mandal et al., 2012). The overburden of topsoil, subsoil, and waste rock excavated during opencut mining operations are ultimately used to construct the postmining landforms and to provide a substrate for revegetation. But as the topsoil remains less compact because of drilling and blasting, soil stockpiles can become a source of fugitive dust during windy dry season (Stovern et al., 2014).

When all the overburden is removed, the underlying coal seam will be exposed. This block of coal may be drilled and blasted (if hard). Drilling of the coal seam releases coal dust, which is also a media of transportation and distribution of pollutants in the surface environment (Stovern et al., 2014). Studies on chemical composition of coal dust have revealed that it contains metals like Fe, Cu, Zn, Mn, Pb, Cd, Cr, Ni, Co, V, Ti, Br, Zr, etc., and organic pollutants (Aneja et al., 2012; Rout et al., 2013). Finkelman and Gross (1999) also reported occurrence of 25 environmentally sensitive elements in the coal.

Besides release of metals, mining operations is also responsible for the release of methane (CH_4) gas. CH_4 is formed in coal during the process of coalification, and the quality and quantity of the gas created and retained is a function of the original organic matter composition and the conditions of burial. Generally, more CH_4 is formed during coalification than can be stored within the coal bed itself, so excess CH_4 migrates into and can be stored in the surrounding strata. CH_4 is retained by the coal bed and surrounding strata as long as it remains under pressure and assuming that no geologic processes breach the reservoir first, mining activity releases this pressure and the CH_4 escapes out (Kirchgessner et al., 2000). Gas drained from fractured formations above mined seams, on the other hand, may contain 30%–95% CH_4 depending on the locations of the boreholes and other operation and completion parameters (Karacan, 2009). Although ventilation air methane from shafts of active mines contributes ~64% of worldwide CH_4 emissions from underground coal mines, CH_4 concentrations in the ventilation air are different for each mine (Su et al., 2008).

After excavation, coal gets loaded on dumpers and carried to the coal preparation (cleaning) plant unit. Coal washing or coal preparation is a generic term that is used to designate the various operations performed on run-of-mine coal to prepare it for specific end uses, without altering its physical identity. During this process, vehicular traffic on the haul road, loading and unloading of mechanized opencast mines have been identified as the most prolific sources of fugitive dust emitted from surface coal mining. Spillage from these vehicles adds to the dust load of the haul roads. Haul roads often show 4–6 in. thick or more of accumulated dust on the surface. Yadav and Jain (2017) estimated that the road dust contains 15%–20% or more of the respirable dust particles, which could easily get airborne during adverse conditions.

There are four basic steps used in cleaning of coal as it contains a mixture of minerals. As Indian coals are believed to be of drift origin (Saini and Srivastava, 2017), they have very high ash content. The demand for coal with ash content within well-defined limits has led to plans for upgrading of coal quality. This has been carried out by the large sizes of coal as mined, being crushed into smaller sizes, and subjected to washing procedures in coal washeries, where dirt is eliminated by gravity separation processes. During the various operations of the washery such as tipping, crushing, screening, loading, unloading, and vehicular movement on dusty roads, a substantial amount of dust is generated. Burning of coal and vehicular movements produce a lot of gaseous pollutants, which not only cause pollution problems in the washery premises but also in surrounding areas.

10.1.2.2 Mine Fires

Coal mine fire is devouring coal seams in major coal-producing countries including China, the United States, India, and Indonesia. Many seams have been burning for decades and some for centuries. Coal fires are a natural phenomenon; however, coal mining by humans has assisted the propagation of these fires and thereby enhancing the environmental pollution (Pone et al., 2007).

The ignition of coal mine fires is of global concern as it may be attributed to significant environmental problems (Stracher and Taylor, 2004; Avila et al., 2014). Environmentally catastrophic effects from coal fires include large amounts of pollutants such as sulfur and nitrogen oxides ((NO_x) acidic gases), CO_2, particulate matter (PM), organic compounds, fly ash, and potentially toxic trace elements such as arsenic, mercury, and selenium (Finkelman, 2004). Coal fires have killed people, affected entire communities to abandon their homes and businesses, damaged floral and faunal habitats, and are accountable for dangerous land subsidence (Pone et al., 2007). CO_2 concentrations exceed those of other gases produced by coal fires; however, CO_2 concentrations are not entirely controlled by the coal itself. CO, hydrogen and the hydrocarbons, ethylene (C_2H_4), propylene (C_3H_6), and acetylene (C_2H_2) are monitored as coal-fire detector gases because these are released sequentially as temperature increases during heating. Consecutive temperatures (°C) at which release begins in medium volatile bituminous coal are 110, 170, 240, and 300°C for CO, H_2, C_2H_4, and C_3H_6, respectively. Combustion occurs between 110 and 170°C, and flames appear at about 200°C (Kong et al., 2017).

Pone et al. (2007) found that gases emitted from coal-fire vents in the Witbank and Sasolburg coalfields consist of a complex mixture of hydrocarbons, halocarbons, greenhouse gases (GHGs), and toxic concentrations of CO, benzene, xylene, and toluene. Based on chemical analyses, they characterized these emissions under four principal groups, such as (1) aromatic compounds mostly volatile organic compounds (VOCs) (Benzene, toluene, ethylbenzene, xylenes, ethyltoluene, and trimethylbenzene), (2) aliphatic hydrocarbons (Ethane, propane, butane, pentane, propene, and ethyne), (3) halogenated hydrocarbons (dichloromethane, chloromethane, bromomethane, iodomethane, and trichloromethane), and (4) greenhouse and other gases (CH_4, CO_2, and CO).

10.1.2.3 Coal Burning

Coal is difficult to burn in simple combustion devices, principally because of difficulty in proper premixing of coal and air during burning, which could be done easily with liquid and gaseous fuels (Smith et al., 2000). Consequently, a substantial fraction of coal carbon is converted to products of incomplete combustion (PICs), compounds other than the ultimate product of CO_2 that results from complete combustion. PICs released from solid fuel combustion are a complex mixture of particulate and gaseous species. Some of the PICs are commonly regulated air pollutants such as SO_2, CO, NO_2, PM, and toxic metals such as arsenic, mercury, and fluorine (Zhang and Smith, 2007).

10.2 TYPES OF AIR POLLUTANTS DUE TO COAL MINING

The air pollutants released by various activities involved in coal mining can be divided into two broad categories:

10.2.1 Nongaseous Pollutants

Nongaseous pollutant is also termed as "particulate pollutant" and is a complex mixture of extremely small particles and liquid droplets and includes all the solid residues of different elements, liquid droplets, aerosols, metallic contaminants, soil dust, fumes, black carbon, oil, and grease. Particulate pollutants are made up of a number of components, including acids (such as nitrates and sulfates), organic chemicals, metals, and soil or dust particles. Atmospheric particles arise either as primary particles—by direct emission from a source or as secondary particles—through in situ formation from the gas phase (nucleation). Particulate air pollution is an assortment of solid, liquid, or solid and liquid particles suspended in the air (Pandey et al., 2014b). These suspended particles vary in size, composition, and origin. It is commodious to assort particles by their aerodynamic attributes because they govern the transport and removal of particles from the air which govern their deposition inside the respiratory system and are linked with the chemical composition and sources of particles (WHO, 2000).

Aerosols with aerodynamic diameters $<2.5\ \mu m$ ($PM_{2.5}$) termed as fine particle are responsible for health hazards. Ultrafine particles with aerodynamic diameters $<1.0\ \mu m$ (PM_1), interfere with visibility and radiation impingement and may get across deeper into the respiratory system (Esposito et al., 2012). PM has broader impact on climate change inducing direct (absorbing, reflecting, and scattering), indirect (clouds formation, clouds albedo, and lifetime), and semidirect (heating and cooling) effects on the global radiative budget (Solomon, 2007). Impacts of particulate deposition on ecosystems are also reported (Chate and Devara, 2009). Combustion of coal and oil can produce coarse particles from the release of noncombustible materials, i.e., fly ash, fine particles from the condensation of materials vaporize during combustion and secondary particles through the atmospheric reactions of sulfur oxide (SO_x) and NO_x initially released as gases.

10.2.2 Fly Ash

Fly ash is a heterogeneous and complex residue resulting from coal combustion, which comprises of fine-grained particles that are preponderantly globular in shape, both solid and hollow and generally glassy (amorphous) in nature. The carbonaceous material in fly ash is composed of angular particles (Ahmaruzaman, 2010). The particle size distribution of most bituminous coal fly ashes is similar to that of silt (<0.075 mm), in general. The specific gravity of fly ash usually ranges from 2.1 to 3.0 (Rai et al., 2013). The color of fly ash may vary from light bronze to gray to black, depending on the amount of unburned carbon in the ash (Ayanda et al., 2012). The color reflects the carbon content in fly ash, the lighter the color of fly ash, the lower the content of carbon. Lignite or subbituminous fly ashes are commonly light tan to buff in color, thus indicating relatively low contents of carbon with some lime or calcium. Bituminous fly ashes usually pose some shades of gray, indicating a higher quality of ash (Ahmaruzaman, 2010).

10.2.3 Gaseous Pollutants

Sulfur oxides (SO_x) are one of a group of highly reactive gases. During standard atmospheric conditions, it is a toxic gas with a pungent, irritating, and rotten smell. SO_2 is a prevalent anthropogenic sulfur-containing air pollutant having a strong inclination to react with O_2 in air (Yavorskyi et al., 2013). The lifetime of SO_2 depends on reaction with OH radical and at typical atmospheric levels of OH, and its lifetime is about 1 week.

Reductions in SO_2 emissions in the United States were particularly large over recent period (USEPA, 2012). Emissions of SO_2 in the European Union decreased since 2005 by about 3700 Gg (Kuenen et al., 2011). Emissions in China were estimated to have peaked around 2006, and then declined toward 2010 as the operation of emission controls increased in the electric power generation sector. In contrast, emissions of SO_2 from India have increased by more than 60% between 2005 and 2012 (Lu et al., 2013; NASA, 2013). Based on the most recent emission estimates (Xing et al.,

2013), India has surpassed the United States, the world's second most prominent SO_2 emitter after China, since 2010.

Nitrogen oxides are produced as a result of chemical reaction between atmospheric N_2 and O_2 in the presence of heat to form NO.

$$N_2(g) + O_2(g) \rightarrow 2\ NO(g) \quad \Delta H = 180.6\ kJ$$

Although N_2 and O_2 are abundant in the air, the formation of NO requires very high temperatures, so it only forms at the time of blasting during coal mining, inside internal combustion engines and fossil fuel power plants, or when lightning strikes. Fume is a combination of postblast gases, which not only contains nitrogen dioxide but also includes nitrous oxide (N_2O). NO into the atmosphere reacts with O_2 to form NO_2 (Seinfeld and Pandis, 2012). The rate of reaction is decided by the temperature of combustion. In atmospheric chemistry, the term "NO_x" means the total concentrations of NO and NO_2. NO_x gases react to form smog and acid rain as well as are central to the formation of tropospheric ozone.

The total NO_x emissions in East Asia were 29.7 Mt in 2010, and the growth rate was 25% during 2005–2010. This trend was because of the increment in emissions from China, which added 82%–88% of total NO_x in Eastern Asia. Monitoring of air quality in Indian cities conducted by Central Pollution Control Board (CPCB) has shown a trend of dramatic increase in NO_x concentrations since 1990 especially in the metropolitan cities. Annual average NO_x concentrations varied from 4.3 to 42.9 ppb in various parts of the country (CPCB, 2009).

A group of chemicals (CH_4 is excluded) that comprises carbon in their molecular structure are "organic". They easily vaporize at room temperature and mostly have no color or smell. Nonmethane volatile organic carbons (NMVOCs) generally include the chemical groups such as alcohols, aldehydes, alkanes, aromatics, ketones, and halogenated derivatives of these substances (Guo et al., 2006). NMVOCs are released from the burning of fossil fuels, in particular, from the burning of petrol in road transport (Streets et al., 2006). Trees and other plants also produce NMVOCs naturally. VOCs (CH_4 too) play a major role in the formation of ozone and photochemical smog.

CH_4 is the most abundant hydrocarbon in the atmosphere. CH_4 stands after CO_2 in its contribution to global warming. CH_4 is stiffer than CO_2 because of the radiative forcing produced per molecule is greater in the former (Yusuf et al., 2012). The energy sector is second to agriculture in contribution to global emissions of CH_4 (Miller et al., 2013). CH_4

undergoes a complex series of reactions, which together constitute the CH_4 cycle in the atmosphere. The principal oxidation reaction of CH_4 is with hydroxyl radicals in the atmosphere.

10.3 COAL MINING CONTRIBUTION TO GLOBAL WARMING

With very significant CO_2, CH_4, and N_2O emissions from different mining activities as fugitive emissions, coal mining is a major contributor to the buildup of GHGs in the atmosphere. These gases allow solar energy to reach the earth's surface but delay that energy's escape into space, and thus effectively trapping heat in the lower atmosphere. The Intergovernmental Panel on Climate Change (IPCC) reported that annual global GHG emissions between 1970 and 2010 increased by 78% (52.0 $GtCO_2e$/year, IPCC, 2014). The International Energy Agency's (IEA) Reference Scenario estimated that worldwide CO_2 emissions will increase by 60% between 2005 and 2050 or 1.6% each year reaching to 1700 $GtCO_2$ (IEA, 2016). In the same period, CO_2 emissions from coal-generated power are projected to increase by 75% (13,544 Mt, IEA, 2016). In 2015, coal was responsible for 93% of the GHG emissions from power generation in the United States (USEPA, 2017). In addition to direct stack emissions, there are CH_4 emissions from coal mines, in the order of 3% of the stack emissions (USEPA, 2017). There are also additional GHG emissions from the other uses of coal ~100 Mt CO_2 (IPCC, 2014).

PM (black carbon or soot) is also a heat-trapping agent, absorbing solar radiation, and even at great distances, decreasing reflectivity (albedo) by settling in snow and ice (Hansen et al., 2007; Ramanathan and Carmichael, 2008). The contribution of particulates (from coal, diesel, and biomass burning) to climate change has been underestimated, until recently. Though short-lived, the global warming potential per volume is 500 times than that of CO_2 (Hansen et al., 2007).

The Energy Information Administration (2010) measured that 1.97 billion tons of CO_2 and 9.3 million tons CO_2e of N_2O were discharged from coal-fired power plants. Cooke et al. (1999) used emissions factor to estimate black carbon emissions based on lignite use, and Hansen et al. (2007) used global temperature potential to convert these emissions to CO_2e. The total costs of climate change damage from coal-derived power, including black carbon, CO_2, and N_2O emissions from combustion, land disturbance in coal mining, and CH_4 leakage from mines are estimated as 63.9 billion dollars with low and high estimates of 21.3–215.9 billion dollars (EIA, 2010). A broad examination of the costs of climate change projects global economic losses between 5% and 20% of global GDP (Stern, 2003). The higher figures of economic losses are based on the potential collapse of ecosystems, such as coral reefs and widespread forest and crop losses. With coal contributing at least one-third of the heat-trapping chemicals, these projections offer a sobering perspective that coal mining companies need to know their own carbon emissions and compare with the other industries. Mining companies need to know where they exist in the emissions curve. Because of that, many countries are now thinking of introducing "Carbon Pollution Reduction Scheme," to control GHGs emissions. The threat of global warming and its increasingly prominent position in the public consciousness demand that every industry need to takes stock of its contribution to the rising atmospheric concentration of GHGs. These concerns have provided tremendous popularity to quantification of the contribution of various activities to global warming usually represented in terms of "carbon footprint."

10.4 CARBON FOOTPRINT FOR COAL MINING

Carbon footprint is a "measure of the impact of human activities left over on the environment" in terms of the amount of GHGs measured in terms of carbon dioxide equivalent (CO_2e). Carbon footprint, being a quantitative expression of GHG emissions from an activity helps in emission management and evaluation of mitigation measures (Pandey et al., 2011). Having quantified the emissions, the important sources of emissions can be identified, and areas of emission reductions and increasing efficiencies can be prioritized. Reporting of carbon footprint to the third party or disclosure to the public is needed in response to legislative requirements, or carbon trading, or as a part of corporate social responsibility, or for improving the brand's image (Carbon Trust, 2007).

10.4.1 Life Cycle Inventory of Coal Mining

For the calculation of carbon footprint, the amount of GHGs emitted/removed or embodied in life cycle of the product has to be determined and summated. A life cycle assessment (LCA), also known as a "cradle-to-grave" (whole life cycle) approach is an ISO 14000 recommended tool used to assess the environmental and social impact of a product throughout its useful life; from its start as a raw material to disposal (ISO, 2006). Mining is the earliest phase in the life cycle of a material, and as such it is markedly important and often sets a material on the path to be used sensibly throughout its life cycle. In its efforts toward sustainability, the mining industry has used several environmental and economic indicators to assess its performance. In recent years, LCA has proved to be one of the excellent tools that can be used to evaluate environmental performance and support decision-making in the mining industry (Durucan et al., 2006). The key elements of LCA are as follows:

- Identification and quantification of the environmental loads involved; e.g., the energy and raw materials consumed and the emissions and wastes generated (life cycle inventory (LCI) analysis).
- Evaluation of the potential environmental impacts of these loads (life cycle impact assessment).
- Assessment of the options available for reducing these environmental impacts (interpretation of LCI analysis and impact assessment).

LCA for GHG estimation are as follows:

- Bottom-up or process analysis
- Top-down or input—output analysis

In bottom-up approach, the emission sources are broken down into different categories for convenient quantification. This method is more accurate for small entities, but it becomes too complex for large firms, which cover more than second-order emissions thus misleading the actual footprint (Wiedmann and Minx, 2007). It is useful in identification of area of process improvement (Green Design Initiative, 2008). Whereas top-down approach makes use of economic input—output model extended to accept and perform operations on environmental variables for calculations of carbon footprint (Green Design Initiative, 2008).

The mining industry first began experimenting with the LCA approach in the mid-1990s; originally utilizing LCIs in metal production processes to help customers with design and choosing products (Durucan et al., 2006). Eventually LCA began to be used in project and process selection with a strong focus on the processes occurring during concentration and refining.

The coal life cycle conceptually consists three major steps: coal mining and processing, transportation, and use/combustion. The system boundaries for LCA should be drawn as broadly as possible (Fig. 10.5). All of the major processes necessary, such as coal mining, equipment manufacturing, transportation, chemicals production, for the mining and power plant operations, are included. The material and energy flows of processes involved in the extraction of raw materials and the production of intermediate feedstocks (e.g., limestone used in the gas cleanup process at the power plant) as well as the disposal of wastes are included (Fig. 10.5).

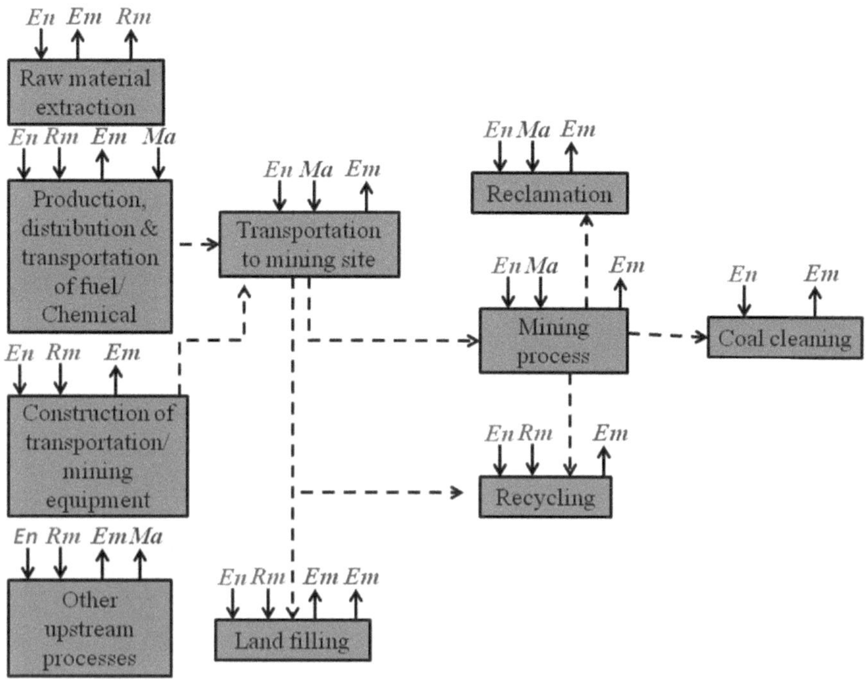

Figure 10.5 Coal mining boundaries for life cycle assessment. *Em*, Emission; *En*, Energy; *Ma*, Materials; *Rm*, Raw material.

The two types of energy basically required: (1) the energy used directly in each process block, and (2) the energy contained in the materials used in each process block. The second type of energy is the sum of combustion and process energies, which is sometimes referred to as the embodied energy of a material. Because of the size and complexities of the complex chain process, it is important to emphasize on the inputs (materials and energy) and outputs (by-products and/or emissions) of the coal mining activities (Tables 10.1 and 10.2). USEPA (2017) reported that CH_4 emission from coal mines in 2015 was 67.3 million tons of CO_2 equivalents, of which 60.9 million tons are from underground mining and 6.4 million tons are from surface mining. CO_2 is also emitted through the combustion of the fuels that provide the energy for mining operations. 43.6 tons of CO_2 equivalents are emitted from rail transportation, 5.89 tons of CO_2 equivalents from H_2O transportation, and 69 tons of CO_2 equivalents from truck transportation, for a million ton per miles of coal transported (Green Design Initiative, 2001).

Table 10.1 Inputs and outputs (by-products and/or emissions) from varying processes involve in coal-mining activities

Processes involved in coal production	Inputs	Outputs	Sources of predominant greenhouse gases
Opening work	Electricity, air, compressed air, explosives, water, cryogenic or petrifying substances (if a shaft is driven), elements of shaft, or mine working lining.	Driven shaft or working, gangue, water of natural inflow or technological processes, land use, outlet air, dust, noise hazardous and nonhazardous wastes, wastewater, and water vapor.	Direct and indirect energy use
Preparation of coal seam for exploitation	Electricity, air, compressed air, explosives, water, and elements of mine working lining.	Driven working, gangue, run-of-mine coal, water of natural inflow or technological processes, outlet air, noise, hazardous and nonhazardous wastes, particulate matter, SO_2, NO_x, methane, mercury (II), water vapor, oil, and grease.	Direct and indirect energy use and fugitive emissions
Coal seam exploitation	Electricity, air, compressed air, water, emulsion, explosives, resins, foams, insulation materials, and elements of mine working lining and timber	Run-of-mine coal, gangue, water of natural inflow or technological processes, outlet air, coal dust, carbon monoxide, NO_x, CO_2, methane, and noise.	Direct and indirect energy use and fugitive emissions
Mechanical coal preparation	Run-of-mine coal, water, electricity, compressed air, various chemicals, flocculants, and other specific exploitation materials.	Commercial coal, coal preparation waste (stone, sludge, slime, sewage, flotation tailings), used chemicals, coal dust, carbon monoxide, NO_x, CO_2, methane, and noise.	Fugitive emissions

Ventilation	Electricity in air	Outlet air and noise.	Fugitive emissions
Prevention of natural hazards	Electricity, compressed air, water, ash–water mixtures, ignition inhibitors, inert gases, explosives, stone dust, and various chemicals.	Outlet air, methane, water natural inflow and technological processes, and noise.	Direct and indirect energy use
Transportation	Electricity and diesel fuel	Noise and coal dust	Direct energy use
Power management	Fuels—methane, gas or coal, and water	Heat energy, ashes, and other wastes.	Direct and indirect energy use, and emissions from AT & C losses
Compressed air management	Electric energy/electricity	Compressed air and noise	Fugitive emissions
Methane drainage	Electric energy/electricity	Methane and noise	Fugitive emissions
Backfilling	Water, electric energy/electricity, compressed air, various backfill materials, and materials for building backfilling installation and dams	Polluted backfill water, slime, noise, and dust	Direct energy use and solid waste disposal
Workshop and workhouse management	Electricity, welding gases, oils, greases and chemicals, and various exploitation materials.	Sewage, worn-out elements of machines and devices, used exploitation materials, and noise.	Direct and indirect energy use

AT & C, Aggregate Technical and Commercial.

Table 10.2 Emissions during different activities from Caval Ridge Coal Mine Project in Moranbah, Australia

Scope	Source	Emissions range (tCO_2e/year)	Emission average (tCO_2e/year)
1	Fugitive emission	120,129–196,359	161,647
1	Diesel combustion	72,450–131,595	105,818
1	Explosive	3348–5011	4430
2	Purchased electricity	56,315–105,572	99,305
3	Coal railing and terminal handling	35,370–48,634	48,192
3	Coal shipping	70,956–97,565	96,678

10.4.2 Greenhouse Gas Accounting

Selection of the set of GHGs covered in calculation depends on the guidelines followed, the need of carbon footprint calculation and on the type of activity for which carbon footprinting is being done. For example, in a thermal power plant, where CO_2 is a predominant emission, and other gases are almost negligibly emitted, only CO_2 emission measurement will be feasible, whereas for a cattle farm, CH_4, CO_2, and N_2O emissions may be significant. Generally for coal mining processes following GHGs emissions measurement are required as follows:

- **CO_2 emissions from fuel combustion:** This refers to emissions of CO_2 produced from combustion of fossil fuels with O_2 in various stationary or mobile combustion plants (e.g., boiler, burner, turbine, heater, incinerator, calcinatory, kiln, and internal combustion engine).
- **CO_2 emissions from torch combustion:** This refers to emissions of CO_2 produced from torch combustion of coal bed gas (coal mine gas).
- **Fugitive emissions of CH_4 and CO_2:** This refers to emissions of CH_4 and CO_2 produced during coal production, including emissions from underground mining, opencast working, and activities after coal mining.
- **CO_2 emissions implied in net purchased electric and heating powers:** This part of CO_2 emissions actually occurs in the power enterprise or heat enterprise that produces the power or heat. However, the emissions are caused by the consumption activity of the reporting entity. Therefore, according to the regulations, the emissions shall also be included into the total emissions of the reporting entity.

10.4.3 Setting Boundary for Coal Mining

Boundary refers to an imaginary line drawn around the activities that will be used for calculating carbon footprint. It depends on the objective of footprinting and characteristics of the entity for which footprinting is to be done. Boundary must be selected so as to represent the organization based on legal, financial, or business control. In case of joint ventures, the organization may take responsibility of the fraction of the emissions for which it is responsible, technically called as "equity share" or may consider all the emission sources, which are under its direct control, depending on the need of carbon footprinting. Once the organizational boundary is established, next step is to select operational boundary. Operational boundary refers to the selection of the direct and indirect emissions.

The criterion for setting the organizational boundary for the inventory is the operational control. The Standard clarifies that a company has operational control over an operation if the former or one of its subsidiaries has the full authority to introduce and implement its operating policies at the operation. According to this interpretation of operational control and based on discussions with the reporting entity's management personnel, contractual activities at the mine are entirely within the operational control of the reporting entity, and hence 100% of the emissions from contractor's activities should be included in report. As an outcome of this approach selection, the operation and emissions-generating activities of the stakeholders/operational entities should be considered as part of the GHG inventory boundary.

Subsequent to organizational boundaries definition in terms of the operations that it owns or controls, the corporate standard requires specifying of operational boundaries, which entail identifying emissions associated with its operations, categorizing them as direct and indirect emissions, and choosing the scope of accounting and reporting for indirect emissions.

- **Direct GHG emissions** are emissions from sources that are owned or controlled by the company.
- **Indirect GHG emissions** are emissions that are a consequence of the activities of the company/industry but occur at sources owned or controlled by another company/industry.

Furthermore, to improve transparency and provide utility for different types of organizations three "Scopes" (Scope 1, Scope 2, and Scope 3) are defined for GHG accounting & reporting.

- **Scope 1:** These are direct GHG emissions from sources that are owned or controlled by the company. For example, emissions from combustion in owned or controlled facilities and vehicles.
- **Scope 2:** These are indirect GHG emissions occurring as a consequence of GHG emissions from the generation of purchased electricity by the company/industry.

- **Scope 3:** These comprise other indirect emissions except those accounted for as Scope 2 emissions. These are a consequence of the activities of the company but occur from sources not owned or controlled by the company and are an optional reporting category. For an instance, embodied carbon emissions from manufacturing of materials used by a company, third party deliveries, business travel activities, and use of sold products and services.

On the basis of GHG emissions, carbon footprint can be primary (calculated from either direct emissions (Tier I) or emissions embodied in energy purchase (Tier II)) and full carbon footprint (calculated from both direct and indirect emissions from the entity) (Pandey et al., 2011).

Fig. 10.6 illustrates the three tiers in carbon footprint estimation. The Tiers 2 and 3 both include indirect emissions, but Tier 2 refers to the emissions embodied in energy production or (and) purchase, transmission, and distribution caused by the entity under consideration, but end user emissions are out of scope of Tier 2. Tier 3 tends to cover all the embodied emissions within the specified boundary. But Tier 3 has vaguely been defined, and the most carbon footprint studies limit up to Tier 2 as it

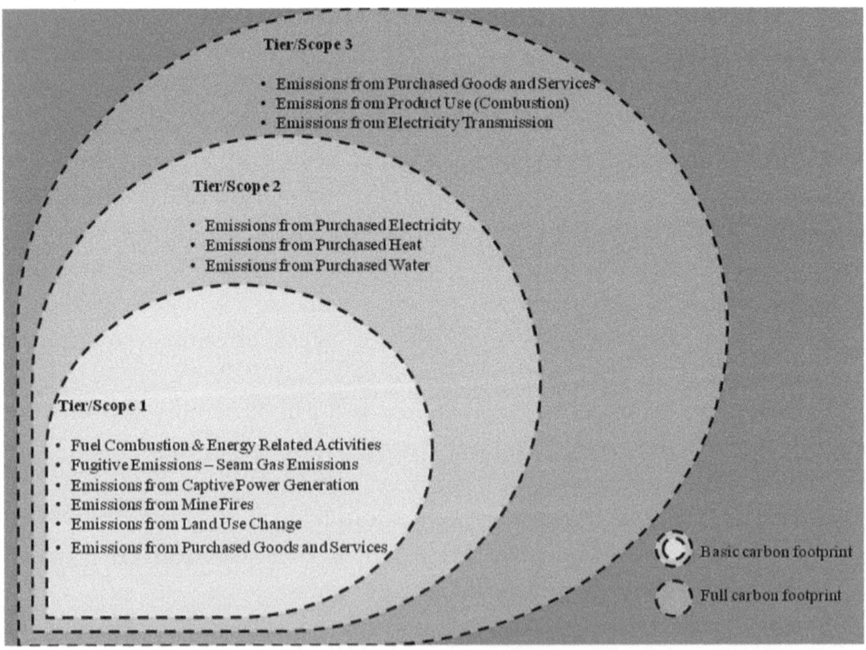

Figure 10.6 Boundaries for calculation of carbon footprint from coal mining industries.

becomes too complex to estimate carbon footprint beyond Tier 2 with accuracy (CDP, 2008; Dada et al., 2008).

For coal mining industries, fuel combustion and energy-related activities, fugitive emissions (seam gas emissions) and emissions from captive power generation, mine fires, land-use change, and from purchased goods and services are included in Scope 1. These activities include the emissions from fossil fuels, biomass fuels, flaring, incineration, mining operations, postmining operations, explosive use, low-temperature oxidation, diesel or residual fuel oil generator, coal/natural, gas/residual fuel oil/diesel-based captive power plant, coal combustion, deforestation, overburden removal, upstream mining operations, and transportation. Scope 2 includes emissions from purchased electricity (grid electricity), heat, and H_2O (municipal H_2O supply), whereas emissions from purchased goods/services (hired vehicles), product use (coal combustion in thermal power plants and other fuel—burning activities), and electricity transmission are included under Scope 3.

10.4.4 Collection of Greenhouse Gas Data and Footprint Calculation

The final step of carbon footprint measurement is collection of GHG data from defined boundaries and calculation. To estimate emitted GHGs data from coal mining can either be collected directly through on-site real time measurement or though emission factors and models (Pandey et al., 2011).

10.4.4.1 Direct On-Site Real Time Measurement

There is a wide range of potential measurement strategies that can be used to measure the concentrations and emission levels. GHGs of interest are measured through several techniques. One of the most common methods for characterizing hydrocarbon seeps is to measure soil gas composition and concentration. This analysis involves inserting a metal probe into the soil and extracting a gas sample for analysis, which may be done either by using in situ analysis equipment or by collecting discrete samples that are transported to a laboratory for analysis. Remote sensing methodologies would overcome a number of the limitations associated with ground-based methodologies such as soil gas and ground-level CH_4 concentration surveys (Van Dijk et al., 2011). Measurement of GHGs emissions is often carried out by the flux chamber method, especially in settings where more complicated experimental approaches appear difficult to install such as in less developed areas

(Hegde et al., 2003), or where power supplies are unavailable (Nakano et al., 2004). Several other techniques, viz. photo acoustic infrared sensors, IR spectroscopy (for CO_2 measurement only), and gas chromatography can be used for measuring all the six Kyoto gases (Berg et al., 2006). Eddy covariance or flux towers are used to continuously monitor CO_2 and CH_4 flux from the entire landscape affected by GHG emissions (Xu et al., 2014), whereas cavity ring-down spectrometers are commonly used for aerial measurements (Lal et al., 2015). Secondary data on GHG emission for carbon footprint estimation can also be obtained from various sources such as Global Trade Analysis Project (Dimaranan, 2006), International Energy Agency, United Nation Development Programme (Brown et al., 2009), "Vulcan" (joint project of NASA and US Department of Energy) (Gurney et al., 2009; Kelly et al., 2009), national GHG inventories, and other government offices. Direct measurement of GHGs at the emission source can give the most accurate and precise assessment of GHG emissions. This is typically not feasible at a coal mine because of the cost ineffectiveness, the disruption to production involved, and the typically large number of trucks and plant equipments involved (USEPA, 2015).

10.4.4.2 Emission Factors and Models

Emission factors take out the need for site-specific assessment of emissions. They are a factor expressed as the amount of GHG emissions per unit of activity during coal mining activities, which can be used to determine inventories for a site. This is much more feasible than testing each source individually, and it is one of the few ways that inventories for proposed sites can be calculated.

10.5 IMPORTANT GREENHOUSE GAS INVENTORY CALCULATIONS FOR COAL MINING

10.5.1 Agriculture Forestry and Other Land Use

Coal mining (especially opencast mining) requires a large scale excavation to extract the coal from the subsurface. This entails commensurately large scale deforestation activity and the associated loss of biomass from the forest area. Forest wood gets transformed to harvested wood products as well as raw material for the manufacturing industry, and fuelwood. GHG emissions and removals per hectare vary according to site factors, forest or plantation types, stages of stand development, and management practices. The forest land is stratified into various subcategories to reduce the variations in growth rate and other forest parameters and to reduce uncertainty. Stratification of forest types into homogeneous subcategories, and if possible at regional or subregional level within a country, reduces the uncertainty of estimates of GHG emissions and removals (IPCC, 2014). In addition to the direct emissions from biomass loss due to deforestation activities from opencast mine excavation, the other significant land-use change impact of the activity is the lost carbon sequestration capacity of the forest area. Because of deforestation, the soil degrades gradually, leading to loss of soil carbon pool and the associated CO_2 emissions. Because of drainage of soil through overburden removal, to extract the coal and mining activities, there is loss of a finite amount of carbon sequestration capacity and the emissions would be accountable year on year indefinitely throughout the mining life. To complete this module, the formulae and data are used following standard methods provided by IPCC (2014).

10.5.2 Direct and Indirect Energy Emissions From Fossil Fuel Combustion

Coal mining operations involve consumption of significant quantities of diesel, captive and grid electricity, lubricant consumption, and explosive used for blasting. These comprise the direct emission sources (from fossils fuels, lubricants, explosives) and indirect emissions from purchased electricity. The methods used for estimating direct and indirect emissions from fossil fuel, electricity consumption, lubricants, and explosives used are (IPCC, 2006) as follows:

- **Total emissions from fuel use**

$$\text{Emissions}_{\text{GHG, fuel}} = \text{Fuel consumption} \times \text{Emission factor}$$

$\text{Emissions}_{\text{GHG, fuel}}$ = Emission of a given GHG by type of fuel

Fuel consumption = Amount of fuel combusted

Emission factor = Default emission factor of a given GHG by type of fuel

- **Total emissions from electricity use**

$$\text{Emissions}_{\text{GHG, purchased electricity}} = \text{Emission factor}_{\text{GHG, electricity}} \times \text{Electricity consumption (kWh)}$$

$\text{Emission factor}_{\text{GHG, electricity}}$ = Grid GHG emission factor for the region

- **Total emissions from Electrical Aggregate Technical and Commercial (AT&C) losses**

$$\text{Emissions}_{\text{GHG, electricity AT\&C losses}} = (\text{Emission factor}_{\text{GHG, electricity}} \times \text{Electricity consumption})/(\text{T\&D losses correction factor})$$

T&D loss correction factor = By auxiliary power consumption.

10.5.3 Emissions From Industrial Processes and Product Use

- **CO_2 emissions from explosive use**
  GHG emission from the use of explosives in coal mining, where the activity level is the mass of explosive used calculated by:

$$\text{Emissions}_{\text{GHG, explosive}} = \text{Explosive consumption} \times \text{Emission factor}$$

$\text{Emissions}_{\text{GHG, explosive}}$ = Emission of a given GHG by type of explosive

Explosive consumption = Amount of explosive blasted

Emission factor = Default emission factor of a given GHG by type of fuel

- **CO_2 emissions from lubricant use**
  The use of lubricants in engines is primarily for their lubricating properties and associated emissions are most GHG inventory calculations considered as noncombustion emissions for and follow the method reported in the Industrial Process and Product Use Sector (IPCC, 2006). During the process of engine operations, small portions of lubricants get oxidized during use. These emissions are estimated as follows:

$$\text{Emissions}_{\text{GHG, lubricant oxidation}} = \text{Lubricant consumption} \times \text{Factor for oxidized during use} \times \text{Carbon content of lubricant} \times \text{Emission factor} \times 44/12$$

10.5.4 Fugitive Emissions

For the period of long geological processes of coal formation, CH_4 and CO_2 are trapped in the coal seam (seam gas). Coal is exposed and broken, and these gases emitted from coal mining and handling as a fugitive emissions (Ju et al., 2016). The key stages for the fugitive emission of GHGs for both underground and surface coal mines are as follows:

- Mining emissions: These GHGs emissions result from the release of stored gas during the breakage of coal during mining operations.

- Postmining emissions: Emissions during handling, processing, and transportation of coal are termed postmining GHGs emissions. Coal normally continues to release GHGs even after it has been mined, although more slowly than during the mining stage.
- Low temperature oxidation: These emissions take place because once coal is exposed to O_2 in air produces CO_2. However, the rate of emission of CO_2 by this process is low.
- Uncontrolled combustion: These emissions are from mine fires because of coal exploitation activities.
- Abandoned coal mines: After closing and sealing of coal mines, CH_4 may still be emitted as a result of gas migrating through natural or man-made conduits.

Fugitive emissions from mining and postmining activities can be estimated using the following approach (IPCC, 2006):

$$\text{Fugitive emissions}_{\text{GHG, coal mining}} = \text{Coal production} \times \text{Emission factor} \times \text{GWP}$$

Coal production = Amount of coal produced

GWP = Global warming potential CO_2 to CH_4 (1:21)

The approach for calculation of emissions from abandoned mines

$$CH_4 \text{ emissions} = \text{Number of abandoned coal mines remaining unflooded} \times \text{Fraction of gassy coal mines} \times \text{Emission factor} \times \text{Conversion factor}$$

Conversion factor is the density of CH_4 and converts volume of CH_4 to mass of CH_4.

Emission factors for a wide range of industrial processes in GHG protocol, PAS-2050, IPCC (2006, 2014) as well as country wise emission factors may be obtained from UN Framework Convention on Climate Change, US Environmental Protection Agency, UK Department for Environment, Food and Rural Affairs, etc. (WRI/WBCSD, 2004; IPCC, 2007).

10.6 CASE STUDIES

10.6.1 Case Study 1

GHGs have been estimated for the operation of the Caval Ridge Coal Mine Project in Moranbah, Australia, which has a project life of 30 years beginning in 2012. Table 10.2 shows a summary of GHG emissions for the project as CO_2e (Alliance, 2011).

Fugitive coal seam gas emissions are on an average ~43% of the sum of Scope 1 and Scope 2 emissions. Combustion of diesel represents ~29% and purchased electricity ~27%. This is consistent with opencut coal mines in Queensland where emissions of coal seam gas have been estimated with the default emission factors rather than from direct measurements.

10.6.2 Case Study 2

A Corporate GHG inventory and product life cycle carbon footprint project report was prepared by cBalance solutions private limited for Central Coalfield Ltd. (CIMFR, 2014) for the period of 2010–13 for coal mines. According to the report a majority of the emissions are attributed to Scope 1 emissions (81%–92% of the total emissions), whereas Scope 2 emissions are the second highest contributor (5.8%–15% of total emissions). Scope 3 emissions represented the lowest contribution to the total GHG emissions. The predominance of Scope 1 emissions in the GHG inventory is because of the significant GHG emissions contributed by fugitive emissions of CH_4 during mining and postmining activities. Fugitive emissions represented the highest contribution (ranging from 71% to 77% of Scope 1 GHG emissions). Emissions from diesel consumption by contractor for overburden removal, coal extraction, and coal transportation contribute ~23%–30% of total Scope 1 GHG emissions. All Scope 2 emissions are the consequence of electricity use. In terms of total magnitude, these emissions have a relatively low impact on the total GHG emissions as the mining process is energy intensive and relies on diesel powered machines or processes. It is also observed that the contribution of office energy consumption is relatively modest (2%–7%) compared with electricity consumption by equipments by mining and coal handling facilities (93%–98%).

10.7 STRATEGIES IN MITIGATING CARBON EMISSIONS REDUCTION FROM COAL MINING

There are many ways to reduce GHG emissions from coal mining, which include basic mitigation strategies and specific tactics/actions within following categories:

10.7.1 Methane Emission Mitigation and Utilization Techniques

CH_4 constitutes about 30%–90% of GHGs emissions from coal-mining activities and poses 25 times higher global warming potential than CO_2, therefore its mitigation through technological innovations are of paramount importance (Jiang et al., 2016). There are several existing techniques to combat with CH_4 emissions from coal-mining activities (Su et al., 2005).

10.7.1.1 Flaring

In flare, collected CH_4 from boreholes during mining is destroyed by combustion and converted to CO_2, a far less potent GHG (Oh et al., 1992; Lee and Trimm, 1995). The combustion of CH_4 may produce CO or CO_2 depending on the air/ CH_4 ratio during the reactions.

10.7.1.2 Methane Purification

Most countries meet the requirement of fuel consumption from coal mine CH_4 (USEPA, 1998; Kruger and Schultz, 2006). However, utilization of mine CH_4 in its mixture form is not possible because of presence of impurities such as N_2, O_2, CO_2, and water vapor. At present there are mainly four basic processes that are commonly used for gas purification:

10.7.2 Solvent Adsorption

Diethylamine and potassium carbonate are used to remove the surplus CO_2 from the raw natural gas, whereas monoethanolamine, diethanolamine, methyl diethanolamine, and diethylene glycol, either individually or in combination are used in different processes based on the same principles. On contrary, these types of processes are not suitable for N_2 removal from gas because it is not readily soluble in aforementioned solvents. It is practicality not possible where air contamination of the gas is likely to result in N_2 concentration over 5% (Beath, 2003).

10.7.3 Pressure Swing Adsorption

The process is similar to solvent adsorption process, but it utilizes solid adsorbent (Molecular Gate adsorbents developed by Engelhard) to adsorb N_2 in favor of CH_4 until the output attains the desired CH_4 proportion (USEPA, 1998).

10.7.4 Cryogenic Separation

It is a stepwise process that involves compression, flash vaporization, heat exchange, and distillation separator to cool the gas stream until it liquefies to separate a N_2-rich gas stream, thereby leaving CH_4-rich liquid stream behind (Su et al., 2005).

10.7.5 Membrane Separation

Utilization of membrane technology is rapidly growing in CH_4 purification, and it is composed of specialized membranes for the separation of different gases emitted from mining activities. Several companies have made a lot of effort to develop membranes for either N_2 or CO_2 removal from gaseous streams. These specific membranes have the capability to efficiently separate CH_4 from all the major contaminant gases found in drainage gas (Lokhandwala et al., 2002). In general, it is a cost-effective and reliable technique but is still under development.

10.7.6 Power Generation Using Methane

Coal mining is an energy-intensive process, which requires high electricity load to run equipments including mining machines, conveyor belts, desalination plants, coal preparation plants, and ventilation fans. In the United States, about 24 kWh of electricity is required per ton of coal extracted for the mine, and 6 kWh of electricity is required per ton of coal processed in the coal preparation plant (Kruger and Schultz, 2006). Mine CH_4-fired stationary power generation technologies such gas engines, gas turbines, and fuel cells can be used in mining of coal. These not only promote the reutilization of generated CH_4 but also reduce the energy consumption during the processes (Su et al., 2005).

10.7.7 Production of Methanol and Carbon Black

Coal mine CH_4 may also be made available as a chemical feedstock for different chemical processes for the production of synthetic fuels and chemicals. Two potential applications in this field are methanol and carbon black production (USEPA, 1998).

10.7.8 Thermal Flow, Catalytic Flow, and Catalytic Monolith Reactor Technologies

Thermal flow (TFRR) and catalytic flow (CFRR) reversal reactors employ the flow reversal principle to transfer heat of combustion of the CH_4 to the incoming air through a solid heat storage medium. This raises the ventilation air temperature to the ignition temperature of CH_4, resulting in emissions of CO_2 and water vapor. Catalytic monolith reactor (CMR) is a honeycomb type monolithic reactor, which performs similar function but at low pressure with high mass flows, high geometrical area, and high mechanical strength (Cimino et al., 2001). Monoliths consist of a structure of parallel channels with walls coated by a porous support containing catalytically active particles. Therefore, compared with the TFRR and CFRR units, the CMR unit should be more compact in terms of processing the same amount of ventilation air. All the three technologies perform similar function; however, they vary in structural pattern and operating behaviors (Su et al., 2005).

10.7.9 Mine Methane Utilization in Gas Turbines

Following TFRR, CFRR, and CMR technologies, mitigation and utilization of mine CH_4 may also be achieved by using various gas turbines, viz. Energy Developments Limited recuperative, Commonwealth Scientific and Industrial Research Organization catalytic, Infrared catalytic microturbines. These turbines utilize heat from CH_4 combustion, which preheats the air in the range of 700–1000°C to drive the turbines (USEPA, 2003).

10.7.10 Concentrators

Concentrators are specialized technique amalgamated with gas turbine to capture VOCs. It is used to enrich CH_4 in mine ventilation air to a level that meets the requirements of lean-burn CH_4 utilization technologies, such as catalytic and recuperative gas turbines (USEPA, 2003).

10.7.11 Underground Sequestration of Carbon Dioxide

Underground sequestration of CO_2 fall into four categories, viz. natural and man-made caverns, unused porous and permeable reservoir rocks, depleted oil and gas fields, and coal beds. Actually, storage of gas in mines cannot make a significant impact on the greenhouse effect. The majority of mines are not leak proof, for an instance abandoned mines when gradually filled with H_2O, it eventually forced the adsorbed gas out into the atmosphere (Holloway, 2005).

10.7.12 Carbon Trading

As per Kyoto Protocol clean development mechanism (CDM), the reduction in GHG emissions from coal mining can be achieved by carbon trading. Carbon trading is an economic activity, which involves buying and selling of environmental services includingGHGs from the atmosphere, which are identified and purchased by eco-consulting firms and then sold to individual or corporate clients to "offset" their polluting emissions (Sada,

2007). Under CDM, GHGs are traded with developed countries, thereby making its mitigation economically viable (Wang et al., 2012).

10.7.13 Miscellaneous Strategies

- Optimization of operational activities and logistics.
- Implementation of a wider fuel management strategy, which encourages the use of more efficient plant and vehicles, planning, logistics, driver education, and maintenance.
- Reduction in the amount of waste disposed to landfill and reuse of waste on site as much as possible, which will subsequently reduce the amount of vehicle movements and therefore fuel usage.
- Procurement of generators, which use biodiesel or natural gas.
- Regular monitoring of fuel and energy consumption during the processes.
- Identification of the significant energy consuming equipments and recognizing opportunities where technical efficiencies in plant and equipment can be applied.
- Regular monitoring and awareness about GHGs emissions from mining activities.
- Afforestation that acts as a sink for emitted carbon from entities.

10.8 CONCLUSION

Coal mining is a significant emitter ofGHGs and is the focus of increasing attention from green groups and regulators. In coal-mining industries, majority of the emissions are attributed to fuel combustion and energy-related activities, fugitive emissions, emissions from captive power generation, mine fires, land-use change, and from purchased goods and services. The predominance of Scope 1 emissions is because of the significant GHG emissions contributed by fugitive emissions of CH_4 (ranging from 71% to 77% of Scope 1 GHG emissions) during mining and postmining activities. Emissions from purchased electricity, heat, and H_2O are the second highest contributors (5.8%–15% of total emissions), whereas emissions from purchased goods/services, product use, and electricity transmission are the lowest contribution to the total GHG emissions. Every coal-mining projects need to know their own carbon emissions and where they fit in the emissions curve. This view seeks to quantify and provide a comprehensive overview of the climate change impacts of coal production based on a comprehensive GHG inventory assessment of the coal mines encompassing all relevant product life cycle stages. This will help end user holistically to understand the GHG emission impact of its production activities, identify key GHG intensity hotspots across life cycle stages, as well as enable benchmarking of coal-mining emissions. The review deals with the key issues and various research outcomes that may help the coal industries and policy makers in mitigating the carbon emission reduction within the coal-mining sector. The review also suggested certain mitigation strategies to be adopted during coal mining to cut down the GHG emissions into the atmosphere.

ACKNOWLEDGMENTS

Authors are thankful to DST-FIST and CAS programme running in the Department of Botany for necessary facilities. Bhanu Pandey and Meenu Gautam are thankful to the University Grants Commission and Council of Scientific and Industrial Research, India, respectively, for financial assistance.

REFERENCES

Ahmaruzaman, M., 2010. A review on the utilization of flyash progress in energy and combustion science, 36 (3), 327–363.

Alliance, B.B.M., 2011. Caval Ridge Mine: Operational Workforce Change Request 4–Submission to Queensland Coordinator-general. BHP Billiton Mitsubishi Alliance, p. 14.

Aneja, V.P., Isherwood, A., Morgan, P., 2012. Characterization of particulate matter (PM_{10}) related to surface coal mining operations in Appalachia. Atmospheric Environment 54, 496–501.

Attalla, M.I., Day, S.J., Lange, T., Lilley, W., Morgan, S., 2008. NO_x emissions from blasting operations in open-cut coal mining. Atmospheric Environment 42 (34), 7874–7883.

Avila, C., Wu, T., Lester, E., 2014. Petrographic characterization of coals as a tool to detect spontaneous combustion potential. Fuel 125, 173–182.

Ayanda, O.S., Fatoki, O.S., Adekola, F.A., Ximba, B.J., 2012. Characterization of fly ash generated from Matla power station in Mpumalanga, South Africa. Journal of Chemistry 9 (4), 1788–1795.

Beath, A.C., 2003. Mine Gas Quality Improvement: Methane Purification, ACARP Project C11025. CSIRO Exploration and Mining Report 1109F (Brisbane, Australia).

Berg, W., Brunsch, R., Hellebrand, H.J., Kern, J., 2006. Methodology for measuring gaseous emissions from agricultural buildings, manure, and soil surfaces. In: Workshop on Agricultural Air Quality, vol. 58.

Birol, F., 2010. World Energy Outlook. International Energy Agency, p. 1.

Brown, M.A., Southworth, F., Sarzynski, A., 2009. The geography of metropolitan carbon footprints. Policy and Society 27, 285–304.

Carbon Disclosure Project (CDP), 2008. Carbon Disclosure Project Report-2008: Global 500, on Behalf of 385 Investors with Assets of $57 Trillion. http://www.cdproject.net/reports.asp.

Carbon Trust, 2007. Carbon Footprint Measurement Methodology. version 1.1. The Carbon Trust, London, UK. http://www.carbontrust.co.uk.

Central Institute of Mining and Fuel Research, Indian Institute of Management and cBalance Solutions Pvt. Ltd., 2014. Corporate GHG Inventory and Product Life Cycle Carbon Footprint Project Report - CCL.

Chate, D.M., Devara, P.C.S., 2009. Acidity of raindrop by uptake of gases and aerosol pollutants. Atmospheric Environment 43 (8), 1571–1577.

Cimino, S., Pirone, R., Russo, G., 2001. Thermal stability of perovskite-based monolithic reactors in the catalytic combustion of methane. Industrial & Engineering Chemistry Research 40 (1), 80–85.

Conti, J., Holtberg, P., Doman, L.E., 2011. International Energy Outlook. US Energy Information Administration, Washington, DC. Paper No. DOE/EIA-0484.

Cooke, W.F., Liousse, C., Cachier, H., Feichter, J., 1999. Construction of a 1×1 fossil fuel emission data set for carbonaceous aerosol and implementation and radiative impact in the ECHAM4 model. Journal of Geophysical Research: Atmospheres 104 (D18), 22137–22162.

CPCB (Central Pollution Control Board), 2009. India, National Ambient Air Quality Standards (NAAQS). Gazette notification, New Delhi.

Dada, A., Staake, T., Fleisch, E., May 2008. The potential of UbiComp technologies to determine the carbon footprints of products. In: Sixth International Conference on Pervasive Computing, Sydney.

Dimaranan, B.V., 2006. Global Trade, Assistance, and Production: The GTAP6 Database. Centre for global trade analysis, Purdue University, West Lafayette, USA.

Durucan, S., Korre, A., Munoz-Melendez, G., 2006. Mining life cycle modelling: a cradle-to-gate approach to environmental management in the minerals industry. Journal of Cleaner Production 14 (12), 1057–1070.

Energy Information Administration, 2010. Energy Information's Outlook Through 2035. From the Annual Energy Outlook 2010, DianeKearney. Energy Information Administration, U.S. Department of Energy, Surface Transportation Board, Washington, DC.

Esposito, V., Lucariello, A., Savarese, L., Cinelli, M.P., Ferraraccio, F., Bianco, A., Mazzarella, G., 2012. Morphology changes in human lung epithelial cells after exposure to diesel exhaust micron sub particles (PM 1.0) and pollen allergens. Environmental Pollution 171, 162–167.

Finkelman, R.B., 2004. Potential health impacts of burning coal beds and waste banks. International Journal of Coal Geology 59 (1), 19–24.

Finkelman, R.B., Gross, P.M., 1999. The types of data needed for assessing the environmental and human health impacts of coal. International Journal of Coal Geology 40 (2), 91–101.

Ghose, M.K., 2007. Generation and quantification of hazardous dusts from coal mining in the Indian context. Environmental Monitoring and Assessment 130 (1–3), 35–45.

Ghose, M.K., Majee, S.R., 2000. Sources of air pollution due to coal mining and their impacts in Jharia coalfield. Environment International 26 (1), 81–85.

Green Design Initiative, 2008. Economic Input–Output Life Cycle Assessment (EIO-LCA). Internet Model available online at: http://www.eiolca.net.

Green Design Initiative, 2001. Economic Input-Output Life-Cycle Assessment Model.

Guo, H., Wang, T., Blake, D.R., Simpson, I.J., Kwok, Y.H., Li, Y.S., 2006. Regional and local contributions to ambient non-methane volatile organic compounds at a polluted rural/coastal site in Pearl River Delta, China. Atmospheric Environment 40 (13), 2345–2359.

Gurney, K.R., Mendoza, D.L., Zhou, Y., Fischcer, M.L., et al., 2009. High resolution fossil fuel combustion CO2 emission fluxes for the United States. Journal of Environmental Science and Technology 43 (14), 5535–5541.

Hansen, J., Sato, M., Kharecha, P., Russell, G., Lea, D.W., Siddall, M., 2007. Climate change and trace gases. Philosophical Transactions of the Royal Society of London A: Mathematical, Physical and Engineering Sciences 365 (1856), 1925–1954.

Hegde, U., Chang, T.C., Yang, S.S., 2003. Methane and carbon dioxide emissions from Shan-Chu-Ku landfill site in northern Taiwan. Chemosphere 52 (8), 1275–1285.

Holloway, S., 2005. Underground sequestration of carbon dioxide—a viable greenhouse gas mitigation option. Energy 30 (11), 2318–2333.

Intergovernmental Panel on Climate Change (IPCC), 2006. National Greenhouse Gas Inventories: Land Use, Land Use Change and Forestry. Institute of Global Environmental Strategies, Hayama, Japan.

International Energy Agency (IEA), 2014. Key World Energy Statistics. OECD/International Energy Agency, Paris.

International Energy Agency (IEA), 2016. International Energy Outlook 2016. With Projections to 2040. International Energy Agency, Paris, France.

IPCC, 2007. Climate Change 2007: Synthesis Report: Contribution of Working Groups I, II and III to the Fourth Assessment Report. Intergovernmental Panel on Climate Change.

IPCC, 2014. In: Hiraishi, T., Krug, T., Tanabe, K., Srivastava, N., Jamsranjav, B., Fukuda, M., Troxler, T. (Eds.), 2013 Revised Supplementary Methods and Good Practice Guidance Arising from the Kyoto Protocol. IPCC, Switzerland.

ISO. Technical Committee ISO/TC 207, Environmental Management. Subcommittee SC 5, Life Cycle Assessment, 2006. Environmental Management: Life Cycle Assessment: Principles and Framework. ISO.

Jiang, X., Mira, D., Cluff, D.L., 2016. The combustion mitigation of methane as a non-CO_2 greenhouse gas. Progress in Energy and Combustion Science. http://dx.doi.org/10.1016/j.pecs.2016.06.002.

Ju, Y., Sun, Y., Sa, Z., Pan, J., Wang, J., Hou, Q., Liu, J., 2016. A new approach to estimate fugitive methane emissions from coal mining in China. Science of the Total Environment 543, 514–523.

Karacan, C.Ö., 2009. Forecasting gob gas venthole production performances using intelligent computing methods for optimum methane control in longwall coal mines. International Journal of Coal Geology 79 (4), 131–144.

Kelly, L.M., Shepson, P.B., Strim, B.P., Karion, A., Sweeney, C., Gurney, K.R., 2009. Aircraft-based measurements of the carbon footprint of Indianapolis. Environmental Science & Technology 43, 7816–7823.

Kirchgessner, D.A., Piccot, S.D., Masemore, S.S., 2000. An improved inventory of methane emissions from coal mining in the United States. Journal of the Air & Waste Management Association 50 (11), 1904–1919.

Kong, B., Wang, E., Li, Z., Niu, Y., 2017. Time-varying characteristics of electromagnetic radiation during the coal-heating process. International Journal of Heat and Mass Transfer 108, 434–442.

Kruger, D., Schultz, K., 2006. A Guide for Methane Mitigation Projects: Gas-to-Energy at Coal Mines (Draft). US Environmental Protection Agency, Office of Air and Radiation, 6202J.

Kuenen, J., Denier van der Gon, H., Visschedijk, A., Van der Brugh, H., Van Gijlswijk, R., 2011. MACC European Emission Inventory for the Years 2003–2007. TNO-report TNO-060-UT-2011-00588, Utrecht.

Kuzu, C., Ergin, H., 2005. An assessment of environmental impacts of quarry-blasting operation: a case study in Istanbul, Turkey. Environmental Geology 48 (2), 211–217.

Lal, S., Chandra, N., Venkataramani, S., 2015. A study of CO_2 and related trace gases using a laser based technique at an urban site in western India. Current Science 109, 2111–2116.

Lee, J.H., Trimm, D.L., 1995. Catalytic combustion of methane. Fuel Processing Technology 42 (2–3), 339–359.

Lokhandwala, K.A., et al., 2002. Membrane Systems for Nitrogen Rejection. GPA Annual Convention, Dallas, USA.

Lu, Z., Streets, D.G., de Foy, B., Krotkov, N.A., 2013. Ozone monitoring instrument observations of interannual increases in SO_2 emissions from Indian coal-fired power plants during 2005–2012. Environmental Science & Technology 47 (24), 13993–14000.

Maiti, S.K., 2013. Ecorestoration of the Coalmine Degraded Lands. Springer Science & Business Media.

Mandal, K., Kumar, A., Tripathi, N., Singh, R.S., Chaulya, S.K., Mishra, P.K., Bandyopadhyay, L.K., 2012. Characterization of different road dusts in opencast coal mining areas of India. Environmental Monitoring and Assessment 184 (6), 3427−3441.

Miller, S.M., Wofsy, S.C., Michalak, A.M., Kort, E.A., Andrews, A.E., Biraud, S.C., Miller, B.R., 2013. Anthropogenic emissions of methane in the United States. Proceedings of the National Academy of Sciences 110 (50), 20018−20022.

Millgate, B., Hagan, P., 2015. Changes in drilling performance and dust suppression with the use of a water separator sub. In: Coal Operators' Conference, pp. 330−336.

Nakano, T., Sawamoto, T., Morishita, T., Inoue, G., Hatano, R., 2004. A comparison of regression methods for estimating soil−atmosphere diffusion gas fluxes by a closed-chamber technique. Soil Biology and Biochemistry 36 (1), 107−113.

NASA, 2013. Fact Sheets. http://nssdc.gsfc.nasa.gov/planetary/planetfact.

Oh, S.H., Mitchell, P.J., Siewert, R.M., 1992. Methane Oxidation Over Noble Metal Catalysts as Related to Controlling Natural Gas Vehicle Exhaust Emissions.

Pandey, B., Agrawal, M., Singh, S., 2014a. Assessment of air pollution around coal mining area: emphasizing on spatial distributions, seasonal variations and heavy metals, using cluster and principal component analysis. Atmospheric Pollution Research 5 (1), 79−86.

Pandey, B., Agrawal, M., Singh, S., 2014b. Coal mining activities change plant community structure due to air pollution and soil degradation. Ecotoxicology 23 (8), 1474−1483.

Pandey, D., Agrawal, M., Pandey, J.S., 2011. Carbon footprint: current methods of estimation. Environmental Monitoring and Assessment 178 (1−4), 135−160.

Pone, J.D.N., Hein, K.A., Stracher, G.B., Annegarn, H.J., Finkleman, R.B., Blake, D.R., Schroeder, P., 2007. The spontaneous combustion of coal and its by-products in the Witbank and Sasolburg coalfields of South Africa. International Journal of Coal Geology 72 (2), 124−140.

Rai, A., Kumar, M., Vajpai, Y.K., 2013. Effects of fly ash and other ingredients to the strength of falgcsd bricks. International Journal of Engineering & Science Research 4, 190−194.

Ramanathan, V., Carmichael, G., 2008. Global and regional climate changes due to black carbon. Nature Geoscience 1 (4), 221−227.

Rout, T.K., Masto, R.E., Ram, L.C., George, J., Padhy, P.K., 2013. Assessment of human health risks from heavy metals in outdoor dust samples in a coal mining area. Environmental Geochemistry and Health 35 (3), 347−356.

Sada, R., 2007. Carbon Trading (Doctoral dissertation). HNB Garhwal University.

Saini, M.K., Srivastava, P.K., 2017. Effect of coal cleaning on ash composition and its fusion characteristics for a high-ash non-coking coal of India. International Journal of Coal Preparation and Utilization 37 (1), 1−11.

Seinfeld, J.H., Pandis, S.N., 2012. Atmospheric Chemistry and Physics: from Air Pollution to Climate Change. John Wiley & Sons.

Smith, K.R., Uma, R., Kishore, V.V.N., Zhang, J., Joshi, V., Khalil, M.A.K., 2000. Greenhouse implications of household stoves: an analysis for India. Annual Review of Energy and the Environment 25 (1), 741−763.

Solomon, S. (Ed.), 2007. Climate Change 2007-The Physical Science Basis. Contribution of Working Group I to the Fourth Assessment Report of the Intergovernmental Panel on Climate Change.

Stern, N., 2003. Stern Review on the Economics of Climate Change. Cambridge University Press, Cambridge, UK.

Stovern, M., Felix, O., Csavina, J., Rine, K.P., Russell, M.R., Jones, R.M., Sáez, A.E., 2014. Simulation of windblown dust transport from a mine tailings impoundment using a computational fluid dynamics model. Aeolian Research 14, 75−83.

Stracher, G.B., Taylor, T.P., 2004. Coal fires burning out of control around the world: thermodynamic recipe for environmental catastrophe. International Journal of Coal Geology 59 (1), 7–17.

Streets, D.G., Zhang, Q., Wang, L., He, K., Hao, J., Wu, Y., Carmichael, G.R., 2006. Revisiting China's CO emissions after the transport and chemical evolution over the Pacific (TRACE-P) mission: synthesis of inventories, atmospheric modeling, and observations. Journal of Geophysical Research: Atmospheres 111 (D14).

Su, S., Beath, A., Guo, H., Mallett, C., 2005. An assessment of mine methane mitigation and utilisation technologies. Progress in Energy and Combustion Science 31 (2), 123–170.

Su, S., Chen, H., Teakle, P., Xue, S., 2008. Characteristics of coal mine ventilation air flows. Journal of Environmental Management 86 (1), 44–62.

U.S. Environmental Protection Agency, 1998. AP42, Compilation of Air Pollutant Emission Factors, Vol. 1: Stationary Point and Area Sources, fifth ed., vol. 1 Chapter 11.9 Western surface coal mining.

U.S. Environmental Protection Agency, 2003. Assessment of the Worldwide Market Potential for Oxidizing Coal Mine Ventilation Air Methane. United States Environmental Protection Agency. EPA 430-R-03–002.

U.S. Environmental Protection Agency, 2012. Regional Screening Level (RSL) Summary Table. United State Environmental Protection Agency, Washington, DC.

U.S. Environmental Protection Agency, 2015. US Environmental Protection Agency, Integrated Risk Information System. http://www.epa.gov/iris/supdocs/0276index.html.

U.S. Environmental Protection Agency, 2017. Inventory of U.S. Greenhouse Gas Emissions and Sinks: 1990–2015 (Washington, DC).

Van Dijk, P., Zhang, J., Jun, W., Kuenzer, C., Wolf, K.H., 2011. Assessment of the contribution of in-situ combustion of coal to greenhouse gas emission; based on a comparison of Chinese mining information to previous remote sensing estimates. International Journal of Coal Geology 86 (1), 108–119.

Wang, F., Ren, T., Tu, S., Hungerford, F., Aziz, N., 2012. Implementation of underground longhole directional drilling technology for greenhouse gas mitigation in Chinese coal mines. International Journal of Greenhouse Gas Control 11, 290–303.

Wiedmann, T., Minx, J., 2007. A Definition of Carbon Footprint. ISAUK Research Report 07-01. ISAUK Research & Consulting, Durham.

World Energy Council, 2013. World Energy Resources, Survey. Available at: http://www.worldenergy.org.

World Health Organization (WHO), 2000. World Health Organization Guidelines for Air Quality. Geneva Xinhua Net (2007) SEPA Together with the Peoples Bank of China and the Baking Regulatory.

World Resource Institute/World Business Council for Sustainable Development (WRI/WBCSD), 2004. The Greenhouse Gas Protocol: A Corporate Accounting and Reporting Standard Revised Edition. World Business Council for Sustainable Development and World Resource Institute, Geneva.

Xing, J., Pleim, J., Mathur, R., Pouliot, G., Hogrefe, C., Gan, C.M., Wei, C., 2013. Historical gaseous and primary aerosol emissions in the United States from 1990 to 2010. Atmospheric Chemistry and Physics 13 (15), 7531–7549.

Xu, L., Lin, X., Amen, J., Welding, K., McDermitt, D., 2014. Impact of changes in barometric pressure on landfill methane emission. Global Biogeochemical Cycles 28 (7), 679–695.

Yadav, S.K., Jain, M.K., 2017. Exposure to particulate matter in different regions along a road network, Jharia coalfield, Dhanbad, Jharkhand, India. Current Science. ISSN: 0011-3891 112 (1).

Yavorskyi, V., Helesh, A., Yavorskyi, I., 2013. Principals for the Creation of Effective and Economically Sound Treating Processes of Industrial Emissions with Sulfur Oxide Low Content.

Yusuf, R.O., Noor, Z.Z., Abba, A.H., Hassan, M.A.A., Din, M.F.M., 2012. Methane emission by sectors: a comprehensive review of emission sources and mitigation methods. Renewable and Sustainable Energy Reviews 16 (7), 5059–5070.

Zhang, J., Smith, K.R., 2007. Household air pollution from coal and biomass fuels in China: measurements, health impacts, and interventions. Environmental Health Perspectives 848–855.

CHAPTER 11

Life Cycle Assessment of an Academic Building: A Case Study

Himanshu Nautiyal[1], Venu Shree[2], Paramvir Singh[2], Sourabh Khurana[3], Varun Goel[2]
[1]THDC Institute of Hydropower Engineering and Technology, Tehri, India; [2]NIT Hamirpur, Hamirpur, India; [3]OM Institute of Technology and Management, Hisar, India

11.1 INTRODUCTION

Building has become one of the basic amenities of human life. It can be expressed as an artificial structure having a roof and walls standing more or less permanently in one place, such as a residential unit or commercial one. There may be several shape, sizes and applications of buildings which can be influenced by variety of factors like weather, available construction materials, type of land etc. The main objective of a building is to provide a suitable indoor environment and shelter due to which the construction rate of buildings is increasing drastically throughout the world. Buildings can be categorized on the basis of their use viz. residential and commercial buildings. Residential buildings include single family house, multi-family house whereas industries, school, office, malls, university are included in the commercial buildings. Buildings consume energy throughout their entire life (Dakwale et al., 2011). The energy consumption associated with a building can be divided into three main phase's viz. construction, operation & maintenance and demolition. Residential buildings include single family house, multi-family house whereas industries, school, office, malls, university are included in commercial buildings. The energy consumption associated with a specific building is influenced by its use and application. The construction of building structure using variety of materials and techniques is associated with the construction phase. It is quite different from manufacturing processes, as it involves production of similar items in large scale without a specified customer whereas construction is typically done on a particular location for a specified buyer. The commencement of construction works is associated with planning, designing, financing etc. and

Environmental Carbon Footprints
ISBN 978-0-12-812849-7
http://dx.doi.org/10.1016/B978-0-12-812849-7.00011-8

continues till the building is commissioned. Cost estimation, design, legal considerations are some common elements in construction projects (Tsai et al., 2011).

Another important phase of the building is its operation and a lot of electrical energy is consumed in lighting, heating, ventilation and air conditioning (HVAC) equipment, and other electrical and economic appliances. Apart from this, maintenance of building is another important factor which consumes energy throughout the life cycle of a building. The process of demolition includes tear-down of a building structure after a long period of time. It is generally done when building is no more useful for the residents or the building land is required for other purposes. The preservation of valuable elements and materials obtained from the building demolition is also important for further recycle and its use. The process of demolition is quite simple for small buildings and can be easily done by using hydraulic equipment, cranes, excavators etc. On the other hand the demolition of the large buildings is a difficult task and it is time consuming and utilizes special techniques. The demolition of a large and multi-stories building in a dense populated area is more difficult since it can affects the other buildings and people around the building. Therefore, proper planning, machineries, skilled workers are essentially required for the demolition process.

Now days, energy consumption in buildings can be considered as an environmental impact and therefore emphasis must be given to quantify the energy use in all phases of a building (Junnila et al., 2006). Over the lifetime, a building can be maintained, refurbished, renovated and demolished and a lot of energy and materials are required to accomplish these processes. The role operation and construction phase of a building is imperative to study the environmental impacts of a building. Today all the materials required in building construction are highly energy intensive as they involve a lot of stages like extraction of raw materials, processing, transportation etc. An important term associated with the production phase of lifecycle energy analysis is embodied energy. The embodied energy associated with a specific material may depends on its abundance, processing transportation etc. For example in a country like India, the use of steel, brick and cement are very common and contribute about 80%–90% of the energy of buildings material (Reddy and Jagadish, 2003). Also, some materials like concrete and reinforced steel contribute the highest embodied energy of a building structure (Buchanan and Honey, 1994). However, many studies concluded that contribution of embodied energy could be decreased significantly using recycled materials and products

(Thormark, 2002; Nautiyal et al., 2015; Debnath et al., 1995). Also, there is a capability to decrease the embodied energy associated with buildings using alternative and local construction materials (Cole and Kernan, 1996). In addition to this, requirement of energy for onsite construction processes like crane operation, space heating, transportation of materials, excavation etc. Further increase in the share of energy use associated with construction phase (Peuportier, 2001).

Also a lot of energy consumption is allied with operation phase of a building. The time period of operation and maintenance phase is the largest among all the phases of a building. So, high amount of energy required for cooling and heating of building depending on the weather situations, running of appliances, machineries etc. Consequently, buildings are associated with the generation of large amount of GHG emissions into the environment (EIA, Annual Energy Outlook, 2008). So, it becomes quite important to find those sources and processes which are mainly liable for generating high amount of GHG emissions (Shree et al., 2015). In addition to this, a high amount of solid waste is also generated from construction and demolition processes. Overall it can be concluded that significant environmental impacts are associated with the buildings from its construction to demolition.

There is a significant share of embodied energy in the total lifecycle energy of a building (Lee et al., 2011). The estimation of embodied energy for various construction materials used in building requires an ample collection of data which can be influenced with time and other financial factors (Varun et al., 2012). On the other hand the estimation of operational energy is usually carried out using electricity, maintenance records etc.

In order to minimize and control environmental effects of buildings it is important to quantify them and explore those factors which have significant contribution in the total environmental impacts. So, an effective methodology is required to perform this type of environmental assessment. LCA is a popular and effective method to perform environmental assessment of buildings. It is useful for approximation of the environmental impacts associated with a product through its life cycle. It is quite appropriate for all phases of building and analyzes the variations required to enhance the performance of a building (Khasreen et al., 2009). Sharma et al. (2011) had carried out a review on life cycle assessment of buildings.

In this chapter discussions begin with the need of estimation of energy consumption and other environmental impacts associated with buildings are discussed in introduction section. In Section 11.2, LCA technique used to estimate environmental impacts is discussed. Section 11.3 describes the methodology used to carry out a case study on estimation of energy consumption and GHG emissions associated with an academic building. The case study is discussed in Section 11.4 followed by results and discussion in Sections 11.5 and 11.6 respectively.

11.2 LIFE CYCLE ANALYSIS

LCA is methodology which is used for the quantitative analysis of input materials, energy flows and environmental impacts associated with a product. It is quite popular due to its systematic approach to study the effects of each process and product. It is quite useful for finding, estimating and evaluating the potential environmental effects throughout the life cycle of product that's why it is popularly known as cradle to grave study (SETAC, 1991; Tshudy, 1994; Basket et al., 1995). There are four main parts in an LCA methodological framework viz. scope and goal definition, life cycle inventory analysis, life cycle impact assessment and elucidation of data. LCA is quite useful to quantify and compare energy consumption and environmental impacts of different buildings (Veerbeck and Hens, 2010). It serves as an important tool to explore areas and inputs where improvements and significant variations can be made in order to reduce the environmental impacts of buildings. Consequently, it can be extended to a good source to provide useful environmental data for environmental policy makers and Government.

There are mainly three types of methodologies which are used to perform LCA viz. process based, input-output and hybrid LCA (Bullard and Herendeen, 1975; Facanha and Horvath, 2006; Guinea, 2002; Heijungs and Suh, 2002; Suh and Huppes, 2005). Process based LCA methodology can be effectively used to estimate the energy consumption for many processes. This method begins with a selection of a specific product (good or service) as objective of the study. Then the resources associated directly and indirectly to produce the product are analyzed. After this, list of data resources is prepared and used to estimate the total energy requirement and associated environmental emissions from the particular product/system is estimated. But in many cases, every stage may consists of several small and large numbers of goods and services which cannot be considered completely used process based analysis. In addition to this the methodology can be suffered from truncation error (Lenzen and Treloar, 2002). Defining a suitable boundary is one of the important issues in process based LCA. It is crucial step to define what to include in the study and what to exclude. Other important issue in process based LCA is the presence of circularity effects. This is because a lot of same product is used to make another product. This requires the LCA of all materials as well as process to perform LCA of a specific product (www.eiolca.net).

In the case of input-output approach, it can be found useful to analyze all the embodied energies/externalities associated with a product/system. In input-output method, a product system consisting of supply chains can be modeled input-output tables or economic flow databases supplied by various statistical agencies of the government. This methodology also provides greater comprehensiveness (Varun, 2010). In economic input-output (EIO) method; the materials, energy requirement and environmental emissions can be evaluated for any activity or process of the system based on the economy of the nation. EIO utilizes the acquaintance about industry dealings - procures of materials by one industry from other industries, and the information about direct environmental emissions from the industries. It is a mathematically defined procedure using economic and environmental data to determine the effect of change in the output of a single sector. This method can be applied to any economy defined by the transactions between sectors (agriculture, heavy machinery etc.). Leontief developed the idea of input-output models of the US economy and theorized about expanding them with non-economic data (Leontief, 1986).

A matrix is created that represents the direct requirements of the inter-sectoral relationships. The rows of matrix A indicate the amount of output from industry i required to produce one dollar of output from industry j. These are considered as the direct requirements that are the output from first tier of suppliers directly to the industry of interest.

The resulting output, (Y_{direct}) from the entire economy can be written as follows (CMU-Green Design Institute):

$$Y_{direct} = (I + A)x \tag{11.1}$$

The relation as is shown in Eq. (11.1) is only for single level of suppliers. Furthermore, the output demand from the main suppliers creates a demand for the secondary suppliers. The requirement of secondary suppliers can be calculated as $A \times A \times x$.

And, the output for the numerous suppliers is calculated as shown in Eq. (2):

$$Y = (I + A + AA + AAA + \ldots)x \tag{11.2}$$

The relation (I + A + AA + AAA + ...) is equal to $(I - A)^{-1}$, which is called the Leontief inverse or total requirements matrix. The relation between demand and output can be expressed compactly as shown in Eq. (11.3):

$$Y = (I - A)^{-1}x \tag{11.3}$$

Another methodology is hybrid LCA which consist the advantage of both process analysis and input-output analysis. However, a demerit associated with it is the deficiency of a reliable and comprehensive database of energy consumption data from industry (Bullard and Herendeen, 1975).

11.3 METHODOLOGY

In the present chapter, total consumption of energy and associated GHG emissions of an academic building starting from all of its components are estimated during its life cycle. The studied building is an academic building of THDC Institute of Hydropower Engineering and Technology located at Tehri, Uttarakhand (India). The life cycle energy consumption is estimated by evaluating the energy used during the complete life cycle of the building. The phases which are considered for calculating the energy used is as follows: construction phase, operation phase, maintenance phase and demolition phase. The construction phase consumes large amount of energy, particularly when the building is situated in a hilly area. The energy is consumed mainly in form of embodied energy of materials. Embodied energy includes energy consumption in extraction, manufacturing, assembly transportation of a particular good and product (www.designingbuildings.co.uk/wiki/home). Embodied energy of some construction materials are presented in Table 11.1 (Gumaste 2006; Reddy 2004; Reddy and Jagadish 2003; Scheuer et al., 2003; www.victoria.ac.nz). Due to the scarcity of data energy consumed in on site construction activities like transportation has not been considered in the case study. A large number of materials were used in the construction of a building. Each material is produced in its respective industry using a particular method which is applicable to its production. Each method consumes certain amount of energy and thus every component is associated with some amount of energy consumption. These energy requirements are fulfilled by some source of energy and each source is associated with its equivalent GHG emissions. These emissions for the concerned building are evaluated in the present analysis.

Operational energy includes energy consumed during the course of the usage phase. These energy includes HVAC (heating, ventilation and air-conditioning), lighting, domestic hot water and other appliances.

Table 11.1 Embodied energy of some common construction materials

S. No.	Material	Embodied energy
1	RCC	730 MJ/m^3
2	Steel	42 MJ/kg
3	Cement mortar	1268 MJ/m^3
4	Red brick masonry	2141 MJ/m^3
5	Timber	10.8 MJ/kg
6	Glass	25.8 MJ/kg
7	Aluminium	207 MJ/kg
8	Cast iron	32.8 MJ/kg
9	Kota stone	2560 MJ/m^3
10	Ceramic tiles	5.5 MJ/kg
11	CGI sheets	32.6 MJ/kg
12	Aggregates	0.083 MJ/kg
13	Sand	232 MJ/m^3
14	Granite	198.1–490.8 MJ/m^2
15	Paint	88.5 MJ/kg

HVAC approximately consumes 40% of the total operational energy used in the building. The share of energy consumed for heating in the winter period and cooling of the particular location. The main consumption during the operational phase is in the form of energy required to run various electrical equipment along with the requirement of HVAC. It is assumed in the study that the operational energy allied with the building is same during its life span. However, its variation due to changes in climatic factors is ignored in the analysis. As the study is carried out in a hilly location in India, it is important to consider the electricity scenario in India in order to carry out the LCA of the selected building which is presented in Table 11.2 (www.cea.nic.in; Varun et al., 2009).

Table 11.2 Current electricity generation scenario in India (CEA, 2017)

S. No.	Source	Installed capacity	Percentage (%)
1	Thermal	214,654.89	68.22
	a. Coal	188,487.88	59.90
	b. Gas	25,329.38	8.05
	c. Diesel	837.63	0.27
2	Hydro	44,189.43	14.04
3	Nuclear	5,780	1.84
4	Renewable sources	50,018	15.90
Total		314,642.32	100

Irrespective of the method used for maintenance, it is pretty clear that maintenance is an essential activity and this activity consumes a lot of energy. Certain standards have been recommended to find out the actual energy consumed in the maintenance phase of a building. Since, the building is relatively new then it so the data for maintenance activities is not available at this particular moment. So, for the sake of calculations it is taken from the standard available data and recommendations as illustrated in the case study section. It is assumed that the energy required for the demolition of the building is less than 2% of the energy used in construction phase (Ramesh et al., 2012). The total life cycle energy of building is calculated as the sum of energy consumed in all the above mentioned phases.

In order to perform the life cycle study, it is required to select a functional unit like rooms, usable area etc. It is quite necessary to relate the inputs and outputs of a system. In this study "m^2 usable floor area" is selected. The total service life of the building is assumed to be 50 years for performing the LCA study. The operational energy is also assumed to be uniform throughout the service life of the building. The total energy use and environmental impacts are estimated with an assumption that no further construction and extension will be made during this life span. The main focus is on the construction and building operation phase. For the construction phase calculation of embodied energy of different building material are carried out using the standard data. Whereas in operation phase the major areas of energy consumption are considered are lighting in the building, HVAC, small machineries and electrical devices.

The methodology used for the present study is process based LCA approach and energy associated with all the phases of the building is calculated separately and associated GHG emissions with each phase has been estimated. There may be variations in operational energy associated with the building in future but those variations are not considered in the present study. Also the result estimated for the buildings are also based on the assumption that there will be no reconstruction and extension of the building will be carried out during the life cycle of the building. There are certain limitations in this study since some simplifications and assumption have been considered. In some cases due to the lack of specific information of manufacturer, common data or data of other equivalent products is considered. Several construction materials are excluded from the analysis due to unavailability of data and only important construction materials are included in construction phase. Also, several human factors like work environment, noise, odor are excluded in the study. In addition to this, furniture, equipment, machineries are not included in the study however,

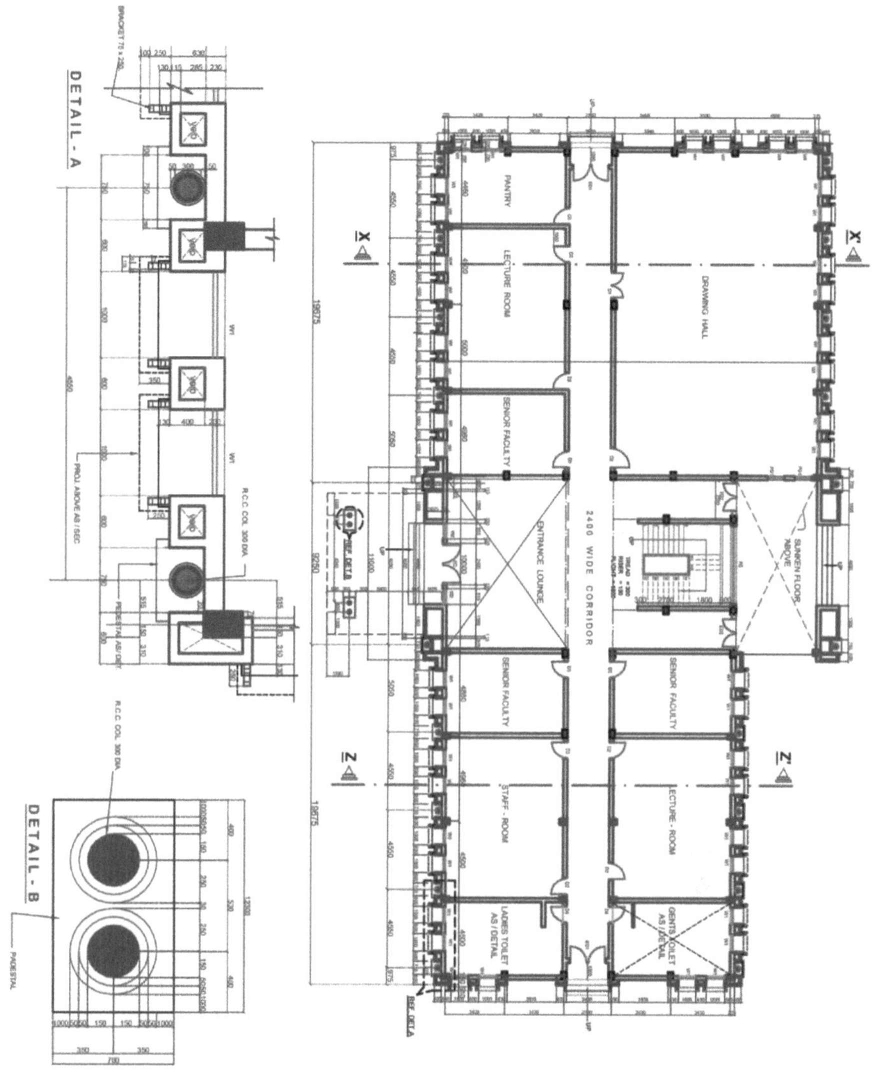

Figure 11.1 Layout of ground floor.

electrical energy consumed by the machineries and equipment are considered in the operational energy of the building.

The analysis begins with the collection and calculation of the inventory data associated with the building. The details of various construction materials consumed during the construction of building are found by using site plans (drawings) as shown in Figs. 11.1 and 11.2 respectively and visual inspection of buildings. After this their corresponding embodied energies are floor wise quantified. Various embodied energy coefficients for different

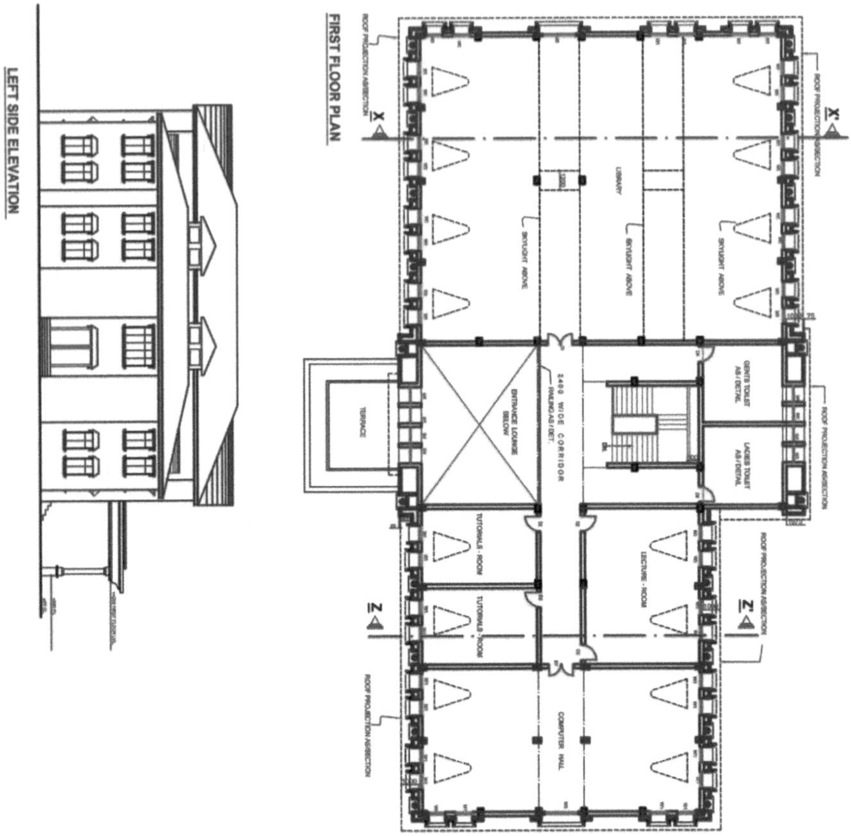

Figure 11.2 Layout of first floor.

materials are used from the previous literature (Gumaste, 2006; Reddy, 2004; Reddy and Jagadish, 2003; Scheuer et al., 2003). In addition to this the associated GHG emissions are estimated with the help of current electricity scenario in India. Since, the studied building is a non-residential building the lighting and other electricity loads are considered as constant with reference to a particular schedule. The operational energy is calculated using the past electricity bills. The life cycle GHG emissions are found by the product of total primary life cycle energy, primary to electrical energy conversion factor and average emission coefficient (Varun et al., 2012).

11.4 CASE STUDY

The subject under consideration is a building of an academic block in THDC Institute of Hydropower Engineering and Technology. The

Figure 11.3 Building for case study.

institute is situated in the hilly terrains of Bhagirathipuram, Tehri in Uttarakhand state (India). It is near to Tehri Dam, which is one of the largest dams of the world with power generation capacity of 2400 MW. The building comes under the category of "Educational buildings." The building was commissioned in 2011. Fig. 11.3 shows the picture of the building under consideration and the detailed specifications of the building are as follows:

Location	30.38 latitude, 78.46 longitude
Floor area	2074 m^2
No. of floors	2
Total number of doors	27
Total number of windows	94
Number of fans	86
Number of tube-lights	131
Number of surface lighting	23
Number of skylights	16

The main objectives in the study are to approximate the total energy consumption and its associated GHG emissions of the building. The building is located in hilly area where a lot of variations in climatic conditions are experienced. It comprises a library section, two lecture rooms etc. in first floor and a seminar of hall of seating capacity is more than 100 persons, two lecture room, a computer lab and four faculty cabins. A lot of electricity is consumed in lightings, fans in summers and electric heating in winters, computers lab equipment etc. The annual electricity consumption of building is estimated around 9224.26 kWh$_e$/year. The detailed analysis is presented in the following sections.

11.4.1 Construction Phase

The main components used in building construction are cement, sand, aggregates, steel, bricks, glass, stones, tiles, granite, putty (wood filler), primer, paint etc. In the construction phase the energy considered is the energy used in production of materials used for construction. In case of a building the embodied energy is calculated by calculating the energy required in producing the different materials used in the construction of building. To estimate the total embodied energy associated with various construction materials used in the building, a survey is conducted on the materials used in the construction to quantify them. The embodied energies are calculated for each floor separately as shown in Figs. 11.4 and 11.5.

11.4.2 Operational Phase

In the operation phase the electrical energy consumed for various reasons is considered. As already discussed that building under consideration is an

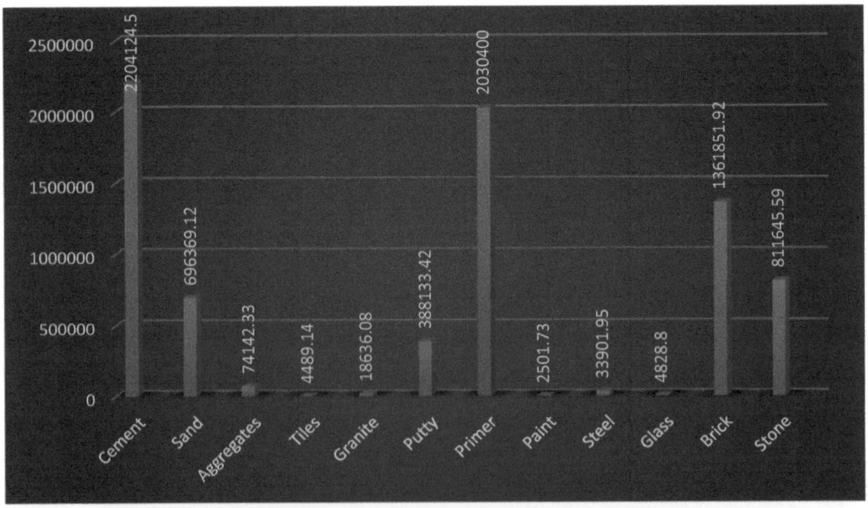

Figure 11.4 Embodied energy of used material in ground floor in MJ.

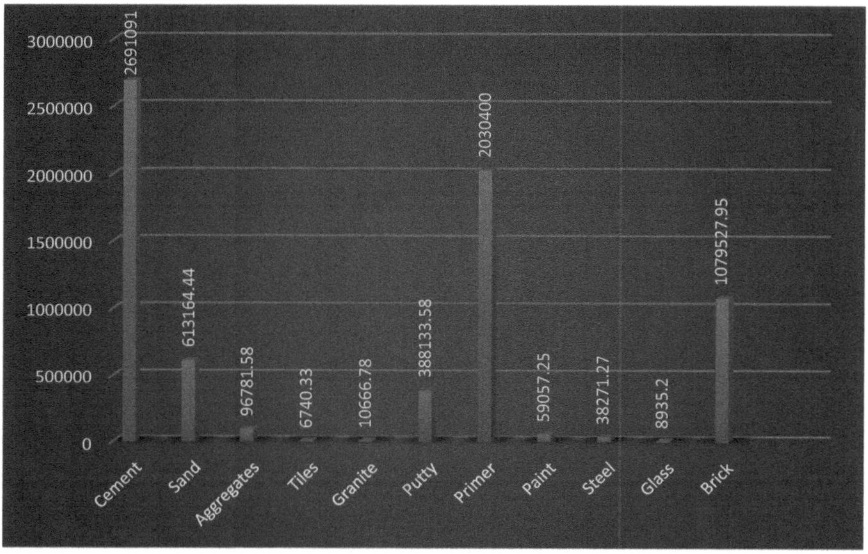

Figure 11.5 Embodied energy of used material in first floor in MJ.

educational building situated in a hilly region. The electricity is used for various purposes are listed as follows:

- Fans (summer)
- Heater (winters)
- Water cooler
- Projectors
- Speakers
- Lighting
- Sanitation equipment
- Computers
- Photocopiers

For calculating overall energy consumed during the operation phase the electricity bills of 12 months are considered. From this annual average energy consumption for each month is estimated. The life time of the building is assumed 50 years.

$$\text{Total operational energy} = \text{Average energy consumed per month} \times \text{No. of year} \times \text{No. of months in the year}$$

11.4.3 Maintenance and Demolition Phase

Similarly, every building consumes energy for its maintenance purpose. The energy for this phase again depends upon the method of maintenance used. Since, the subject under consideration is only 4 years old, hence no considerable maintenance activity has been placed so far. So, for the sake of calculating the energy for maintenance throughout the life span of building the energy consumed for maintenance is assumed to be 3% of initial embodied energy. Every building is supposed to be demolished after a certain period of time. Demolition of a building consumes certain amount of energy as per the medium used for doing this task. Since, building is relatively new, the energy consumed in demolition is again assumed to be 3% of the total operational energy (Cooper and Fava, 2006). The energy required in demolition is calculated with taking an assumption that the building will be demolished after its service life.

11.4.4 Life Cycle Energy

The total life cycle energy of the building is calculated as the sum of total energy associated with different phases of the building throughout its lifetime.

$$\text{LCE} = \text{CE} + (\text{OE}/\text{year} \times \text{lifetime of building}) + \text{ME} + \text{DE}$$

where, LCE, Life cycle energy; CE, Energy consumed in construction phase; OE, Operational energy; ME, Energy consumed in maintenance; DE, Energy consumed in demolition.

The life cycle energy of the building for a lifetime of 50 years is calculated as 19,412,038.04 MJ.The contribution of various phases in the life cycle energy is shown in Fig. 11.6.

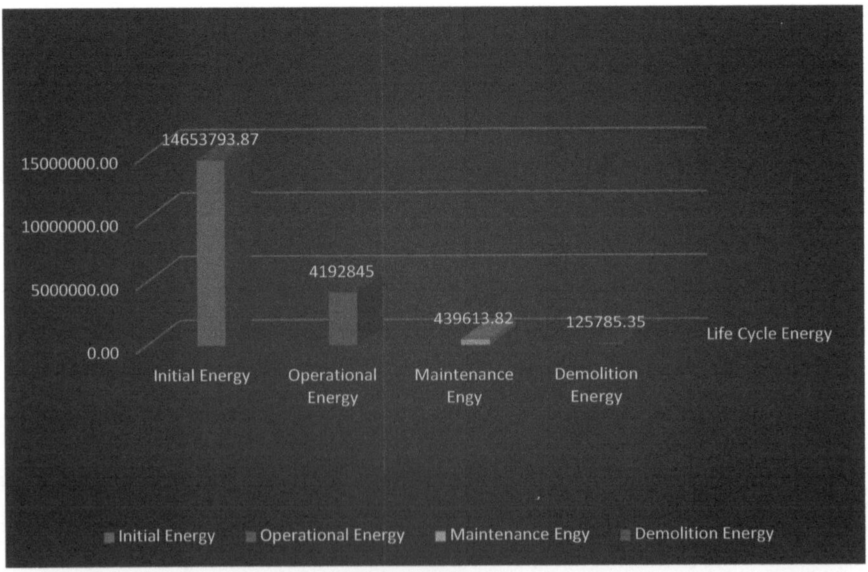

Figure 11.6 Contributions of various phases in the life cycle energy in MJ.

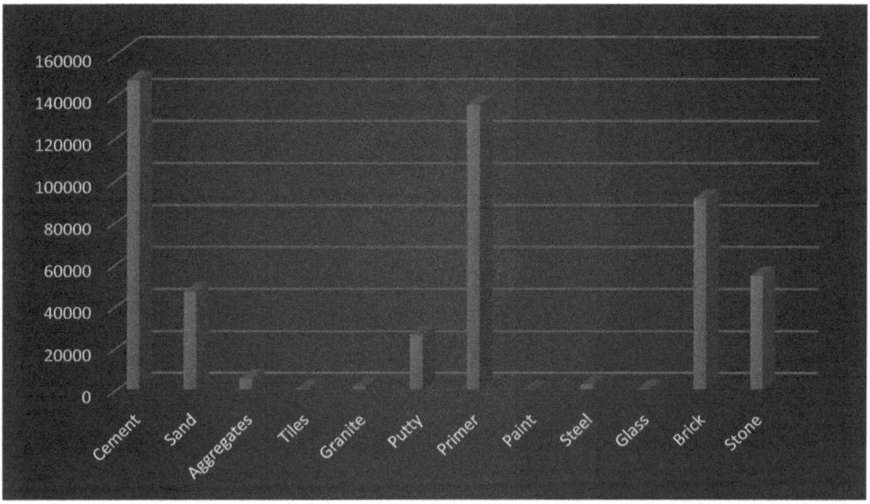

Figure 11.7 GHG emissions associated with ground floor.

11.4.5 Green House Gases Emission

In previous section total life cycle energy of the building has been quantified and now GHG emissions are estimated. For this the methodology of estimating the carbon di-oxide equivalent (CO_{2eq}) of electrical energy is adopted. The GHG emissions associated with the different phases of the building are estimated using current electricity scenario of India. As the electricity consumed in operational phase is being supplied from the national grid. The primary energy in MJ is converted into kWh_{eq} by multiplying it with an electrical equivalent factor of 0.11. Also the average electricity conversion efficiency for thermal energy is assumed 0.40 for India (Varun et al., 2010). The electrical energy is then converted into CO_{2eq} using average emission coefficient in terms of CO_{2eq}/kWh and shown in Figs. 11.7 and 11.8.

$$\text{GHG emissions} = \text{Primary life cycle energy} \times \text{Conversion factor} \times \text{average emission coefficient}$$

11.5 DISCUSSIONS

Buildings have a notable share in total energy consumption and with the prognosis for forthcoming development there is the expectation of further escalation in consumption. Based upon attempts to achieve sustainable constructions, attention is focused on reducing overall consumption. From the calculations it can be seen that different materials contribute different percentages in total embodied energy of the building (shown in Fig. 11.9).

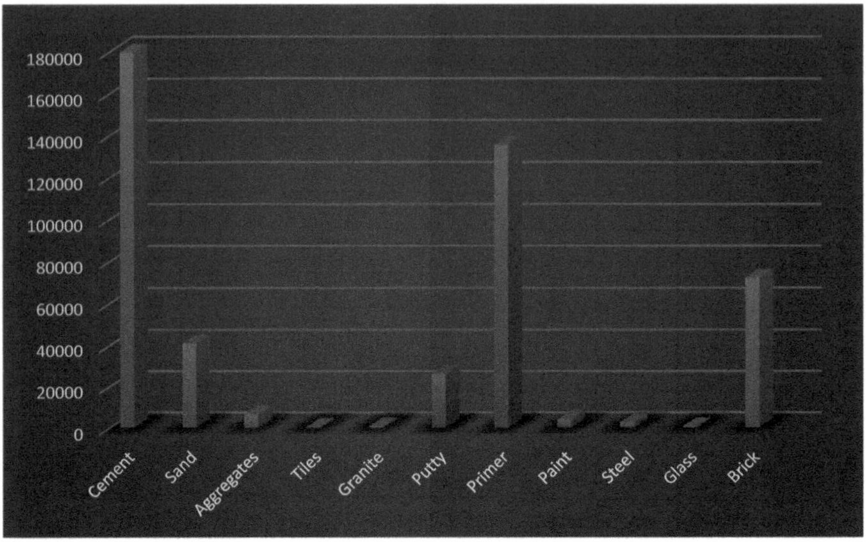

Figure 11.8 GHG emissions associated with first floor.

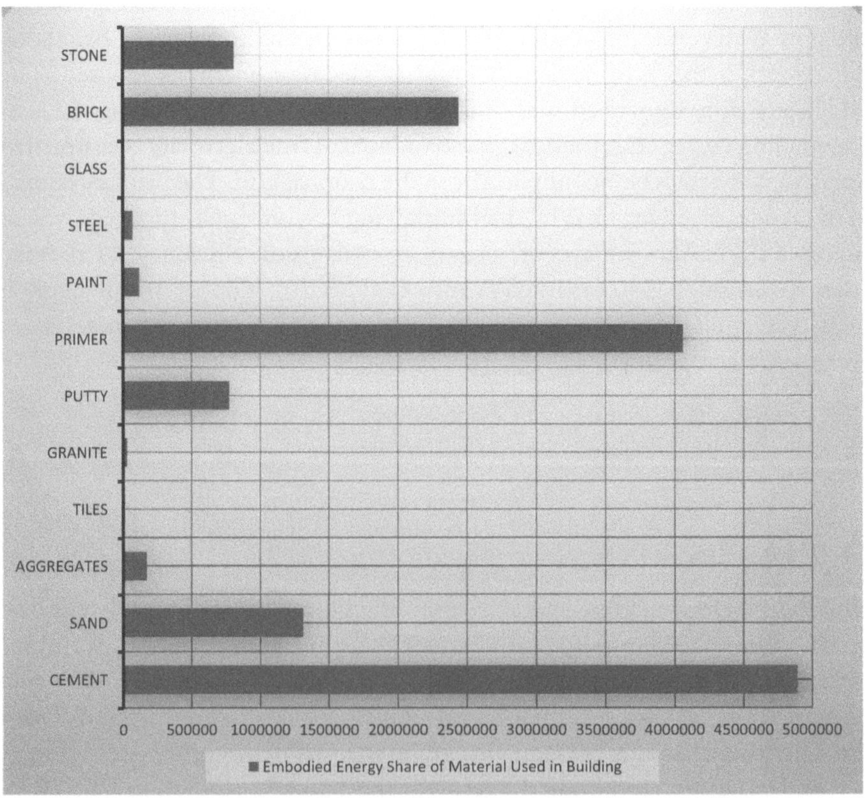

Figure 11.9 Embodied energy share of material used in building.

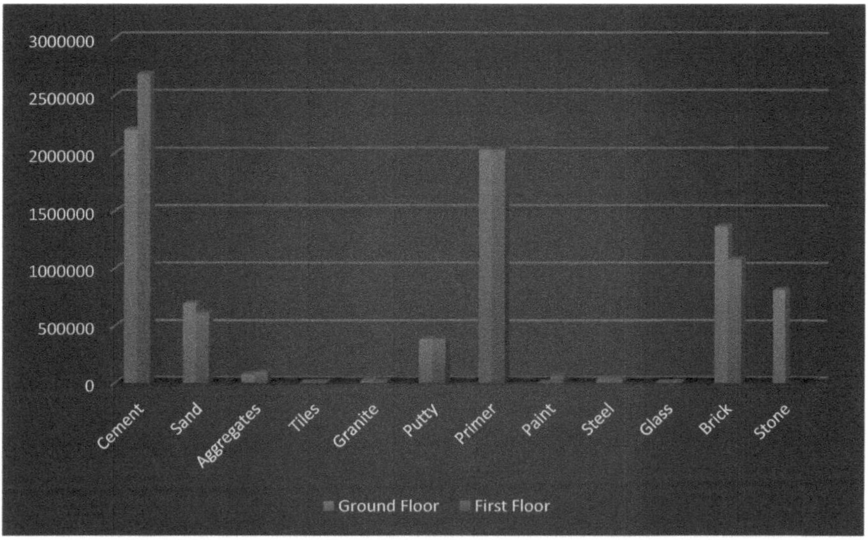

Figure 11.10 Comparison of embodied energy.

The total embodied energy for all construction materials turns out is 14,653,793.87 MJ and the associated GHG equivalent of this comes out as 978,873.40 kg-CO_{2eq}. The total operational energy for the entire life cycle is 4,192,845 MJ and associated GHG equivalent 18,909 kg-CO_{2eq}. Fig. 11.10 shows the comparison of embodied energy and in construction phase of ground and first floors. It is found that the energy and emissions associated with construction phase in both the floors are approximately same expect for bricks & cement. This is due to the difference in their quantity utilized in the construction of each floor. For the operational energy the GHG equivalent is kg-CO_{2eq}. In general operational energy of a building is dominated by HVAC requirements. But in the present case the building is located at a height of 1750 m from the sea level. Thus, the HVAC requirements of building are considerably small. Hence, the operational energy is considerably less when compared to a similar building in general. Also the building is relatively isolated and hence it further reduces the energy requirements of the building due to a phenomenon known as urban heat isolation effect. The energy associated with maintenance and demolition is found to be 439,613.82 and 125,785.35 MJ respectively.

In the present study it is visible that for building in hilly terrain the main energy consumption is in the construction phase. The main reason for this is the

Table 11.3 Comparison of life cycle studies on academic buildings

Study	Varun et al. (2012)	Present study
Building	Academic building NIT Hamirpur	THDC-IHET Tehri
Year of commissioning	1994	2011
Location	Hamirpur, Himachal Pradesh	Tehri, Uttarakhand
Region	North India	North India
Climate	Sub tropical	Warm & moderate
Altitude from sea level (in m)	785	1750
No. of floors	03	02
Floor area (m^2)	3,960	2,074
Service life time (years)	50	50
Life cycle energy (MJ)	263,98,334.7	168,60,950.51
Life cycle energy (MJ/m^2)	6,666.25	8,129.68
GHG emissions (ton CO_{2eq}/m^2)	0.45	0.55

fact that the heating and ventilation energy requirements are reduced drastically thus leading to low operational energy in the building life cycle. Also, due to the hilly location of building the energy required in transportation must be greater than the building located in plane or populated urban area. Therefore, it is important to increase our focus to evaluate the embodied energies of materials used in construction. The use of alternative techniques for building construction like passive solar building design (Chel and Tiwari, 2009) can be helpful to reduce life cycle energy used in building construction.

The present study can be compared with the other life cycle study carried out by Varun et al. (2012) on an academic building. The detailed comparison of both the studies is shown in Table 11.3. The size of the building in present study is smaller than the building studied by Varun et al. (2012) but due to the difference in their locations the results are highly varied. The building studied in present analysis is located in high hill areas which justify the demand of more resources and energy in construction and operation of a building.

11.6 CONCLUSIONS

Buildings are one of the major sources of energy consumption in present day scenario. A building consumes energy in four phases: construction phase, operational phase, maintenance phase, demolition phase. It is important to understand that the energy consumption pattern of any particular building is very different from other buildings depending on the geography, demography etc. of the location. Life cycle energy analysis is a tool which is very useful in calculating the energy consumptions in afore mentioned phases and also in quantifying their environmental impacts in the form of GHGs emissions. After applying the tool on an educational building situated in a hilly terrain, the energy consumption and greenhouse gases emissions of the building have been studied. It is observed that a building located in a hilly terrain does not consume maximum energy in the operational phase, which is very opposite of the case with a building situated in plains. It is also due to the fact that the subject building is a day time educational building, further concatenate in the reduction of operational energy. It is observed that the embodied energy of the construction materials accounts for maximum energy consumption and subsequently for maximum GHGs emissions. Thus it is recommended that in order to formulate any reduction strategies, main focus should be on using the materials with lesser embodied energy without compromising on building safety standards.

ACKNOWLEDGMENTS

The authorities of THDC Institute of Hydropower Engineering & Technology, Tehri (India) are gratefully acknowledged for providing valuable data required for this research work.

REFERENCES

Basket, J.S., Lacke, C.J., Weizt, K.A., Warren, J.L., 1995. Guidelines for Assessing the Quality of Life Cycle Inventory Analysis. Research Triangle Institute, National Technical Information Service, Springfield.

Buchanan, A.H., Honey, B.G., 1994. Energy and carbon dioxide implications of building construction. Energy and Buildings 20, 205–217.

Bullard, C., Herendeen, R., 1975. The energy cost of goods and services. Energy Policy 3, 268–278.

Carnegie Mellon University, Green Design Institute (CMU-Green Design Institute), US. Economic Input-Output Life Cycle Assessment (EIO-LCA) Model. www.eiolca.net.

Central Electricity Authority (CEA), 2017. Ministry of Power Govt. of India. www.cea.nic.in.

Chel, A., Tiwari, G.N., 2009. Thermal performance and embodied energy analysis of a passive house – case study of vault roof mud-house in India. Applied Energy 86, 1956–1969.

Cole, R.J., Kernan, P.C., 1996. Life cycle energy use in office buildings. Building and Environment 31 (4), 307–317.

Cooper, J.S., Fava, J., 2006. Life cycle assessment practitioner survey: summary of results. Journal of Industrial Ecology 10 (4), 12–14.

Dakwale, V.A., Ralegaonkar, R.V., Mandavgane, S., 2011. Improving environmental performance of building through increased energy efficiency: a review. Sustainable Cities and Society 1, 211–218.

Debnath, A., Singh, S.V., Singh, Y.P., 1995. Comparative assessment or energy requirements for different types of residential buildings in India. Energy and Buildings 23 (2), 141–146.

Designing Buildings Wiki. www.designingbuildings.co.uk/wiki/home.

EIA, 2008. Annual Energy Outlook. Energy Information Administration, U.S. Department of Energy.

Facanha, C., Horvath, A., 2006. Environmental assessment of freight in the US. International Journal for Life Cycle Assessment 11 (4), 229–239.

Guinea, J.B., 2002. LCA-an Operational Guide to the ISO-Standards.

Gumaste, K.S., 2006. Embodied energy computations in buildings. Advances in Building Research 404–409.

Heijungs, R., Suh, S., 2002. The Computational Structures of Life Cycle Assessment. Kluwer Academic Publisher, Dordrecht.

Junnila, S., Horvath, A., Guggemos, A.A., 2006. Life-cycle assessment of office buildings in Europe and the United States. Journal of Infrastructure System 12, 10–17.

Khasreen, M.M., Banfill, P.F.G., Menzies, G.F., 2009. Life-cycle assessment and the environmental impact of buildings: a review. Sustainability 1, 674–701.

Lee, B., Trcka, M., Hensen, J.L.M., 2011. Embodied energy of building materials and green building rating systems-a case study for industrial halls. Sustainable Cities and Society 1, 67–71.

Lenzen, M., Treloar, G., 2002. Embodied energy in buildings: wood versus concrete-Freply to Borjesson and Gustavsson. Energy Policy 30, 249–255.

Leontief, W., 1986. Input-Output Economics. Oxford University Press, England.

Nautiyal, H., Shree, V., Khurana, S., Kumar, N., Varun, 2015. Recycling potential of building materials: a review, environmental implications of recycling and recycled products. In: Environmental Footprints and Eco-Design of Products and Processes, pp. 31–50.

Peuportier, B.L., 2001. Life cycle assessment applied to the comparative evaluation in single family houses in the French context. Energy and Buildings 33 (5), 443–450.

Ramesh, T., Prakash, R., Shukla, K.K., 2012. Life cycle energy analysis of a residential building with different envelopes and climates in Indian context. Applied Energy 89, 193–202.

Reddy, B.V.V., 2004. Sustainable building technologies. Current Science 87 (7), 899–907.

Reddy, B.V.V., Jagadish, K.S., 2003. Embodied energy of common and alternative building materials and technologies. Energy and Buildings 35, 129–137.

Scheuer, C., Keoleian, G.A., Reppe, P., 2003. Life cycle energy and environmental performance of new university building: modelling challenges and design implications. Energy and Buildings 35, 1049–1064.

SETAC (Society for Environmental Toxicology and Chemistry), 1991. A technical framework for life-cycle assessment. In: Workshop Rep., SETAC, Washington, DC.

Sharma, A., Saxena, A., Sethi, M., Shree, V., Varun, 2011. Life cycle assessment of buildings: a review. Renewable and Sustainable Energy Reviews 15, 871–875.

Shree, V., Varun, Nautiyal, H., 2015. Carbon footprint estimation from a building sector in India. In: The Carbon Foot Print Handbook, pp. 239–258.

Suh, S., Huppes, G., 2005. Methods for life cycle inventory of a product. Journal of Clean Production 13, 687–697.

Thormark, C., 2002. A low energy building in a life cycle embodied energy, energy need for operation and recycling potential: International Journal of Building and Environment 37 (4), 429–435.

Tsai, W.H., Lin, S.J., Liu, J.Y., Lin, W.R., Lee, K.C., 2011. Incorporating life cycle assessments into building project decision-making: an energy consumption and CO_2 emission perspective. Energy 36, 3022–3029.
Tshudy, J.A., 1994. Environmental life cycle analysis—the foundation for understanding environmental issues and concerns. In: Proceedings, International Symposium on Resource, Conservation and Environmental Technologies in Metallurgical Industries, 20–25 August 1994, Lancaster USA, pp. 229–237.
University of Wellington Victoria. www.victoria.ac.nz.
Varun, 2010. Life Cycle Assessment of Small Hydropower Systems in India (Ph.D. Thesis). National Institute of Technology, Hamirpur, India.
Varun, Bhat, I.K., Prakash, R., 2009. LCA of renewable energy for electricity generation systems – a review. Renewable and Sustainable Energy Reviews 13 (5), 1067–1073.
Varun, Prakash, R., Bhat, I.K., 2010. Life cycle energy and GHG analysis of hydro electric power development in India. International Journal of Green Energy 7 (4), 361–375.
Varun, Sharma, A., Shree, V., Nautiyal, H., 2012. Life cycle environmental assessment of an educational building in northern India: a case study. Sustainable Cities and Societies 4, 22–28.
Veerbeck, G., Hens, H., 2010. Life cycle inventory of buildings: a calculation method. Building and Environment 45, 1037–1041.

FURTHER READING

Smith, M.G., Smiley, L., 2003. The Hand-Sculpted House: A Practical and Philosophical Guide to Building a Cob Cottage. Chelsea Green Publishing.

CHAPTER 12

The Role of Demand-Side Management in Carbon Footprint Reduction in Modern Energy Services for Rural Health Clinics

Olubayo M. Babatunde, Peter O. Oluseyi, Tolulope O. Akinbulire, Henry I. Denwigwe, Tolulope J. Akin-Adeniyi
Department of Electrical Electronic Engineering, University of Lagos, Akoka, Nigeria

12.1 DEMAND-SIDE MANAGEMENT

12.1.1 Introduction

Because of globalization, industrialization, and development due to technology, the demand for electrical energy is on the increase. There is therefore a need for efficient energy measures to ensure conservation, thereby saving costs. Demand-side management (DSM) deals with conversion of energy demand of consumers into activities/programs/tactics (e.g., financial incentives and public awareness/education), which brings about less use of energy by the consumers.

Gellings and Parmenter gave a history of DSM in the United States and its influence on energy resources. They also explained the role of DSM in integrated resource planning, the main elements of DSM programs and summarized the key best practices for program design and delivery.

Palensky and Dietrich (2011) described DSM as using measures such as sophisticated real-time control of distributed energy resources, better materials, smart energy tariffs with incentives for certain consumption patterns to improve energy efficiency. Various types of DSM were analyzed, and an overview of modern DSM projects was given.

Haney et al. (2010) highlighted how integrated government DSM policies, targeting residential demand for electricity and heat are more likely

Environmental Carbon Footprints
ISBN 978-0-12-812849-7
http://dx.doi.org/10.1016/B978-0-12-812849-7.00012-X

to be successful than single policies. DSM was also used to show how large untapped potentials could be uncovered through barriers to energy demand reduction.

A study on sustainable energy regulation and policymaking for Africa (Demand-side management sustainable energy regulation and policymaking for africa) examined some of the challenges facing the implementation of DSM programs and also gave a detailed review of DSM measures (this includes review of housekeeping and preventive maintenance, which are the simplest and most effective ways of reducing demand, and marketing of DSM programs) managing and controlling loads from the utility side, converting unsustainable energy practices into more efficient and sustainable energy use, thereby reducing energy demand for the end user. Worldwide, the consumption of energy is massively on the rise, and as a result, the carbon footprint is also increasing in an exponential manner. The rise in energy consumption and carbon footprint are the significant components responsible for "Global Warming." Currently, climate change a consequence of global warming is a global challenge with severe penalties for our socioeconomic infrastructure as well as the natural environment, and future generation. Long-terms planning will be required to slow down the grave effect of climate change. One of the ways of reducing the energy consumption is the adoption of sustainable DSM programs. Sustainability is also imperative in DSM programs. Sustainability means "meeting the needs of the present without compromising the ability of future generations to meet their own needs."

The rest of this chapter gives an in-depth background into the importance of DSM, DSM program types, and the various benefits of its adoption. It further presents the opportunities of adopting DSM in rural healthcare centres (RHC). Finally, a case study validating the effect of adopting DSM in sizing of hybrid renewable energy systems for RHC is analyzed.

12.1.2 Importance of Demand-Side Management

The importance of DSM can be summarized into the following (Demand-side management sustainable energy regulation and policymaking for Africa):

- Reduction in cost of producing energy by generating companies and purchasing energy by consumers.
- Provision for modern technologies and innovation, which create job and bring about economic development.
- Air pollution is reduced, therefore having a positive impact on the environment and social life.

12.1.3 Motivation Behind Demand-Side Management

According to S. Saini, the major motives behind DSM are the need for cost to be reduced and continued sustenance of the environment, also the need for reliability in energy efficiency through the network infrastructure is a motive that encourages DSM.

As a result of the rising costs of power generations and environmental hazard potentials available in power-generating stations because of generation of electricity, DSM therefore not only provides integrated resource planning for reducing costs of generation but also meeting the demands for generation (Course Module). This can be achieved by setting up energy commissions or agencies, which set targets and goals toward energy conservation, load management, and environmental protection. These agencies are to work with public energy utility companies that oversee generation, distribution, and transmission of electricity to ensure the success of these programs. It is very important that, good policies are put in place to sustain DSM programs. Some of these policies include a standard program for cost-benefit analysis of DSM activities, a predetermined target for energy savings, which is generally accepted, a system of funding, which promotes the competitive nature of energy companies practicing DSM, and finally regulation of prices of energy sold by the energy companies to customers who are noneligible for DSM financial incentives to increase sales, save incentives for eligible customers and therefore save energy (Course Module). The implementation of DSM programs can be incurred in cost, either by the government through taxes or by the electricity companies through higher tariffs whose burden on the consumers can be eased by the government through subsidies and loans. The energy commissions and agencies are to also ensure that, no planning of theirs affects environmental and infrastructural development, even as they ensure that the natural environment is not polluted, and it is therefore sustained.

Because of increasing demand for electricity and inadequate infrastructure for transmission and distribution networks in developing and developed countries, the need for DSM program is therefore of priority. Planning through DSM can reduce the need and possibly eliminate the cost of constructing transmission and distribution network infrastructure expansion (Course Module).

12.1.4 Types of Demand-Side Management Measures

DSM measures are classified into three (Demand-side management sustainable energy regulation and policymaking for Africa), namely: energy reduction programs, load management programs, and load growth and conservation programs.

12.1.4.1 Energy Reduction Programs

This is also referred to as "energy-saving tips". Some of the programs require capital, whereas others are capital free.

Energy-saving tips for steaming and heating systems include the following:

1. It should be ensured that there are adequate control for adjusting the quantity of combustion air.
2. It should be ensured that, insulation and refractory pipes are in good condition, their thicknesses should also be appropriate for good modern practice.
3. Combustion conditions should be routinely monitored, and the efficiency should always be kept as high as possible.
4. It should be ensured that, the water treatment system is always in good working order, and there should be constant and regular monitoring of the boiler feed water quality.
5. It should always be ensured that, there are no areas for steam and water leaks in the equipment.
6. Vessels, return lines, and fittings should always be condensed, and any steam present should be insulated
7. Steam leaks and steam traps should always be maintained and repaired when faulty.
8. Flash steam in the plant should always be considered during use.
9. Automatic temperature controls should be placed correctly in equipment wherever waste of steam that can overheat equipment or processes is minimized.

Energy-saving tips for lighting include the following:

1. Efficient energy fluorescent tubes, CFLs, and other low-energy light sources should be used.
2. Luminaries should be cleaned regularly.
3. Appropriate lighting levels at different periods and different zones should be used.
4. Natural light should always be encouraged in areas where possible, e.g., roof panels and skylights, etc.
5. Walls and ceilings should be painted with white or bright colors to improve reflection of light.

Energy-saving tips for motors include the following:

1. Use of highly efficient motors should be encouraged.
2. Improved bearings should be installed, and there should be regular and constant lubrication.
3. Motors should be properly sized and should be used only when needed.
4. All equipment should be maintained regularly.
5. Controls having electronic variable speed where motor loads vary in normal operation should be used.
6. Power factor should be checked regularly and should also be improved using capacitor banks, preferably closely installed to the running equipment.

Energy-saving tips for compressed air systems include the following:

1. Wrong use of compressed air should be discouraged and eliminated.
2. Leaks should be checked out when workshops are typically quiet and are not supposed to be using air.
3. Important parts of the system should be checked regularly to prevent the effect of early damage.
4. Overall system efficiency should be ensured by checking compressor running times to improve the system.
5. System pressure should always be optimized.
6. Heat recovery systems should be installed.

In households, some energy-saving tips include the following:

1. Air infiltration or hot air escapes, and inadequate or failed insulation should be eliminated in fans and vents, heating, cooling and ventilating ducts, and fireplaces, electric outlets, plumbing penetrations through walls, floors, walls, ceilings, doors, and windows.
2. Insulating blankets, insulating pipes, and thermostat set to 50°C should be encouraged for use in hot water cylinders.
3. Incandescent lightings should be replaced with CFLs and LEDs of appropriate and equivalent illumination.
4. Curtains should be used as insulators at night and as a blind against the sun to cool when hot; the curtain should also be open for sun to enter to heat when cold.
5. Dryers, dishwashers, and washing machine should only run with full load.

Some energy-saving measures that require investments include the following:

1. Replacing appliances that are old and worn-out.
2. Making use of a solar water heater.
3. Making use of double-glazing windows.

The measures discussed above are not generally applicable. Some of the generally applicable measures include the following:

1. *Efficient Lighting*: Efficient lighting measures which require investments, are not so large, and the investments are eased through paybacks and subsidies. Changing light bulbs, fittings, switches, and increased use of natural light are some of the tips involved in efficient lighting. Incandescent bulbs lose energy to heat and are therefore inefficient, compact fluorescent lamps with electronic ballasts are more efficient than incandescent bulbs with conventional ballasts, although they are more expensive, they last longer and consume less energy than the required energy for transmitting the light energy output. Some of the most common opportunities involved in efficient lighting include
   a. Long-term benefits by replacing existing lamps with more efficient light sources. This is known as light retrofitting.
   b. Safety in lighting by removing selected lamps, which do not support safety of the lighting zones from existing light fixtures in a uniform pattern throughout specific zones.
   c. Switching off selected areas of lighting with absence of people, whereas adjacent areas remain switched on to save energy, which can be wasted because of little or nonusage.
2. *Energy-efficient Motors:* Energy-efficient motors are needed to sustain manufacturing and mining industries and the natural environment. Many opportunities exist through energy-efficient motors. Air-conditioning for air circulation and compressor motors is a major source of power usage in industries to remove heat, dust, and gases, which can cause loss in energy. High efficiency motors can reduce electrical loads, thereby saving energy and ensuring the ventilation of mines (Pitis (Femco) and Livingstone (Anglo Gold Mining Motors), 2004). Per feasibility studies carried out in 1999 for establishment of motor repair and sales centers in Ghana based on sustainable energy regulation and policymaking training manual (Demand-side management sustainable energy regulation and policymaking for africa). It was concluded that, repeated motor rewinding and refurbishment lead to significant efficiency in losses. The study also made a recommendation on developing and implementing procedures on motor testing, labels, and standards for minimum efficiency and also setting up a facility in Ghana for the manufacture of small electric fan and pump motors (Ahenkorah).

12.1.4.1.1 Energy Management Practices in Organizations

Energy management practices through DSM are very important in industries/organizations for cost reduction of energy. Energy management in organizations depends on organizational budget, skills of staff available, and the nature of the organization. Energy management programs include: capital investment management (including equipment procurement), energy purchasing, performance measurement, energy policy development, metering and billing, energy surveying and auditing, awareness-raising, and training and education. An energy manager in any organization is usually tasked with the following responsibilities:

- Development and evaluation of projects to save energy.
- Proper and regular identification of energy-saving opportunities.
- Regular collection and analysis of energy-related data.
- Constant supervision of energy purchases and equipment procurement.
- Project implementation and future monitoring of implementation performance.
- Effective communication and public relation skills.

Cost-management activities that aid DSM programs include (Martel, 2000) the following:

- Accounting of energy and base-lining analysis.
- Billing of tenants for multiple occupancy buildings.
- Verification of utility bills and tracking of budget.
- Production of load profiles of individual and multiple facilities to ease energy decisions.
- Reporting of management of facility to the senior managers.
- Benchmarking of internal and external energy performance.

12.1.4.1.2 Energy Management in Households

Some good housekeeping tips for energy savings in households include:

- Switching off loads that are not in use: e.g., lightings, computers, monitors, sockets, etc.
- Removing lighting fixtures and pipings that are redundant and contribute to unnecessary heat losses.
- Faulty electricity lines should be replaced to prevent losses and distribution systems should be improved regularly for efficient output.
- Steam leaks and faulty equipment should be replaced or prevented to prevent losses.

12.1.4.1.3 Precautions to Prevent Energy Losses in Building

Some preventive measures for ensuring that energy losses do not occur in industries include the following:

- Filter cleaning on air compressors, pumps, upstream of steam traps in ventilation ducts, and others should be undertaken regularly.

- Hot spots on boilers and furnaces should be monitored regularly and checked for refractory failure.
- Transformer temperatures should be monitored regularly to prevent abnormalities.
- Noise and vibration of bearings should be monitored regularly to prevent failure of bearings.
- Having a routine schedule for lubrication of parts and equipment.
- Regular replacement of worn-out equipment.

Building regulations and standards are also important practices of DSM. Building regulations has an impact on energy efficiency of buildings. Some of these regulations include the following:

- Passive lighting should be encouraged.
- Use of quality building construction materials to reduce heat losses.
- Roof spaces should be insulated.

It is also important as part of customer awareness that, equipment and buildings indicate their expected energy consumption. Customer education and labeling of appliance help customers make good judgment when buying appliances and equipment, thereby reducing energy consumption.

12.1.4.1.4 Energy Auditing

Energy auditing deals with analyzing and surveying the flow of energy to conserve energy in a building/plant. This can be done by using energy record keeping and measuring equipment like energy meters. Energy auditing is an essential component of energy management program, whereby a complete review of energy consumption activities, such as energy consumed in manufacturing and energy consumed in heating or cooling. Energy audits then collate all relevant data to carry out a detailed analysis of the performance of the building or system and to identify deficiencies and make recommendations for improvement.

Kumar et al. (Sameeullah et al., 2014) identified energy-saving methods experienced during an energy audit of a building in India and also gave a detailed explanation on how these methods could be implemented. This was done with the aim of showing the importance of energy audit in energy conservation.

Lamba and Sanghi (2015) gave a complete emphasis and explanation as to how energy audit is a continuous process toward achieving energy efficiency. They gave a detailed analysis on how to conserve and efficiently utilize scarce resources and how to identify and implement energy saving potentials through energy auditing. Opportunities using renewable energy technologies were also discussed.

The following steps explain how energy audit can be carried out in a building:

- Information on the processes employed, machinery characteristics, plant equipment and physical facilities, design data, and production capacities are obtained and collated.
- Historical records, the emissions, energy consumptions, and production levels for machines and energy consuming equipment (over a period of say 2–3 years) are determined.
- The actual operating parameters and performance of equipment and processes should be determined.
- The data obtained and the observations made should be properly established in efficiency of energy utilization by key equipment.
- Constraints to improving performance, including organizational, technical, and financial constraints should be identified and characterized.
- Potential measures for improvement should be identified and financial evaluation, where investment is needed should be carried out.
- A comprehensive action plan, which is very logical to address the constraints, should be developed, including specific recommendations and priorities for the different measures.

Audits can be divided into preliminary audit and detailed audit, which are both discussed, respectively.

1. Preliminary audit: This is an exercise in form of a fieldwork to gather initial data at the primary stages of the auditing program. It is also known as walk-through or short audit. In this type of survey, there is no need for the use of equipment because data from the building are sufficient. Data is collected via a "walk-through" of the building during which the general condition of equipment, the standard of maintenance, the level of operations control exercised by management, and the reporting procedures in effect are observed.

Dongellini et al. (2014) gave a good explanation on how "walk-through" audit is important in reducing energy consumptions for sustainable and energy-efficient manufacturing, continuous energy audit, and process tracking of industrial machines. This was carried out on eight large industrial buildings of a famous car manufacturing holding in Italy. Preliminary audit is very easy and less stressful as few measurement and easy calculations are used. It is therefore quick and can be completed within a short period. The information obtained from a preliminary audit is used for a thorough analysis of the energy performance of a building/plant.

2. Detailed Audit: This is a more comprehensive form of energy auditing. It deals with a survey on home energy by obtaining more detailed information on the home's energy usage, as well as a more proper financial analysis of its energy costs. Portable instruments are commonly used for accessing parameters on equipment and processes, the auditors must be well experienced with a good sense of judgment when collecting and interpreting data. Half of the effort put into a detailed audit should be spent on collecting data on-site, whereas the other half on proper analysis of the data and preparing the report. Detailed energy audit takes a longer time to complete, and it is used for a longer-term performance monitoring.

Singh et al. (2012) defined detailed audit as one, which provides a dynamic model of energy use characteristics of both existing facilities and all energy conservation measures identified, and therefore calibrates the building model against actual utility data to provide a realistic baseline and to compute operating savings for the proposed measures. They also presented a detailed energy audit, for design and implementation of a physically based model for industrial load management, thereby improving the plant efficiency and reducing the energy wastages.

Srinath and Uday Kumar (2014) explained detailed audit in the industrial sector as one requiring a comprehensive recording and analysis of energy consumption data, which is split into various sectors (steam/hot water production, compressed air, electricity, and heating, ventilation, and air conditioning). This is done to present and analyze different parameters that determine each type of energy use (e.g., production capacity, climate conditions, raw materials, and others). A list of potential energy—saving measures, which requires investment capital are presented together with a cost-benefit analysis for each proposed measure.

12.1.4.2 Load Management Program

Load management programs include: load leveling, load control, tariff incentives, and penalties.

12.1.4.2.1 Load Leveling

Load leveling optimizes the generating base load, operating at the present time without the need for reserve capacity to meet the periods of high demand. It therefore reduces fluctuations in customer demand. Classic forms of load leveling include the following:

- Peak clipping: Here the peaks of demand are clipped to reduce the load at peak times and intervals. The diagram of peak clipping is represented in the first diagram of Fig. 12.1.
- Valley filling: Here the low-demand periods are "filled" by building up off-peak capacities achieved by using thermal energy storage to displace fossil fuel loads. The diagram of valley filling is represented in the second diagram of Fig. 12.1.

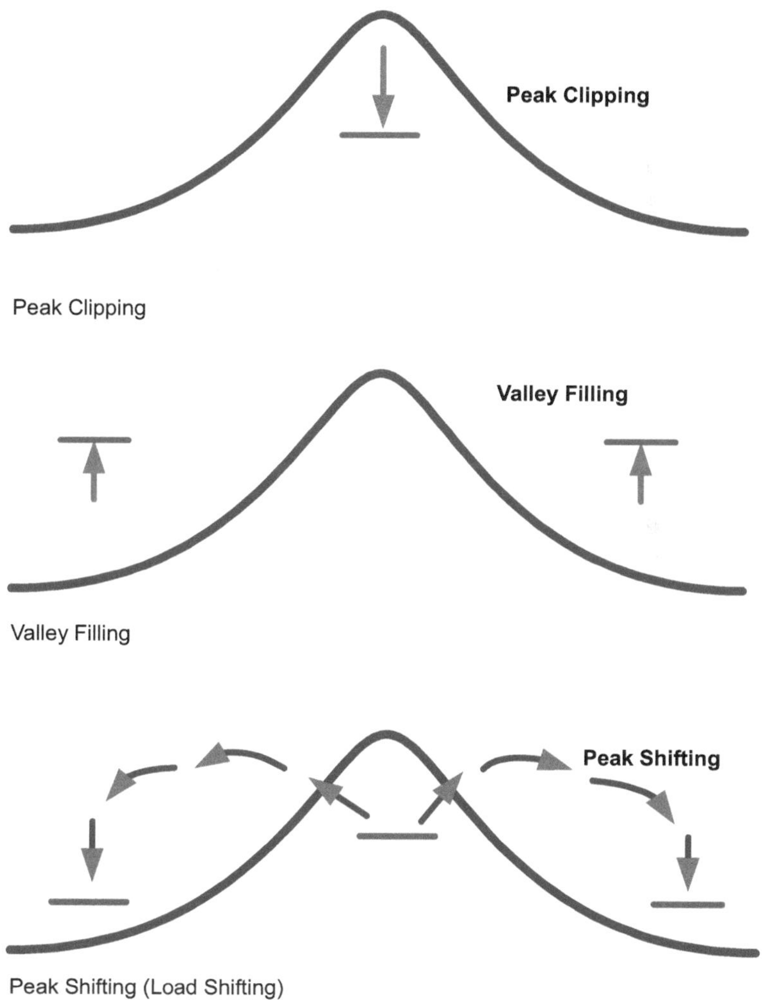

Figure 12.1 Classic forms of load leveling. *(From https://image.slidesharecdn.com/presentazionecordoba-160319093812/95/residential-demand-response-operation-in-a-microgrid-8—638.jpg?cb=1458380410.)*

- Load shifting: Here loads are "shifted" from high demand to low demand to achieve clipping and filling. Some applications of load shifting include: storage space heating, storage water heating, customer load shifting, and coolness storage. The diagram of load shifting is represented in the third diagram of Fig. 12.1.

12.1.4.2.2 Load Control

This is the automatic regulation of load because of heating, cooling, ventilation, and lighting by the utility or electricity company. Here, agreements are made between the customers and utility company to meet peak demand on the grid. This is done by customers using energy storage equipment and generators for backup. Electricity companies can organize a publicized schedule for a systematic switching off supply to different areas within a particular region at different intervals, whereby different areas take turns in losing supply. This is done to enable businesses and homes to plan their use of energy for that period.

12.1.4.2.3 Tariff Incentives and Penalties

Whenever customers use energy at some certain periods of time to ensure a better-priced rate for their energy use, the electricity companies encourages tariff incentives as a reward for these customers. These include the following:

- Providing different charges for power use at different periods, whereby high peak time charges encourage users to carry out their activities requiring high load in an off-peak period when the rates are lower. This is known as time-of-use rates.
- Penalizing users for going below a fixed power factor threshold, usually between 0.90 and 0.95. This is also known as power factor charges.
- Ensuring that, energy rates vary based on the energy provided by the electricity company. This is known as real-time pricing.

12.1.4.3 Load Growth and Conservation Programs

These programs are carried out with the aim of ensuring productivity and environmental compliance on the part of customers while making sure that, electricity companies increase their sale of energy. These programs can eliminate energy practices, which are unsustainable and bringing about more efficient practices such as: reducing the use of fossil fuels and also raw materials. Some of these programs include the following:

- *Strategic load growth:* This is the change of shape in load supplied to the customers by the electricity company which means that, there is a general increase in sales of electricity. This is represented in Fig. 12.2.

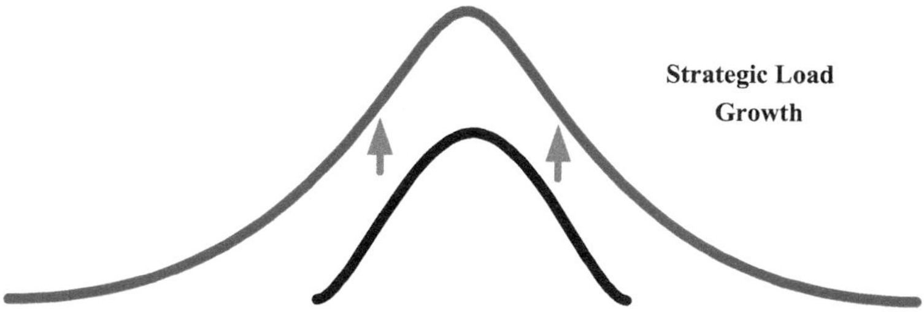

Figure 12.2 Diagram representing strategic load growth. *(From https://image.slidesharecdn.com/presentazionecordoba-160319093812/95/residential-demand-response-operation-in-a-microgrid-8—638.jpg?cb=1458380410.)*

Increase in market share of loads that are or can be served by competing fuels, as well as new areas development encourages load growth. Electrification involving electric technologies, industrial process, heating, and automation are modern day techniques for encouraging load growth.

- *Strategic load conservation:* These are programs targeted at the end use customers, whereby reducing sales as well as a change in the pattern of use is highly encouraged. Some of these programs include: insulation, sealing, and double-glazed windows (also known as weatherization). This is done for improvement in efficiency, in homes, and appliance. Strategic conservation is the change of shape in load supplied to the customers by the electricity company, which is directed at the end use consumption. This is represented in Fig. 12.3.

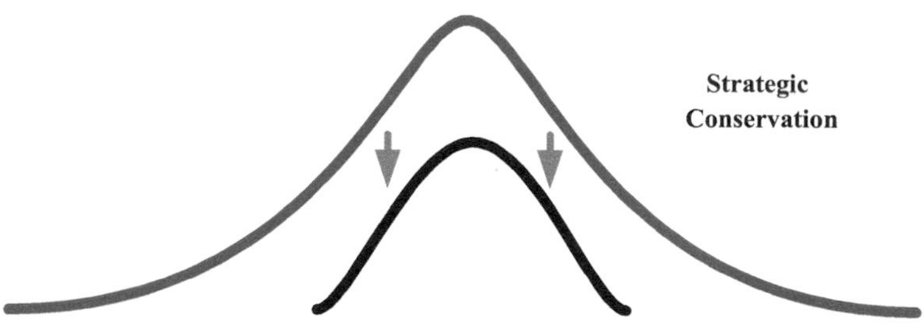

Figure 12.3 Diagram representing strategic load conservation. *(From https://image.slidesharecdn.com/presentazionecordoba-160319093812/95/residential-demand-response-operation-in-a-microgrid-8—638.jpg?cb=1458380410.)*

12.1.5 Information Dissemination on Demand-Side Management

This involves a collective effort from different stakeholders in developing, and implementing energy efficiency policies that are necessary for uninterrupted energy efficiency improvement. Information can be disseminated by awareness through campaigns promoting energy efficiency options and specific DSM techniques. This can be done through visual media (e.g., leaflets, brochures, fliers, posters, and video clips). Advertising and conducting energy audits are modern day forms of information dissemination.

12.1.6 Challenges of Implementing Demand-Side Management Programs

Challenges, ranging from insufficient provision, low awareness of energy, and efficiency/DSM programs to inability of industrial and commercial companies to carry out energy audits to obtain important information on their current operations are current trends affecting developing nations today. This can be as a result of failure by the company's management to explore the potential benefits of energy efficiency and also lack of skilled personnel for performing audits. Some measures that organizations can use in overcoming these challenges include the following:

- Ensuring that, competence and comprehensiveness of the assessment of DSM programs and audits are accomplished.
- Ensuring that, DSM systems and opportunities are well known and understood.
- Ensuring that, accuracy of assumptions is always considered.
- Ensuring that, there is proper awareness on production and safety constraints of involved plants/companies.

Factors that load management programs need to consider increasing energy efficiency include the following:

- Varying prices of electricity and other fuels.
- Cost burden on the customer.
- The value of losses prevented by ensuring an improved and reliable electricity system.
- Losses that can occur in production when implementing DSM programs.

Thorough financial analysis of the benefits of energy efficiency improvement needs to be undertaken when planning to set up DSM activities. Every DSM investments need to be properly accounted for to ensure easy assessment of funds for DSM projects because, if a project is not properly evaluated, approval of funds would be difficult, and this is the most vital challenge of implementing DSM projects.

12.1.7 Case Study—Energy Efficiency Opportunities for Rural Health-Care Facilities

Energy efficiency and conservation are a well-established practice in developed country hospitals and few developing countries. Energy efficiency measures have the tendency to reduce energy consumption as well as health facility's energy bill (Oluseyi et al., 2016). Apart from these, it can reduce emissions from conventional captive diesel/gasoline generators used in powering the facility and also allow backup systems to work more effectively during power outages or failure. Energy-efficient equipment is important for any facility, which is intended to operate off grid or on intermittent power sources or which must rely on battery storage.

Like in other power-consuming sectors, energy efficiency measures should be an essential factor in all existing and new health-care facilities in developing countries. Hence, the use of energy-efficient modern and functional medical equipment should be encouraged. This at times is usually a complex issue as agreement needs to be reached between medical personnel, design engineers, and other stakeholders. For example, in most developing countries with hot climate, there is usually the need for air-conditioning, which usually consumes significant amount of energy. Also, air-conditioning units are particularly sensitive to power fluctuations and, as such, are difficult to keep operational in developing countries. On the other hand, air-conditioned rooms are essential for sensitive laboratory equipment as well as sanitary purposes. In addition to sizing and designing, the operation of the health clinic can also affect its energy efficiency and power demand. The type of equipment purchased is important in reducing power consumption at health facilities. Energy star—rated equipment perform better in energy consumption when compared with conventional and obsolete equipment. All equipment and lighting connected to the storage device (backup) should be energy star—rated. If air-conditioning must be used, units with a high ratio of British thermal unit to watts should be procured. Training of both medical and maintenance staff on the importance of utilizing and ensuring efficient, low-energy operation of a health facility should also be ensured.

Typically, energy-efficient equipment and appliances are more expensive than standard-efficiency models. Conversely, this high cost may be

Table 12.1 Energy efficiency opportunities in health-care centers

Appliance	**Energy efficiency/conservation measures**
Lights	Lighting retrofits with energy-efficient light bulbs such as CFL or LED lamps. Switch off redundant lamps.
Vaccine refrigerator and ice pack freezer	Energy-efficient refrigerators should be encouraged because these refrigerators use 20% less energy.
Centrifuge	Turn off redundant centrifuge. Select centrifuge models that have relatively low rotor friction, which generates less heat, specifying smaller energy.
Autoclave	The equipment is fitted with a timer to cycle off when not in use. This will save maintenance by replacing the element as well as save energy.
Air-conditioner	Air-conditioner energy efficiency is measured by the energy-efficient ratio (EER) of the equipment. When buying new systems, buy the model with the highest EER rating. For operation, ensure coils are kept clean and that the thermostat is set at a comfortable, but not overly cool, temperature.
Computer	Energy star computers with flat screens and enabled with power management software can consume 40% less electricity than the conventional counterparts.
Printer	60% energy can be saved if printers that enable power management and duplexing features are used.

From http://www.poweringhealth.org/index.php/topics/technology/energyefficiency.

recovered through reduced operational costs of a smaller electricity generation system. Table 12.1 shows energy efficiency opportunities for some appliances/devices used in health clinics.

12.1.7.1 Need for Reliable Electricity in Rural Healthcare Centres

The need for a constant and efficient electrical power supply in our rural and local health-care centers all over the world is now a major concern. This is because of the adverse effect; the lack of a reliable electrical power system exposes the operators of these rural health-care systems and their patients, thereby posing a risk when saving human lives (Jimenez and Olson, 1999). Today, the distribution of vaccines and other temperature sensitive drugs and substances such as: antiretroviral drugs in our health-care centers globally. This has placed its own demand for electricity in health centers where there are limited or even no access to reliable power supply (Olatomiwa et al., 2016). The emergence of highly communicable diseases in modern time such as: the HIV/AIDS and Ebola virus, which were predominant in western Africa, and the Zika virus, which was and is still intense in South American countries, has necessitated the need for an efficient and reliable electricity power supply to fight these diseases to a standstill. For instance, unreliable power supply can hinder the treatment process of these diseases, if there are no refrigerators to store most of the vaccines, which are temperature sensitive, needed to fight these diseases. Also, a health personnel, attending to a patient or trying to take samples for medical tests in an enclosure or room without proper illumination stands the risk of being infected with these diseases easily as a result of poor illumination to comply with the laid down ethics and procedures for fighting the disease. The chances of carrying out the required medical tests could be impeded even during the day because electricity is required to power most medical equipment required to carry out these tests. Furthermore, a patient that arrives a health center that lacks electric power at night for medical attention will have to wait till the next day before proper examination can be carried out, which could be fatal.

Nigeria as a country has persistently been facing the challenge of lack of access to constant modern energy services over the years with the high dependence of its citizens on petroleum fuel for its individual electricity production. This challenge has its tremendous effect in other facets of the economy. In a country where about 40% of its citizens have access to electricity and 80% of this percentage lives in the urban areas (Nwulu and Agboola, 2011), this means that, most rural areas in Nigeria are not connected to national grid and therefore do not have access to electricity. These rural areas depend mostly on their traditionally domestic sources of energy for their livelihood. In a country with an electricity installed capacity of ~6538.3 MW, not more than 4500 MW is ever produced (Nwulu and Agboola, 2011), and with the ever growing demand for electricity all over the country in both urban cities and rural areas, especially in sectors that need critical attention such as: the health sector, there is the need to provide electricity that is reliable, efficient, and constant. As at present, Nigeria has not utilized or tapped fully the renewable energy technology to solve its electricity problems even though it is highly endowed with renewable energy. This form of energy will provide an affordable, reliable, efficient as well as constant source of electrical power, especially to critical sectors of the economy such as: health and in the rural areas that face utmost lack of these basic social amenities for the provision of good, reliable, and stable healthcare system.

Only a few know that, Nigeria is endowed with abundant renewable energy resources such as: solar energy, hydroelectric power resources, wind, biomass with potentials for hydrogen utilization and development of geothermal, and ocean energy (Ohumakin, 2010). As a result of the different resources of renewable energy, various approaches have been applied to determine and meet the optimal size of the system and components (Zhou et al., 2010; Bajpai and Dash, 2012). Renewable energy have been proved to provide quality and reliable electricity for different applications in rural areas (Dihrab and Sopian, 2010; Hiendro et al., 2013).

12.1.8 Methodology

12.1.8.1 Sizing Hybrid Renewable Energy Systems

12.1.8.1.1 Hybrid Optimization Model for Electric Renewables

A typical hybrid system will comprise a photovoltaic (PV) array, wind turbines, possibly a diesel generator, inverters, and battery bank with associated control devices (Lal and Raturi, 2012). Software tools such as Hybrid Optimization Model for Electric Renewables (HOMER) can be used to carry out simulation and optimization of hybrid systems. There are three main tasks that can be carried out by HOMER, which are simulation, optimization, and sensitivity analysis (HOMER Energy LLC: Hybrid optimization model for electric renewable, 2009). The HOMER is designed to perform simulation on long-term operation of a combination of micropower system configurations, with the inclusion of components like PV, wind, small hydro, diesel generators, and storage devices that can be battery banks. The HOMER can also be used for the modeling of grid-connected systems. After the simulation of quite a number of combinations, HOMER will make suggestions based on the net present cost (NPC) optimal configuration of the system. The sensitivity analysis can be performed using the sensitivity variables, i.e., wind, speed, and diesel price, which will enable the designer to determine the best combination of system components under different conditions.

12.1.8.1.2 Researches for Renewable Energy Access for Rural Healthcare Centres

Several researches have been carried out to study the hybrid systems in many locations around the world (Bekele and Palm, 2010; Akinbulire et al., 2014;

Hassan et al., 2011; Nandi and Ghosh, 2010; Rahman and Al-Hadhrami, 2010), Bekele and Palm researched on the feasibility study for a stand-alone solar wind based on hybrid energy system for application in Ethiopia and determined the optimal system for supplying electricity to a neighborhood of 200 families in Ethiopia and discovered that the price for diesel in 2009 was the most cost effective for the diesel generator/battery/converter set up (Bekele and Palm, 2010). The study of the optimization and life cycle cost of a hybrid system for a rural area health clinic in southern Iraq using HOMER software was carried out by Al-Karaghouli and Kazmerski. It was discovered that, the system when consisting the PV/battery inverter, was the most economic system (Al-Karaghouli and Kazmerski, 2010). The study of the technoeconomic evaluation of PV/diesel/battery systems for rural electrification in Saudi Arabia was carried out by Akinbulire et al. (2014). They researched on the effect of the increase in PV/battery on the cost of energy, operational hours of diesel generators, and the reduction in greenhouse gas (GHG) emissions. The study of the renewable energy technologies in the Maldives was performed by Van Alphen et al. (2007), and it was discovered that, 10% of the total electricity needed by Maldives can be supplied using renewable energy resources. Benghamem performed a study of the optimization of tilt angle for solar panel to increase the energy production, using Madinah, Saudi Arabia as case study (Benghanem, 2011) and revealed a gain of 8% at a tilt of solar collector in a particular angle on monthly basis when compared with tilting it annually. Besarati et al. (2013) studied the potential of harnessing solar radiation in different parts of Iran was analyzed and the results revealed that, there is greater potential for renewable energy application in the central and southern part of Iran.

12.1.8.2 Site Description

The sites used for the research were six towns selected across the six geopolitical zone/regions of Nigeria. The towns were selected across the six geopolitical zones to enable the study covers all the characteristics of the weather and climatic conditions of all the zones in Nigeria. Also, the towns selected are not the popular urban cities well known in Nigeria but are towns that are rural in nature and close to rural areas. The towns include: Okrika (south south zone), Ihiala (south east zone), Igbeti (south west zone), Kabo (north east zone), Abadan (North West zone), and Gboko (north central zone). Analysis was performed on these selected sites on the prospects of PV to power RHC.

12.1.8.3 Solar Resource

The NASA website provided the metrological data used in the analysis of this research work. The agency provided the average solar radiation for a stipulated range of time used for the analysis for the sake of the study.

12.1.8.4 Proposed Hybrid System

Based on the amount of solar resources present at these selected sites, the proposed hybrid system will be a combination of PV modules, diesel generators, power converters, and storage batteries. HOMER is the software to be used for this study. It is a general purpose system design tool for electric power systems. The software has several inputs, as well as component's technical and economic details (Olatomiwa et al., 2016). To arrive at an optimal value for system components, HOMER accepts more than one value to be entered to a tune of multiple values. Reliability and efficiency can be strengthened when the right merging of different sources of energy is done. The energy storage requirements are also reduced when compared with systems that have only one renewable energy source (Zhou et al., 2010; Olatomiwa et al., 2014). When solar energy resources and diesel generators are combined, a hybrid solar/diesel generator configuration is formed with an energy storage system to make the system a more reliable source of electrical power. The diesel generator in the hybrid system acts as a backup energy source, when there is redundancy or insufficient renewable energy resources.

The hybrid system design is better because there is more than one source of energy for alternation, and compliments the efforts of one another to achieve a reliable and efficient system in producing the much-needed electrical power for rural health centers, unlike the one source of renewable system that does not have these benefits. Also, in the hybrid system, different components need to be sized so as to achieve the required load demand, while in a single source of renewable system, the system to accommodate the load demand, the becomes oversized.

12.1.8.5 Data Requirements

To achieve the aim of this study, the data required for this analysis are categorized into three, namely: (1) metrological (solar irradiance), (2) estimated load, and (3) the component data. These are discussed in the next section.

12.1.8.6 Metrological Data

For the purpose of this study, a town each falling within a geopolitical zone in Nigeria is picked as a pilot study for the entire geopolitical zone.

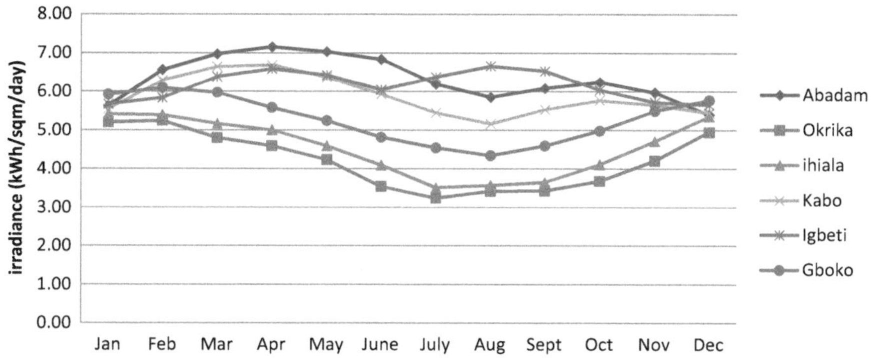

Figure 12.4 The solar irradiance for the chosen cities.

Fig. 12.4 shows the solar irradiance for the chosen cities as obtained from NASA's global satellite database. This serves as input into HOMER to ascertain if the load demand of the clinic can be met.

12.1.8.6.1 Load Data

12.1.8.6.1.1 Categorization of Health Clinics To adequately capture the energy demand of a health facility, it is essential to categorize health facilities based on size and energy consumption. This section describes classes of health facilities and their corresponding energy demand. The energy demands of a health facility will be a critical component in the selection of the most appropriate electrification technology. The descriptions provided are according to United States Agency for International Development documents available in open literature (http://www.poweringhealth.org/index.php/topics/technology/energyefficiency).

12.1.8.6.1.2 Health Post Health post is the smallest, most basic health facility in terms of size and services provided. These facilities usually do not have a permanent doctor or nurse. Most times, the health post may have a full- or part-time primary health-care provider and trained health officers. Services provided by health post facilities include: the treatment of minor illnesses, the nursing of minor injuries and, and the provision of basic immunization services. Because of the limited medical services provided, the medical equipments are few, and as a consequence, the overall energy consumption is moderately low. The energy demands of a health post will be satisfied through category of health clinic electrification options, while taking into account the reduced daily demand for energy.

12.1.8.6.1.3 Health Clinics Health clinics are generally larger than health posts and engage one or more full-time nurses. Clinics may also employ part-time medical doctors, depending on the services provided, size and location. Health clinics offer comprehensive category of services when compared with health post. Therefore, it is expected that it acquires more medical equipment so as to enable more sophisticated and advanced diagnoses. They can be subdivided into the three categories (Categories I, II, and III). Rural health clinics generally fall into one of the three categories, based on the type and number of medical devices used in the facility and the frequency with which they are used on a daily basis (http://www.poweringhealth.org/index.php/topics/technology/energyefficiency).

Category I Health Clinic (low-energy requirements, 5–10 kWh/day)

1. Typically located in a remote setting with limited services and a small staff.
2. Approximately 0–60 beds.
3. Electric power is required for:
   - lighting the facility during evening hours and to support limited surgical procedures (e.g., suturing)
   - maintaining the cold chain for vaccines, blood, and other medical supplies—one or two refrigerators may be used
   - utilizing basic lab equipment—a centrifuge, hematology mixer, microscope, incubator, and hand-powered aspirator.

Category II Health Clinic (moderate energy requirements, 10–20 kWh/day)

1. Approximately 60–120 beds.
2. Medical equipment similar to Category I Health Clinic; frequency of use and number of devices are key factors of differentiation between Category I and II health clinics.
3. Separate refrigerators may be used for food storage and cold chain.
4. Communication device, such as a radio, may be utilized.
5. May accommodate more sophisticated diagnostic medical equipment and perform more complex surgical procedures.

Category III Health Clinic (high energy requirements, 20–30 kWh/day)

1. Approximately 120 beds or more.
2. May serve as a regional referral center and coordinate communication between several smaller facilities and hospitals in large cities.
3. May need to communicate with remote health centers and hospitals by way of telephone, fax, computer, and Internet.

4. May contain sophisticated diagnostic devices (X-ray machine, CD4 counters, blood typing equipment, others) requiring additional power.

12.1.8.6.1.4 Energy Assessment of a Rural Health Center Because health-care facilities offer a comprehensive variety of health services, it is essential to specify energy needs based on equipment required for such services. In literature and practice, this has enjoyed limited attention (http://www.who.int/hia/green_economy/modern-energy-services/en/). Therefore, this area needs more attention by health and energy professional at the international, regional and national level, and grassroots. Nevertheless, infrastructural studies such as Service Availability and Readiness Assessment (SARA) itemized a detailed list of medical equipment accessible in health facilities, and the outcome of these reviews can be adopted as a basis for evaluating prevailing energy needs. In the survey, SARA classified health services, essential electric equipment and its indicative power requirements as: (1) infrastructure (lighting, communication, water supply, and waste management); (2) medical devices; and (3) support appliances for specific health services, which include vaccination, infectious and non-communicable disease treatment, emergency care such as blood transfusions and surgical services. An extract of this is presented in Table 12.2.

12.1.8.6.1.5 Criteria for Selection of the Appropriate Energy System for the Load For the selection of appropriate hybrid power system for health-care facilities, a list of factors needs to be considered. These include the following:

- *Peak power capacity*: This is defined as the maximum capacity the system can supply.

  Different appliances consume different amount of electricity, which ranges from a few watts to several kilowatts. Similarly, the various power-generating systems generate different quantities of electricity. The generated electricity should adequately meet the expected electrical load demand. Hence, the use of energy-efficient equipment is of great emphasis, so as to reduce the available peak power needs of the health-care facilities.
- *Daily energy capacity*: Generally, energy use by various appliances in health-care facility could be classified as continuous (e.g., refrigerators and space heating) and intermittent such as in laboratory equipment. The intermittent loads are load that are operated periodically, either in the evening or early morning (radio, laptop, and others) or some

Table 12.2 Health equipment by power load level

Health services	Electrical devices	Indicative power rating (W)	AC power supply	DC power supply
Infrastructure				
Basic amenities	Lightings:			
	Incandescent bulb	10.8 W/m^2	110/220 V	—
	Halogen bulb	1.8 W/m^2	110/220 V	12 V
	CFL bulb	2.16 W/m^2	110/220 V	—
	LED bulb	1.8–2.14 W/m^2	110/220 V	10–30 V
	Security lighting/outdoors CFL/LED	40–160 W	110/220 V	10–30 V
	Laptop desktop incandescent	200–600 W	110/220 V	10–30 V
		20–60 W	110/220 V	12–20 V
	Mobile phone battery (charging)	5–20 W	110/220 V	5–16.5 V
	Desktop computer	15–200 W	110/220 V	8–20 V
	Printer (ink jet)	65–100 W	110/220 V	12–20 V
	VHF radio receiver: standby	2 W	110/220 V	12 V
	transmitting	30 W	110/220 V	12 V
	Ceiling fan	50–100 W	110/220 V	—
	Refrigerator (for food and water)	150–200 W	110/220 V	—
	Portable air-conditioner (AC)	1000–1500 W	110/220 V	48 V

Specific services				
General outpatient services	Nebulizer	80–90 W	110/220 V	–
	Oxygen concentrator	270–310 W	110/220 V	–
	Pulse oximeter	50 W	110/220 V	–
Antenatal child and adolescent health	Vaccine refrigerator	60–115 W	110/220 V	N/A
Obstetric and new born	LED lighting (phototherapy)	440 W	110/220 V	–
	Suction apparatus	90–200 W	110/220 V	–
	Vacuum aspirator	36–96 W	110/220 V	3–6 V
	Neonatal incubator	800–1035 W	110/220 V	–
	Ultrasound	800–1000 W	110/220 V	–
General diagnostics, blood analysis, and laboratory equipment	Laboratory refrigerator	60–160 W	110/220 V	–
	Centrifuge	250–400 W	110/220 V	–
	Hematology analyzer	230–400 W	110/220 V	–
	Blood chemistry analyzer	45–88 W	110/220 V	–
	CD4 counter	200 W	110/220 V	12 V
	Microscope (with LED light)	20–30 W	110/220 V	3–6 V
	X-ray machine (portable)	3–4 kW	110/220 V	–
	Laboratory incubator	200 W	110/220 V	12 V
Basic surgical services	Suction apparatus	90–200 W	110/220 V	–
	Anesthesia machine	1440 W	110/220 V	

From http://www.who.int/hia/green_economy/modern-energy-services/en/.

medical equipment operated once or twice a day (e.g., autoclaves for instrument sterilization). The breakdown for expected daily energy capacity needed for the health-care facilities considered for this chapter is presented in Table 12.3. This is an extract from the SARA requirement (Table 12.2). The proposed energy system must be capable of supplying the daily energy requirements of a typical rural health clinic.

- *Evening peak hours' supply*: In a typical RHC, some patients, including pregnant women may show up for treatment in the evening time. To avoid or minimize wasting time during their daytime jobs. In case of first aid emergencies, the availability of power supply in health-care facility in the evening and during the night hours is considered imperative for quality health-care delivery, especially when there are no referral clinics nearby. In RHC that have enough personnel to do night duties, electric power supply in the evening is vital to enable them feel secured and improve their productivity during night shifts. Therefore, the power supply is designed to suit the expected nighttime load demand.
- *Duration of supply*: The duration of power supply to health-care facility is dependent on size as well as types of health-care services rendered. It is expected that, electricity supply be available to power all equipment required to enhance quality medical services delivery at all times RHCs, with maternal health care and childbirth services or other emergency services are expected to have power supply for basic applications. Some equipment, such as laboratory equipment, for testing of samples collected during the day and analyzed in a batch is used intermittently. However, other appliances such as of refrigerators are run as base load and may require continuous electricity supply.
- *Reliability:* Reliability in this context deals with the probability of power failure because of unmet load. Unmet load is electrical load that the power system is unable to serve. It occurs when the electrical demand exceeds the supply. Because of the sensitive nature of services rendered by RHCs during power disruptions, health facility is expected to be provided with backup power generators. This will improve the reliability of the system and as such, reduce the loss of load. Backup generators may not provide sufficient electricity for all the equipment; nevertheless, it should serve as support to priority applications such as: lighting, refrigerators, and other critical equipments. Consequently, reliability of the power supply is given significant attention, while sizing and designing the power system.

Table 12.3 Clinic appliances and their power ratings

S/ no	Power consumption	Qty	Power (CME) watts	Power (EEME) watts	Total (watts)	Total (watts)	Daytime hours 7 a.m.–6 p.m.	Evening hours (6–10p.m.)	Night hours (10p.m.–7a.m.)	Total hours/day	Total energy (CME) (kWh/day)	Total energy (EEME) (kWh/day)
1	Refrigerator-vaccine (RV)	1	60	40	60	40	5	3	2	10	0.6	0.4
2	Refrigerator-nonmedical (RNM)	1	300	125	300	125	2	2	1	5	1.5	0.625
3	Centrifuge	2	242	242	484	484	4			4	1.936	1.936
4	Microscope	2	20	20	40	40	6			6	0.24	0.24
5	Blood chemical analyzer (BCA)	1	88	45	88	45	4			4	0.352	0.18
6	Hematology analyzer (HA)	1	230	230	230	230	4			4	0.92	0.92
7	CD4 Machine	1	200	200	200	200	4			4	0.8	0.8
8	Radio	1	30	15	30	15	10			10	0.3	0.15
9	Tube fluorescent lights (TFL)	4	40	18	160	72	8			8	1.28	0.576
10	Wall fan	5	65	65	325	325	8			8	2.6	2.6
11	Halogen lamp (security)	1	100	50	100	50	0	3	8	11	1.1	0.55
12	Desktop computer	1	230	65	230	65	4			4	0.92	0.26
	Total energy (kWh/day)										12.548	9.237
	% Energy reduction										26%	
	Oil equivalent of energy saved										0.0003	tonnes
	CO_2 emission reduction										4.3043	kg/day

CME, Conventional medical equipment; *EEME*, energy efficient medical equipment.

- *Environmental health and sustainability*: It is expected that, energy supply to any health-care facility should not lead to environmental hazards. Environmental impart such as air pollution, water pollution, and noise pollution in any quantities that could further deteriorate the health of patients, affect the medical staff or people living at the health-facility premises, should be avoided. This can be achieved by making sure high penetration of renewable energy sources. This will ensure high value of renewable energy fraction. Consequently, reduction of GHG emission per kWh of power generated will lead to environmental sustainability. To this aim more priority should be given renewable energy sources than captive gasoline or diesel generator source in providing power supply to the rural health-care facilities.

A 24-h energy demand is required, so as to determine the sizes of the components. Because of the nonavailability of a typical load demand for RHC in Nigeria, load profile based on SARA as indicated by UNICEF for a standard rural health center is developed for the purpose of this study. This was centered on the estimates of the medical equipment, power ratings at different times of the day and night. It is assumed that, the power will vary across the hour of the day, and that the clinic is only operational between 8a.m. and 5p.m. daily. Essential appliances such as refrigerator, operated intermittently during the night. Two kinds of medical devices and appliances are modeled for the load-conventional (inefficient) and energy-efficient equipment. The existing devices in most RHC visited are conventional and obsolete technologies. The various appliances used in this study and the presumed distribution of the daily energy consumption by appliance are shown in Table 12.3 and Fig. 12.5A and B, respectively. The fan and the centrifuge consume bulk of the power during the daytime.

As shown in Fig. 12.6, the peak load of 1.557 kW/day and 1.411 kW/day occur for both types on load between 12p.m. and 2p.m. To model the change in pattern of electricity usage during the year, a day-to-day variability of 10% and hour-to-hour random variability of 15% was specified in HOMER. The clinic utilizing conventional equipment will require 12.55 kWh/day to run its services, whereas the clinic deploying energy-efficient appliance and equipment will require 9.24 kW/day-26% less than the former. The oil equivalent of energy saved and the CO_2 avoided with use of energy efficient medical equipment (EEME) as against conventional medical equipment (CME) is 0.0003 tonnes/day and 4.3043 kg/day. The hourly load distribution analysis for pre-and post-DSM is given in Tables 12.4 and 12.5.

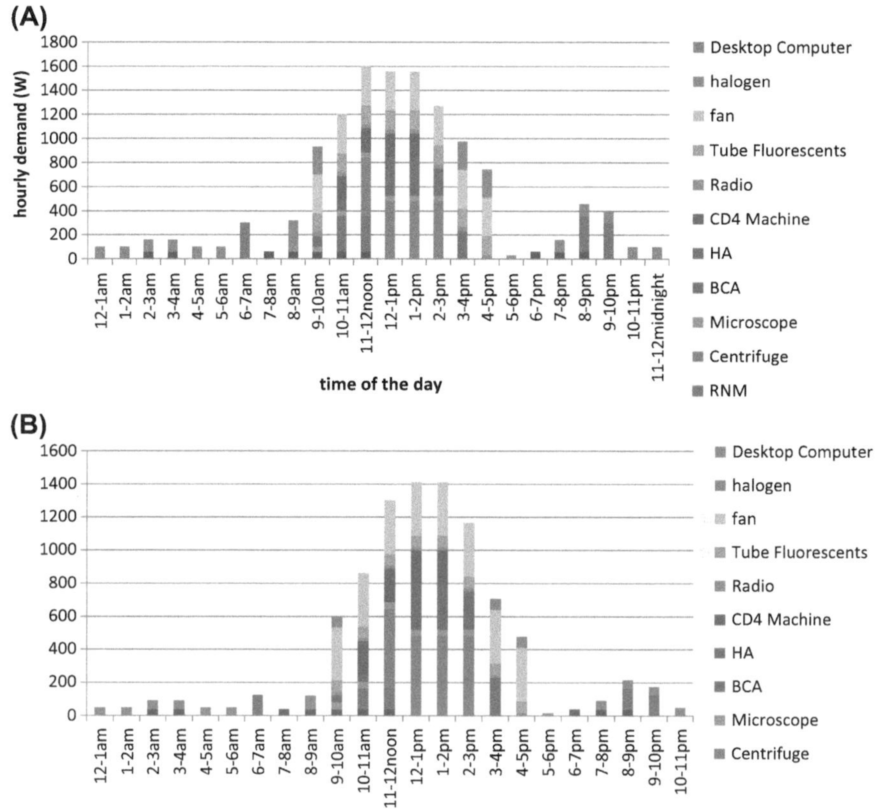

Figure 12.5 (A) Daily energy demand contributions of conventional appliances. (B) Daily energy demand contributions of energy-efficient appliances. *BCA*, blood chemical analyzer; *HA*, Hematology analyzer; *RNM*, Refrigerator-nonmedical.

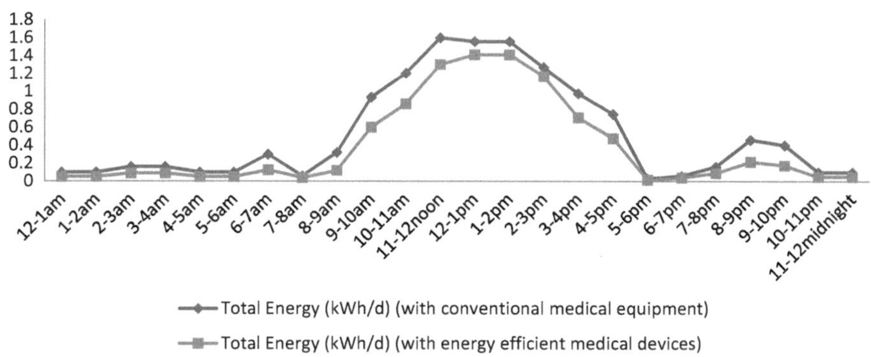

Figure 12.6 Electric load profile for both scenarios (conventional and energy-efficient devices).

Table 12.4 Hourly distribution of load demands without demand-side management

Time of the day	RV	RNM	Centrifuge	Microscope	BCA	HA	CD4 machine	Radio	Tube fluorescent	Fan	Halogen	Desktop computer	Total energy (Wh/d)
12–1a.m.											100		100
1–2a.m.											100		100
2–3a.m.	60										100		160
3–4a.m.	60										100		160
4–5a.m.											100		100
5–6a.m.											100		100
6–7a.m.		300											300
7–8a.m.	60												60
8–9a.m.	60							30				230	320
9–10a.m.	60			40	88			30	160	325		230	933
10-11a.m.	60	300		40	88		200	30	160	325			1203
11–12noon	60	300	484	40			200	30	160	325			1599
12–1p.m.			484	40	88	230	200	30	160	325			1557
1–2p.m.			484	40	88	230	200	30	160	325			1557
2–3p.m.			484	40		230		30	160	325			1269
3–4p.m.						230		30	160	325		230	975
4–5p.m.								30	160	325		230	745
5–6p.m.								30					30
6–7p.m.	60												60
7–8p.m.	60										100		160
8–9p.m.	60	300									100		460
9–10p.m.		300									100		400
10–11p.m.											100		100
11–12midnight											100		100
Total	600	1500	1936	240	352	920	800	300	1280	2600	1100	920	12548

BCA, blood chemical analyzer; *HA*, Hematology analyzer; *RNM*, Refrigerator-nonmedical; *RV*, refrigerator-vaccine.

Table 12.5 Hourly distribution of load demands with demand-side management

Time of the day	RV	RNM	Centrifuge	Microscope	BCA	HA	CD4 machine	Radio	Tube fluorescent	Fan	Halogen	Desktop computer	Total energy (Wh/d)
12–1a.m.											50		50
1–2a.m.											50		50
2–3a.m.	40										50		90
3–4a.m.	40										50		90
4–5a.m.											50		50
5–6a.m.											50		50
6–7a.m.		125											125
7–8a.m.	40												40
8–9a.m.	40							15				65	120
9–10a.m.	40			40	45			15	72	325		65	602
10–11a.m.	40	125		40	45		200	15	72	325			862
11–12noon	40	125	484	40			200	15	72	325			1301
12–1p.m.			484	40	45	230	200	15	72	325			1411
1–2p.m.			484	40	45	230	200	15	72	325			1411
2–3p.m.			484	40		230		15	72	325			1166
3–4p.m.						230		15	72	325		65	707
4–5p.m.								15	72	325		65	477
5–6p.m.								15					15
6–7p.m.	40												40
7–8p.m.	40										50		90
8–9p.m.	40	125									50		215
9–10p.m.		125									50		175
10–11p.m.											50		50
11–12midnight											50		50
Total	400	625	1936	240	180	920	800	150	576	2600	550	260	9237

BCA, blood chemical analyzer; *HA*, Hematology analyzer; *RNM*, Refrigerator-nonmedical; *RV*, refrigerator-vaccine.

Table 12.6 Assumption regarding component sizing and cost

Component parameter	Value	Component parameter	Value
PV		**Converter**	
Rated capacity	1 kW	Rated power	1 kW
Derating factor	80%	Efficiency	90%
Capital cost	\$4250	Capital cost	\$621.80
replacement cost	\$4200	Replacement cost	\$569
Operational life	20 years	O & M cost	\$3/year
Ground reflectance	20%	Operational life	15 years
Battery		**Diesel**	
Rating	4 V, 1900 Ah	Rated power	1 kW
Round-trip efficiency	85%	Capital cost	\$280
Min. state of charge	40%	Replacement cost	\$280
Capital cost	\$269	O & M cost	0.5\$/h
Replacement cost	\$260	Operational life	15000 h
O & M cost	\$5/year	Minimum load ratio	30%
Operational life	4 years		

O & M cost, operation and maintenance cost; *PV*, photovoltaic.

12.1.8.6.1.6 Component Data The initial selections of components are based on the load profile in Fig. 12.6. The details of components and other assumption regarding them are specified in Table 12.6. Prices are specified in US dollar. HOMER models two types of dispatch strategies—"load following," where the generator supplies exactly enough power to serve the load in the event of insufficient renewable energy and the "cycle charging," where the generators run at full load and excess electricity charges the batteries. In this study, both dispatch strategies were specified.

12.1.8.7 Emission Factors

An emission factor is a characteristic value that relates to the quantity of a pollutant released into the atmosphere with an activity associated with the release of that pollutant. These factors are usually calculated by dividing the weight of a pollutant by a unit weight, volume, distance, or interval of the activity emitting the pollutant. These factors are used to calculate emissions from different sources of air pollution.

These factors are averages of all reliable available data and are assumed to be typical for long-term means of the source category. The emission factor is used to compute the total emission from a source. Table 12.7 shows the state-by-state average emission factors for all the pollutants that HOMER

Table 12.7 Average emissions factors for the year 2000 for each us state

		Average grid emissions factors					Average grid emissions factors					Average grid emissions factors		
Serial number	State	CO_2 g/kWh	SO_2 g/kWh	NO_x g/kWh	SN	State	CO_2 g/kWh	SO_2 g/kWh	NO_x g/kWh	SN	State	CO_2 g/kWh	SO_2 g/kWh	NO_x g/kWh
1	Alabama	656	3.76	1.38	18	Kentucky	1011	5.7	2.41	35	North Dakota	1086	4.41	2.27
2	Alaska	586	0.61	2.21	19	Louisiana	629	1.6	1.15	36	Ohio	836	7.5	2.33
3	Arizona	533	0.74	1.06	20	Maine	297	0.96	0.65	37	Oklahoma	833	1.57	1.66
4	Arkansas	659	1.57	1.1	21	Maryland	623	4.58	1.58	38	Oregon	149	0.26	0.25
5	California	287	0.08	0.26	22	Massachusetts	587	2.53	0.91	39	Pennsylvania	560	4.32	1.23
6	Colorado	913	1.85	1.59	23	Michigan	710	3.2	1.52	40	Rhode Island	454	0.09	0.24
7	Connecticut	335	1.02	0.62	24	Minnesota	744	2.26	1.78	41	South Carolina	405	2	0.89
8	Delaware	894	5.93	1.59	25	Mississippi	597	3.2	1.55	42	South Dakota	378	1.25	1.61
9	District of Columbia	1205	6.18	2.36	26	Missouri	898	2.79	1.96	43	Tennessee	621	3.96	1.51
10	Florida	644	2.74	1.53	27	Montana	662	0.79	1.29	44	Texas	666	1.38	1.05
11	Georgia	641	3.87	1.43	28	Nebraska	702	1.95	1.42	45	Utah	950	0.71	2.01
12	Hawaii	779	2.08	2.38	29	Nevada	704	1.35	1.34	46	Vermont	26	0.02	0.14
13	Idaho	42	0.04	0.07	30	New Hampshire	321	3.12	0.64	47	Virginia	559	2.64	1.17
14	Illinois	503	2.24	1.23	31	New Jersey	332	0.96	0.62	48	Washington	130	0.72	0.25
15	Indiana	977	6.01	2.37	32	New Mexico	969	1.83	2.32	49	West Virginia	920	5.84	2.62
16	Iowa	894	3.13	1.84	33	New York	444	1.88	0.66	50	Wisconsin	799	3.01	1.7
17	Kansas	848	2.36	1.93	34	North Carolina	586	3.45	1.33	51	Wyoming	1044	1.69	1.84

From https://www.epa.gov/energy.

models. Prior to simulating the energy system, HOMER defines the emission factors (kg of pollutant emitted divided by the unit of fuel consumed) for each pollutant. Subsequently, it computes the annual emission of that pollutant by multiplying the emission factors by the total annual fuel consumption. Emissions factors for four of the six pollutants: carbon monoxide, unburned hydrocarbons, particulate matter, and nitrogen oxides are directly specified. Using these values and the carbon and sulfur content of the fuel, HOMER estimates the emissions factors for the two-remaining pollutants: carbon dioxide and sulfur dioxide. To achieve these, HOMER uses three primary assumptions following:

- Any carbon in the fuel that does not get emitted as carbon monoxide or unburned hydrocarbons get emitted as carbon dioxide.
- The carbon fraction of the unburned hydrocarbon emission is the same as that of the fuel.
- Any sulfur in the burned fuel that does not get emitted as particulate matter gets emitted as sulfur dioxide.

Eq. (12.1) is generally used in the estimation of the emission.

$$E = A_r \times E_f \times \left(1 - \frac{EF_\eta}{100}\right) \tag{12.1}$$

where E = emissions; A_r = activity rate; E_f = emission factor; and EF_η = overall emission reduction efficiency (%).

12.1.8.8 Results and Discussion

Systems constraints along with input parameters described in Table 12.3 were used in the simulations of the PV battery system to obtain the optimal configuration. Optimal system configuration is obtained by choosing suitable system components, depending on parameters such as: solar irradiance, diesel price, and maximum capacity shortage. The most viable system depends on the total net present cost (TNPC), as well as the hourly performance. This section presents the result obtained from the simulations, both in graphical and tabular forms.

12.1.9 System Configuration

Displayed in Table 12.8 is the optimal system architecture for the selected geopolitical zones, with both CME and EEME. It is worth noting that, during component sizing and optimization, HOMER priorities continuous satisfaction of load. It also makes sure, there is minimum diesel consumption, and the PV array is not oversized. In Table 12.8, for both case of

Table 12.8 System configuration for the six geopolitical zones

City	**Gboko**		**Igbeti**		**Kabo**	
Load type	**EEME**	**CME**	**EEME**	**CME**	**EEME**	**CME**
PV array (kW)	3	3	3	3	3	3
Generator (kW)	2	2	1	2	2	2
Number of batteries	2	3	2	3	2	3
Converter (kW)	2	2	2	2	2	2
Dispatch strategy	LF	CC	CC	CC	LF	CC
	Ihiala		**Abadam**		**Okrika**	
PV array (kW)	3	3	3	3	3	3
Generator (kW)	2	2	1	2	2	2
Number of batteries	3	3	2	3	3	3
Converter (kW)	2	2	2	2	2	2
Dispatch strategy	LF	LF	CC	CC	LF	LF

CC, Cycle charging; *CME*, conventional medical equipment; *EEME*, energy efficient medical equipment; *LF*, load following; *PV*, photovoltaic.

CME and EEME, the PV array size across the sampled locations is the same (3 kW). Except for Gboko and Abadam with a smaller size of diesel generator of 1 kW for EEME loads, all other locations require a 2 kW generator for both EEME and CME loads. For all locations, the number of batteries required with the use of CME is three each, whereas with the use of CME; all locations returned two batteries each except for Ihiala and Okrika with three batteries each. The latter is because of the fact that, Ihiala and Okrika, receives the lowest average solar irradiance. The size of the converter for both categories of medical equipment for all locations is 2 kW.

12.1.10 Economic Analysis

Table 12.9 shows the economic analysis of the optimal configuration of the system in the six geopolitical zones. It gives the breakdown of the initial capital cost (ICC), operating cost, TNPC, and the levelized cost of energy (LCOE). The percentage reduction that accrues with the use of EEME instead of CME is also shown in Table 12.9. With 17% reduction, Igbeti and Abadam had the highest percentage reduction in TNPC and operating cost. Gboko recorded the least percentage difference of 15%. If installed, clinics in Igbeti will have the highest TNPC of $17,611 and $20,948 for EEME and CME, respectively. The initial cost of system installation using CME is the same across all locations that were considered ($15,252),

Table 12.9 Economic analysis of optimal system architecture

City	Gboko			Igbeti			Kabo		
Load type	**EEME**	**CME**	**% Reduction**	**EEME**	**CME**	**% Reduction**	**EEME**	**CME**	**% Reduction**
Initial capital cost	14,983	15,252	2%	14,757	15,252	3%	14,983	15,252	2%
Operating cost	262	628	58%	198	584	66%	213	599	64%
Total net present cost	16,937	19,940	15%	16,238	19,611	17%	16,573	19,724	16%
Levelized cost of energy	0.672	0.585	−15%	0.645	0.575	−12%	0.658	0.579	−14%
	Ihiala			**Abadam**			**Okrika**		
Initial capital cost	15,252	15,252	0%	14,757	15,252	3%	15,252	15,252	0%
Operating cost	290	728	60%	200	585	66%	316	763	59%
Total net present cost	17,414	20,690	16%	16,249	19,621	17%	17,611	20,948	16%
Levelized cost of energy	0.691	0.607	−14%	0.645	0.576	−12%	0.699	0.615	−14%

CME, Conventional medical equipment; *EEME*, energy efficient medical equipment.

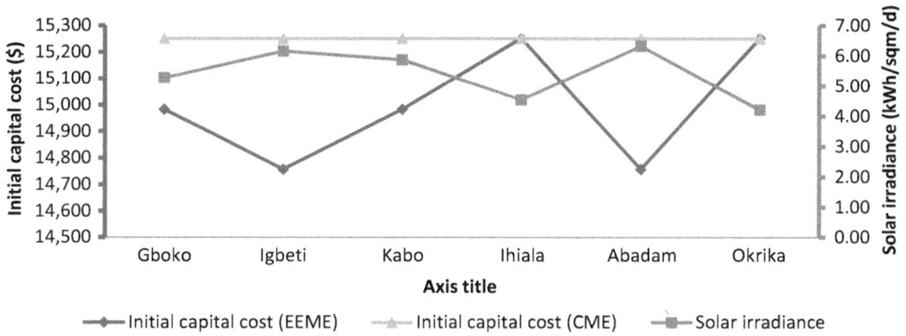

Figure 12.7 Relationship between initial capital cost for both energy efficient medical equipment (EEME) and conventional medical equipment (CME) and solar irradiance.

whereas that of the EEME varies directly. In the solar irradiance as could be seen in Fig. 12.7. At higher solar irradiance, the ICC is lower and high at lower, solar irradiance. The LCOE was higher for PV—DG battery system with EEME loads and lower for CME. In all scenarios, all capital cost for EEME and CME are close with just a maximum of 3% difference.

Fig. 12.8 shows the monthly contributions of the PV and diesel generator for the six locations across the six geopolitical zones. The contribution pattern of the PV array follows the same pattern for both types of loads across all sites except Igbeti. Generally, PV contribution falls to its lowest in the month of August, except in Igbeti, which experiences its lowest in the months of April and August. It can also be observed that, the diesel generator contribution peaks when the PV contribution is at its lowest across all locations.

Table 12.10 shows the breakdown of annual electricity production across all sites. Okrika recorded the highest annual electricity production of 4611 kWh/years and 5468 kWh/years, for EEME and CME, respectively. With the use of EEME, Okrika has the highest electricity production of 680 kWh/years, whereas Gboko produce the highest excess electricity of 1360 kWh/years with the use of CME. Unmet load and shortage capacity is mostly low and negligible as compared to the load served. The highest renewable energy fraction of 91.8% and 73.2% can be observed in Kabo. This is expected because it receives the highest solar irradiance out of all the locations considered. Okrika has the lowest renewable faction of 86.3% and 71% for EEME and CME, respectively. This location receives the lowest average solar irradiance. Across all locations, the value of the renewable fraction for EEME is higher that, CME with the highest increment of 28% in Kabo and least in Ihiala (21%).

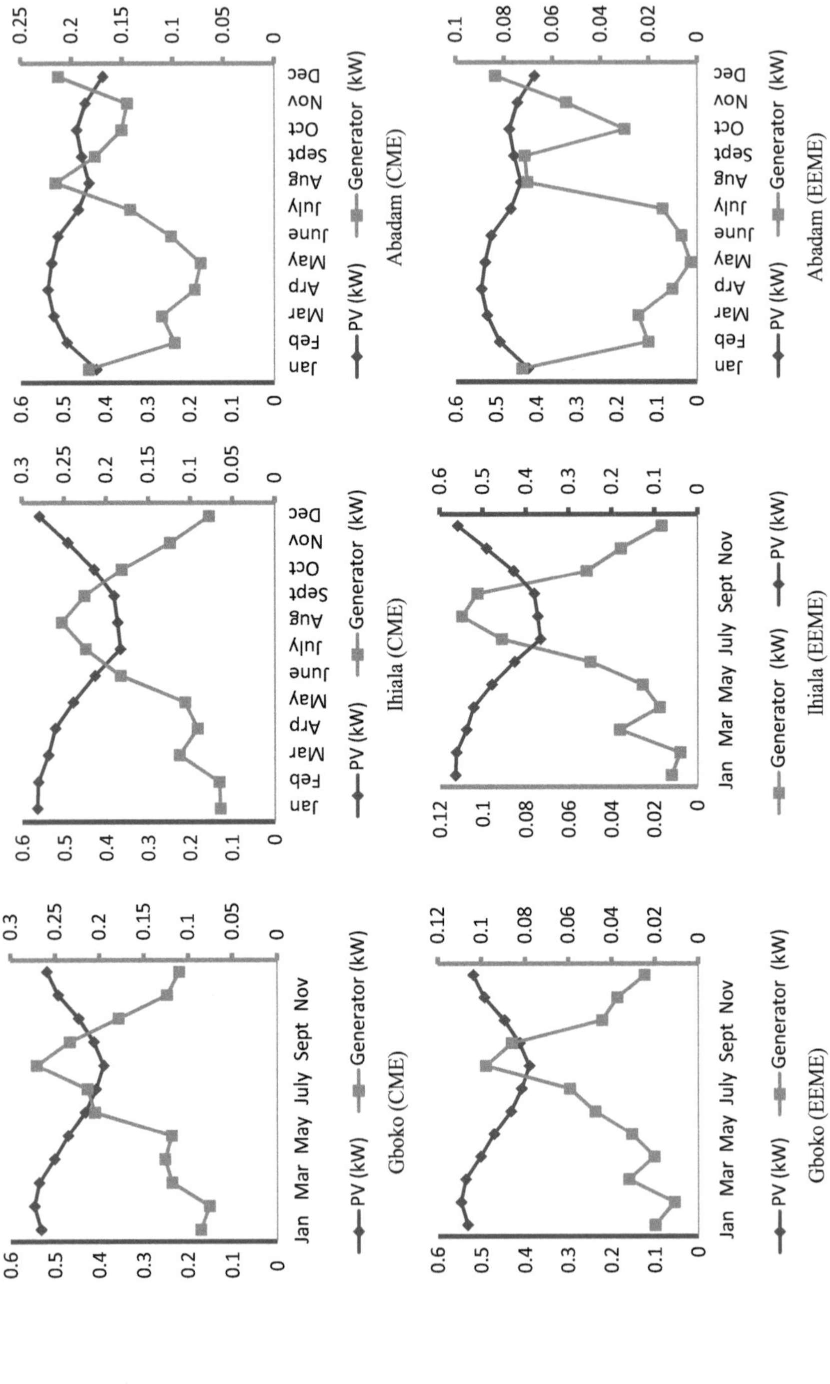
PV (kW)
Generator (kW)
Jan Mar May July Sept Nov
Jan Feb Mar Arp May June July Aug Sept Oct Nov Dec
Gboko (CME)
Ihiala (CME)
Abadam (CME)
Gboko (EEME)
Ihiala (EEME)
Abadam (EEME)

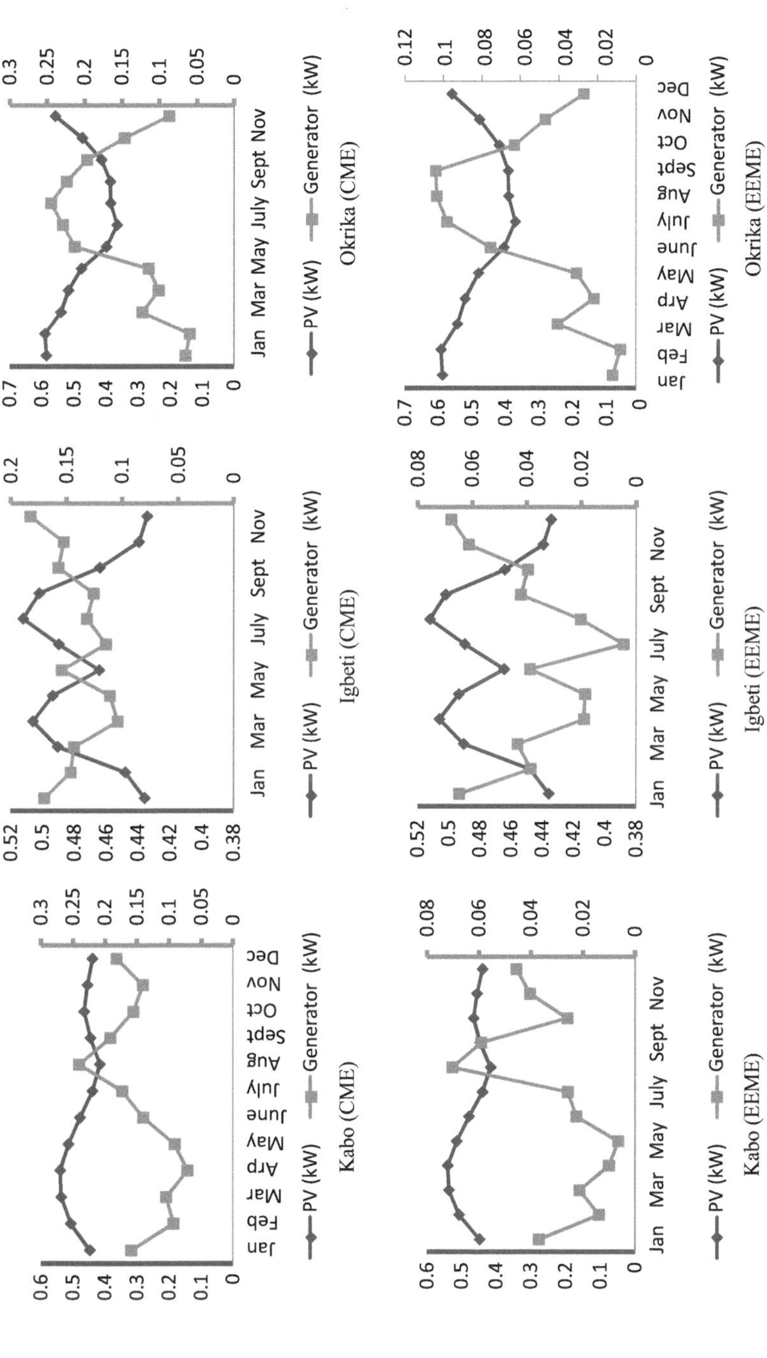

Figure 12.8 Monthly average electricity production. *CME*, Conventional medical equipment; *EEME*, energy efficient medical equipment; *PV*, photovoltaic.

Table 12.10 Energy production for optimal system

City	Gboko		Igbeti		Kabo	
Load type	**EEME**	**CME**	**EEME**	**CME**	**EEME**	**CME**
Annual electricity Production (kWh/year)	4522	4148	4482	5383	4425	5425
Excess electricity (kWh/year)	613	1360	545	36	518	53.6
Unmet load (kWh/year)	0.00000775	2.2	0.0427	0.000012	0.00000696	0.000011
Shortage capacity (kWh/year)	0.0185	0	1.33	0	0	0
Renewable fraction	0.889	0.702	0.901	0.729	0.918	0.732
	Ihiala		**Abadam**		**Okrika**	
Annual electricity production (kWh/year)	4556	5399	4485	5383	4611	5468
Excess electricity (kWh/year)	626	147	548	34.1	680	218
Unmet load kWh/year	0.00000771	0.49	1.24	0.0000128	0.00000753	0.598
Shortage capacity (kWh/year)	0.108	3.27	2.95	0	0.118	3.97
Renewable fraction	0.879	0.726	0.9	0.729	0.863	0.71

CME, Conventional medical equipment; *EEME*, energy efficient medical equipment.

12.1.11 Emission Results

The environmental benefit analysis of the proposed hybrid photovoltaic system is first considered in terms of the amount of fossil fuel consumed by the generator. This information is then employed to estimate the quantity of emissions produced by the generator. The emissions by the generator are assumed to be the quantity of emissions avoided by the solar PV system. As indicated in (U. S, 1998), various values of CO_2 emission factors are reported in the literature; 3.2, 3.15, and 3.0 kg CO_2/L of diesel, respectively. This work adopted an emission factor of 2.66 kg CO_2/L of diesel to evaluate the quantity of carbon dioxide. All other factors used by homer is given in Appendices A.1. All emissions are estimated by HOMER software and results are given in Table 12.11. Emissions are directly proportional to the quantity of diesel consumed by the generator. The diesel consumption will depend on the annual running hours, total load severed annually, which also depends on whether the solar resource can adequately meet the load or not. Table 12.12 shows the diesel consumption reduction that is achievable with the use of EEME as against CME. Fig. 12.9A and B show the relationship between CO_2 emission and diesel consumption for both CME and EEME. Ihiala consumed the highest liter of fuel and consequently the highest level of emission, whereas as expected, Kabo had the least fuel consumption as well as emission. There was a 72% reduction in Kabo, 74% in Igbeti and Abadam, whereas Okrika recorded the least reduction of 64%. The relationship between the running hours of diesel generator for both load types (EEME and CME) is shown in Fig. 12.10. The RHC located in Okrika has the highest diesel generator running hours for both equipment types, whereas Abadam recorded the least diesel generator running hour.

12.1.12 Conclusion

Various aspects of DSM are very crucial in ensuring that, energy resources available to a country is efficiently used. So many benefits of DSM include: provision of relief to power grids and generation plants, mitigation of emergencies in electrical systems, minimization of blackouts to increase reliability of energy systems, reduction in energy prices, mitigation of emergencies in electrical systems, to reduce or partially eliminate investments in generation of energy, reduction of dependency on expensive imports, transmission and distribution networks, and reduction of environmental emissions.

Table 12.11 Emission for both conventional medical equipment (CME) and energy efficient medical equipment (EEME)

City	Gboko			Igbeti			Kabo		
Load type	**EEME**	**CME**	**% Reduction**	**EEME**	**CME**	**% Reduction**	**EEME**	**CME**	**% Reduction**
Carbon dioxide	488	1391	65%	328	1278	74%	364	1317	72%
Carbon monoxide	1.2	3.43	65%	0.81	3.16	74%	0.899	3.25	72%
Unburned hydrocarbons	0.133	0.38	65%	0.0898	0.35	74%	0.0996	0.36	72%
Particulate matter	0.0908	0.259	65%	0.0611	0.238	74%	0.0678	0.245	72%
Sulfur dioxide	0.98	2.79	65%	0.659	2.57	74%	0.732	2.64	72%
Nitrogen oxides	10.7	30.6	65%	7.23	28.2	74%	8.03	29	72%
	Ihiala			**Abadam**			**Okrika**		
Carbon dioxide	526	1586	67%	332	1281	74%	593	1669	64%
Carbon monoxide	1.3	3.91	67%	0.819	3.16	74%	1.46	4.12	65%
Unburned hydrocarbons	0.144	0.434	67%	0.0908	0.35	74%	0.162	0.456	64%
Particulate matter	0.0979	0.295	67%	0.0618	0.238	74%	0.11	0.311	65%
Sulfur dioxide	1.06	3.19	67%	0.667	2.57	74%	1.19	3.35	64%
Nitrogen oxides	11.6	34.9	67%	7.31	28.2	74%	13.1	36.8	64%

Table 12.12 Diesel fuel consumption analysis

Location	Gboko	Igbeti	Kabo	Ihiala	Abadam	Okrika	Base case
Fuel consumption (EEME)L/yr	185	125	138	200	126	225	20364
Fuel consumption (CME)L/yr	528	485	500	602	486	634	20364
% reduction	65%	74%	72%	67%	74%	65%	0

CME, Conventional medical equipment; *EEME*, energy efficient medical equipment.

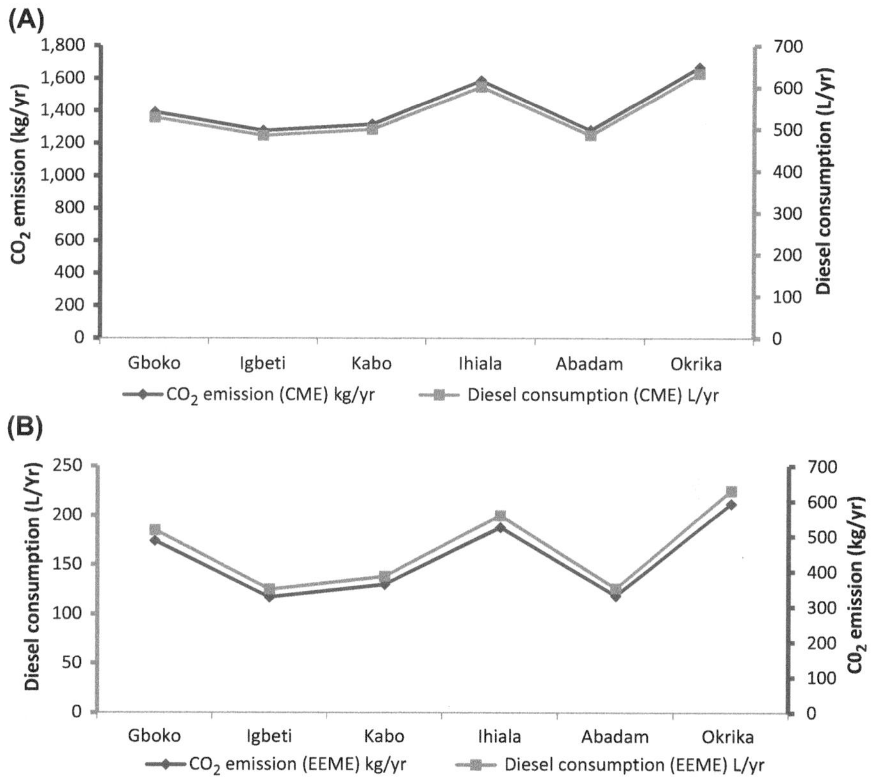

Figure 12.9 (A) The relationship between CO_2 emission and diesel consumption for conventional medical equipment (CME). (B) The relationship between CO_2 emission and diesel consumption for energy efficient medical equipment (EEME).

The technoeconomic and emission comparison of deploying energy efficiency as a DSM tool in sizing alternative energy for a typical health-care center has been presented in this chapter. In conclusion, this study validates the effectiveness of utilizing energy efficient-equipment as compared with

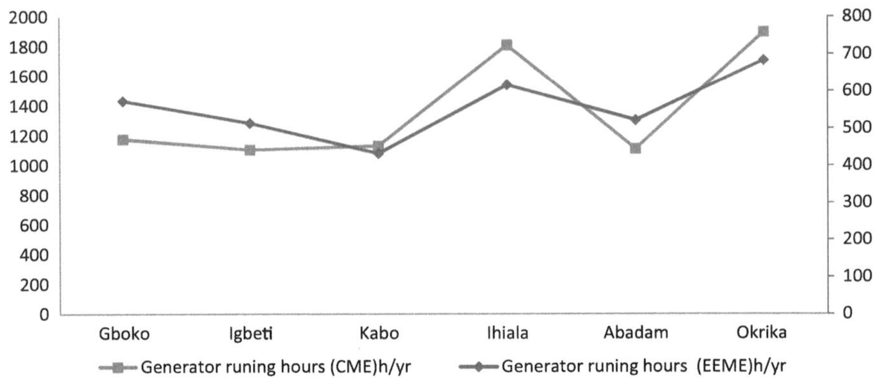

Figure 12.10 Generator running hours. *CME*, conventional medical equipment; *EEME*, energy efficient medical equipment.

conventional ones. It shows that, investments in more energy-efficient medical devices, though at higher costs, has the tendency to reduce the capital investment in hybrid renewable energy system for a rural health clinic, and as such, removes major barrier to the consumption of greener energy systems at lower costs over its life span.

For newly proposed clinics of similar status in rural communities of developing countries like the ones described in this study, a similar load profile is expected as the cautious selection of energy-efficient medical devices and energy conservation practices along with a hybrid PV/generator/battery system architecture would most possibly ensure the lowest NPC without compromising quality and reliability energy services delivery. The application of energy-efficient equipment is seen to increase the renewable energy penetration, across the six geopolitical zones in Nigeria. This in turn reduces the use operational hours of use of captive generators and consequently, the carbon footprint.

In terms of retrofitting, older health centers, the cost of procurement of energy-efficient equipment may initially appear expensive. Conversely, the NPC of the system over its life span makes investments in more energy-efficient devices reduce generator fuel costs considerably over time. Apart from this, cost of offsetting the effect of emission is also an advantage.

It is worth noting that, this model did not fully quantify the initial cost and maintenance costs of procuring newer, and more energy efficient medical equipment but only the approximate estimate. In future study, a search light can be beamed to explore these costs, the effect of inflation, carbon tax, as well as the variation in diesel fuel price, and solar irradiance so as to create a robust cost-benefit and environmental assessment model. Furthermore, health centers with constrained budgets, and facing challenges, choosing between investing in hybrid energy system or efficient medical devices, models that fully consider both supply and demand side, management options can help generate a precise assessment of costs and benefits, based on the options available locally.

APPENDICES

A.1 Emission Factors

Carbon monoxide	6.5	g/L of fuel
Unburned hydrocarbons	0.72	g/L of fuel
Particulate matter	0.49	g/L of fuel
Portion of fuel sulfur converted to particulate matter	2.2	%
Nitrogen oxides	58	g/L of fuel

A.2 Destination of Fuel Carbon

Carbon dioxide	99.5%
Carbon monoxide	0.4
Unburned carbon	0.1
Total	100

A.3 Conversion Factor (Abolarin et al., 2013)

1 kW = 100 W
12,000 kWh = 1 tonne of oil equivalent
1 L = 2.68 × 10−3 tonnes of CO_2
1 kWh = 1.25 × 10−3 tonnes of CO_2

REFERENCES

Abolarin, S.M., Gbadegesin, A.O., Shitta, M.B., Yussuff, A., Eguma, C.A., Ehwerhemuepha, L., Adegbenro, O., 2013. A collective approach to reducing carbon dioxide emission: a case study of four University of Lagos Halls of residence. Energy Build. 61, 318–322.

Ahenkorah, A.K.O., Capacity Building in Energy Efficiency and Renewable Energy Regulation and Policy-Making in Africa, Ghana – Energy Efficiency Country Profile.

Akinbulire, T.O., Oluseyi, P.O., Babatunde, O.M., 2014. Techno-economic and environmental evaluation of demand side management techniques for rural electrification in Ibadan, Nigeria. International Journal of Energy and Environmental Engineering 5 (4), 375–385.

Al-Karaghouli, A., Kazmerski, L., 2010. Optimization and life cycle cost of health clinic PV system for a rural area in southern Iraq using HOMER software. Solar Energy 84, 710–714.

Bajpai, P., Dash, V., 2012. Hybrid renewable energy systems for power generation in stand-alone applications a review. Renewable & Sustainable Energy Reviews 16 (1), 2926–2939.

Bekele, G., Palm, B., 2010. Feasibility study for a stand-alone-solar wind based hybrid energy system for application in Ethopia. Applied Energy 87, 487–495.

Benghanem, M., 2011. Optimization of tilt angle for solar panel: case study for Madinah, Saudi-Arabia. Applied Energy 88 (4), 1427–1433.

Besarati, S.M., Padilla, R.V., Goswami, D.Y., Stefamakos, E., 2013. The potential of harnessing solar radiation in Iran: generating solar maps and viability study of PV power plants. Renewable Energy 53, 193–199.

Course Module, Energy Efficiency and Demand Side Management. University of Warwick, REEEP.

Demand-Side Management Sustainable Energy Regulation and Policymaking for Africa.

Dihrab, S.S., Sopian, K., 2010. Electricity generation of hybrid PV/wind systems in Iraq. Renewable Energy 35 (6), 1303–1307.

Dongellini, M., Marinosci, C., Morini, G.C., 2014. Energy audit of an industrial site: a case Study. Energy Procedia 45, 424–443.

https://www.epa.gov/energy.

Gellings, C.W., Parmenter, K.E., Energy Efficiency and Renewable Energy Handbook.

Haney, A.B., Tooraj, J., Platchkov, L.M., Pollitt, M.G., 2010. Demand-Side Management Strategies and the Residential Sector: Lessons from International Experience.

Hassan, K., Faitma, K., Mahmood, H.S., 2011. Feasibility of hybrid power generation over wind and solar standalone system. In: 5th Power Engineering and Optimization Conference (PENCO), 7, pp. 6–7.

Hiendro, A., Kumianto, R., Rajagukguk, M., Simarajuntak, Y.M., 2013. Techno-economic analysis of photovoltaic/wind hybrid system for onshore remote area in Indonesia. Energy 59, 652–657.

HOMER Energy LLC: Hybrid Optimization Model for Electric Renewable, 2009. http://www.homerenergy.com/.

https://image.slidesharecdn.com/presentazionecordoba-160319093812/95/residential-demand-response-operation-in-a-microgrid-8–638.jpg?cb=1458380410.

Jimenez, T., Olson, K., 1999. Renewable Energy for Rural Health Clinics.

Lal, S., Raturi, A., 2012. Techno-economic analysis of a hybrid mini-grid system for Fiji Islands. International Journal of Energy and Environmental Engineering 3, 10.

Lamba, M.K., Sanghi, A., 2015. Energy audit on academic building. International Journal of Engineering Research and General Science 3 (4).

Martel, C.M., May 2000. Lessons in MRP Can Help Control Deregulated Energy Costs, 25 (5), p. 14.

Nandi, S.K., Ghosh, H.R., 2010. Prospect of wind-PV battery hybrid power system as an alternative to grid extension in Bangledash. Energy 35, 3040–3047.

Nwulu, N.I., Agboola, O.P., 2011. Utilizing renewable energy resources to solve Nigeria's electricity generation problem. International Journal of Thermal and Environmental Engineering 3 (1), 15–20.

Ohumakin, O.S., 2010. Energy utilization and renewable sources in Nigeria. Journal of Engineering and Applied Sciences 5 (2), 171–177.

Olatomiwa, L., Mekhilef, S., Huda, A., 2014. Optimal sizing of hybrid energy system for a remote telecom tower: a case study in Nigeria. In: IEEE Conference on Energy Conversion (CENCON), pp. 243–247.

Olatomiwa, L., Mekhilef, S., Ohunakin, O.S., 2016. Hybrid renewable power supply for rural health clinics (RHC) in six geo-political zones of Nigeria. Sustainable Energy Technologies and Assessments 13, 1–12.

Oluseyi, P.O., Babatunde, O.M., Babatunde, O.A., 2016. Assessment of energy consumption and carbon footprint from the hotel sector within Lagos, Nigeria. Energy & Buildings 118, 106–113. http://dx.doi.org/10.1016/j.enbuild.2016.02.046.

Palensky, P., Dietrich, D., 2011. Demand side management: demand response, intelligent energy systems, and smart loads. IEEE Transactions on Industrial Informatics 7 (3), 381–388.

Pitis (Femco), C.D., Livingstone (Anglo Gold Mining Motors), A., 2004. Energy efficient fans in underground auxiliary ventilation systems. In: Paper Presented at ICUE Conference (Cape Town, South Africa).

http://www.poweringhealth.org/index.php/topics/technology/energyefficiency.

Rahman, S., Al-Hadhrami, M.L., 2010. Study of a solar PV-diesel battery hybrid power system for a remotely located population near Rafha Saudi-Arabia. Energy 36, 4986–4995.

Saini, S., The Growing Need of Demand-Side Management, Energy Program, Sanc oGlobals

Sameeullah, M., Kumar, J., Lal, K., Chander, J., 2014. Energy audit: a case study of hostel building. International Journal of Research in Management, Science & Technology 2 (2).

Singh, H., Singh, M., Singh, G., 2012. Energy audit: a case study to reduce lighting cost. Asian Journal of Computer Science and Information Technology 2 (5), 119–122.

Srinath, G., Uday Kumar, N., 2014. Energy audit as a tool for improving system efficiency in industrial sector. International Journal of Engineering Research and Applications 4 (6), 6–11. ISSN: 2248-9622.

Van Alphen, K., Wilfred, G.J.H.M., Herkert, M.P., 2007. Renewable energy technologies in the Maldives – determining the potential. Renewable & Sustainable Energy Reviews 11, 1650–1674.

http://www.who.int/hia/green_economy/modern-energy-services/en/.

Zhou, W., Lou, C., Li, Z., Lu, L., Yang, H., 2010. Current status of research on optimum sizing of stand-alone hybrid solar-wind power generation systems. Applied Energy 87, 380–389.

CHAPTER 13

Carbon Footprint Analysis of Printed Circuit Board

Winco K.C. Yung[1], Subramanian Senthilkannan Muthu[2], Karpagam Subramanian[1]
[1]The Hong Kong Polytechnic University, Kowloon, Hong Kong SAR; [2]Bestseller, Kowloon, Hong Kong SAR

13.1 INTRODUCTION

Printed circuit board (PCB) is one of the major subcomponents of almost all electronic devices. It supports the electronic components such as capacitors, resistors, etc., which are soldered on the board. Generally, electronic products are often criticized for their contribution toward carbon emissions. Manufacturers also thrive to produce low-carbon products to achieve a sustainable production. Simultaneously the urge among consumers for green products is also increasing. Life cycle assessment (LCA) determines the entire environmental impacts of a product, whereas product carbon footprint (PCF) focuses on climate change and global warming potential (GWP) only. Though PCF focuses on single environmental indicator, it is still a very significant one, and it is considered as a threat to natural resources and livelihood of people globally. Hence PCF analysis has gained so much attention in recent years. Though carbon foot printing of products is gaining more attention, the focus is more on large-scale information and communication technology devices and not much attention is given to small-scale products and intermediate products. PCF analysis of a PCB, an intermediate product that is used as one of the components in many larger electronic products performed and reported here is the first of its kind in this area of research. The results will definitely help manufacturers in reducing the carbon emissions of their intermediate products.

The goal of this PCF study is to calculate the potential contribution of the chosen product, PCB, to global warming expressed as grams or kilograms of CO_2 eq., by quantifying all significant greenhouse gas (GHG) emissions and removals over the product's life cycle (only "cradle-to-gate"

Environmental Carbon Footprints
ISBN 978-0-12-812849-7
http://dx.doi.org/10.1016/B978-0-12-812849-7.00013-1

stages are considered in this study). This report follows the standard of ISO/TS 14067:2013 "Greenhouse gases- Carbon footprint of products- Requirements and guidelines for quantification and communication" (International Organization for Standardization (ISO), 2013). The results, data, methods, assumptions, and life cycle interpretations are presented in sufficient detail to allow the reader to comprehend the complexities and trade-offs inherent in the PCF study. The PCF value is measured by quantifying the GHG emissions for the entire product life cycle expressed in carbon dioxide equivalents value (kg CO_2 eq.), as per ISO/TS 14067.

$$CF_{total} = CF_{RM} + CF_{M} + CF_{D} + CF_{U} + CF_{EOL} \tag{13.1}$$

where CF_{RM} is the carbon footprint of the raw material stage (kg CO_2 eq.); CF_M is the carbon footprint of the manufacturing stage (kg CO_2 eq.); CF_D is the carbon footprint of the distribution stage (kg CO_2 eq.); CF_U is the carbon footprint of the use stage (kg CO_2 eq.); CF_{EOL} is the carbon footprint of the end-of-life stage (kg CO_2 eq.) (International Organization for Standardization (ISO), 2013).

13.2 METHODOLOGY

13.2.1 Scope

13.2.1.1 Product System and Its Function(s)

The product selected for this study is a PCB, a semifinished product used as a component in ignition systems in automobiles. Fig. 13.1 shows the outline of the selected PCB.

13.2.1.2 Functional Unit

As there is no specific product category rule (PCR) for an LCD panel, the functional unit (f.u.) of the LCD The functional unit (f.u.) of the PCB is referenced to the PCR of PCB in Korea (EDP Korea, 2005). The functional unit is defined as 1 m^2 of PCB (weight: 4.2725 kg/m^2 excluding packaging) with packaging (packaging weight: 0.1463 kg/m^2, including primary packaging: packaging film and bubble wrap; and secondary packaging: silica gel, humidity card, carton box, stickers, and poly strap). The reference flow is defined as 1 m^2 of PCB.

13.2.1.3 Data and Data Quality

13.2.1.3.1 Description of Data

The report covers "cradle-to-gate" stages of product life cycle of the PCB. Data of the unit processes collected for raw material, manufacturing, and

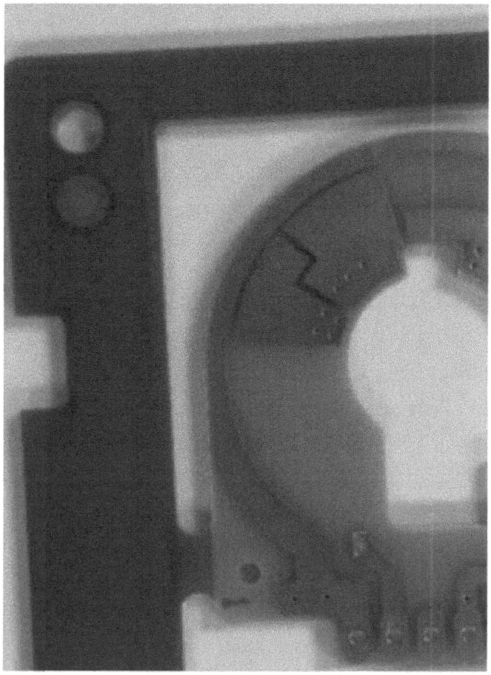

Figure 13.1 Selected printed circuit board (PCB) model.

transportation stages were primarily collected by Shenzhen Sunshine Circuits Technology Co., Ltd., which had financial and operational control of the unit processes. The emission factors used were localized to the extent possible with geographic, technology, and time-related considerations. Table 13.1 shows the details of individual data.

Table 13.1 Description of individual data

Life cycle stage	Type of data	Data
Raw material stage	Primary data (site specific)	Amount of materials and components used
	Secondary data	Emission factors of materials and components
Manufacturing stage	Primary data (site specific)	Electricity and water consumption, amount of waste generated in the manufacturing process
	Secondary data	Emission factors of electricity and water usage
Transportation stage	Primary data (site specific)	Transportation tools used
	Secondary data	Emission factors of transportation route Distance of transportation routes

13.2.1.3.2 Characterization of Data Quality

The best quality data were used to reduce bias and uncertainty as far as practically possible. Primary and secondary data were selected to meet the goal and the scope of the PCF study.

13.2.1.4 Cutoff Criteria and Cutoff

No materials or processes have been omitted in this study, thus, no cutoff criteria were established.

13.2.1.5 Allocation Procedures

Allocation of emissions is based on the ratio of the manufactured quantities of the studied product and the total production in the factory in Shenzhen from Feb 2014 to Aug 2014. The allocation was applied to electricity, water, consumables usage, and waste amount generated from the production (Section 13.4.2).

13.2.1.6 Geographical and Time Boundary of Data

The PCBs were manufactured in Shenzhen from Feb 2014 to Aug 2014 and were sold to an American customer.

13.2.1.7 System Boundary

As a manufacturer, Shenzhen Sunshine Circuits Technology Co., Ltd. provides products to other manufacturers (B2B business model). The system boundary for this study covers "cradle to gate" of the products life cycle stages of the PCB, from raw material extraction and manufacturing, to product distribution stage. However, in this study, emission at the distribution stage is not included in the final PCF, though it is reported for reference. Regarding the all the waste generated at the material and manufacturing stages, the following activities are included in the total PCF calculation: (1) transportation of the waste to recycling sites; (2) transportation of sludge and hazardous liquid; (3) amount and transportation of consumables used in the treatment inside the factory of gas emissions and wastewater; (4) electricity and water consumption for wastewater and emitted gas treatment; and (5) the municipal sewage wastewater treatment of factory end effluent; but the following emissions are excluded from the calculation: (1) environmental burdens from the recycling processes; and (2) treatments of sludge and hazardous liquid. Fig. 13.2 shows the system boundary of this study.

All processes included in this study directly contributed to the end result (GHG emissions) from "cradle-to-gate" life cycle stages. The overheads of the factory and equipment maintenance in all life cycle stages were excluded in this study because they were not significant and not directly related to the GHG emissions and removal of the PCB.

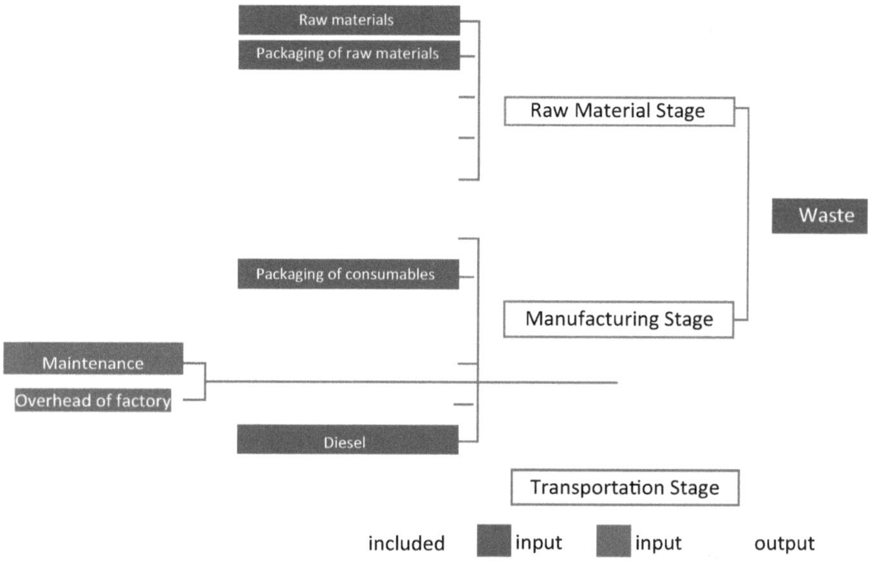

Figure 13.2 System boundary of carbon footprint analysis of "the printed circuit board."

13.2.1.8 Relevant Assumptions in This Study

13.2.1.8.1 Assumptions in "Raw Material" Stage

- There were differences in terms of weights among some of the components and consumables; the variances may be because of different suppliers of the components and consumables.
- Mean weight of materials (for example, silica gels, product stickers, humidity cards, etc.) used for three samples was taken as the actual weight used for the production.
- The emission factor of the unknown organic chemicals was assumed on the basis of database from Swiss ecoinvent data version 2.2—chemicals organic, at plant [GLO].
- When there was not exact emission factor for an organic chemical in the database, its emission factor was assumed on the basis of database from Swiss ecoinvent data version 2.2—chemicals organic, at plant [GLO].
- The emission factor of the unknown inorganic chemicals was assumed on the basis of database from Swiss ecoinvent data version 2.2—chemicals inorganic, at plant [GLO].
- When there was not exact emission factor for an inorganic chemical in the database, its emission factor was assumed on the basis of database from Swiss ecoinvent data version 2.2—chemicals inorganic, at plant [GLO].

- The closest and the most similar emission factor has been chosen for different materials, for example, potassium dicyanoaurate, milling cutter, backup board (Syndyne Industrial Co. Ltd., 2014), diazo film (Wikipedia, 2013), degreasing agent (Su Zhou He-Chuan Chemical Technology Service Co. Ltd., (n.d.)), defoamer (PCBcity, (n.d.)), and so on.
- Weights of drilling needles and milling cutters were obtained based on the factory internal production data and calculation methods (Tables A1.1 and A1.2) (Appendix 1).
- The weight and usage of etching liquid and tin striping solution transported by mobile tank were calculated based on the factory internal calculation methods, which cannot be disclosed because of confidential concerns.
- 68.3% of potassium dicyanoaurate usage was assumed to be left on the PCB product, and the rest was assumed to be consumables based on the information provided by the factory personnel.
- The packaging of final products was assumed to be the same for each delivery, which included a printed carton box containing 150 sets of PCBs (0.034 m^2), and each set of PCB consists of 20 pieces of single PCB units (0.0017 m^2). The quantity of PCBs in one box was assumed to be 5.1 m^2 for every delivery.
- Data estimation and calculation were made by the engineers of the company.
- Supplementary data for materials and transportation activities were provided by the factory.
- All distances in each material transportation route taken were obtained from Google Map.

13.2.1.8.2 Assumptions in "Manufacturing" Stage

- The energy consumption of each PCB module manufactured in the factory was assumed to be the same.
- The consumables consumption of each PCB module manufactured in the factory was assumed to be the same.
- The amount of consumables used was calculated based on the factory internal production data and calculation methods, which cannot be disclosed because of confidential concerns.
- The EF of packaging bag for consumables only includes packaging film material and production; the bag production process was excluded because relevant EF is unavailable in the database.

- The energy consumption data were for 7 months as the selected PCB was only manufactured from Feb 2014 to Aug 2014. No seasonal factor of production is included in this study.
- Electricity consumption for lift usage was included in the production.
- Electricity consumption of the canteen and residential hall for workers was excluded from electricity consumption for production of the PCB module.
- All solid wastes (for example, material cutoffs, packaging materials, scraps, etc.) generated from manufacturing processes are transported to three officially certified recycling plants according to different waste types. Emissions attributable to transportation of solid wastes were included, but the environmental burdens from the recycling processes were excluded from the calculation.
- Hazardous liquids generated from manufacturing processes are transported to a certified treatment plant. Emissions attributable to transportation of these hazardous liquids were included, but the environmental burdens from the treatment processes were excluded from the calculation because there was no relevant emission factor for the treatment available in the database.
- All wastewater is treated inside the factory before being discharged into public sewage. The end effluent is monitored on-site by a 24-h online system that is connected to the environmental monitoring station.
- All the end effluent from the factory was assumed to be discharged into municipal sewage water system, and the emission of sewage wastewater treatment is included in the calculation.
- Sludge generated from wastewater treatment processes at manufacturing stage is transported to a certified treatment plant. Emissions attributable to transportation of the sludge were included, but the environmental burdens from the treatment processes were excluded from the calculation because there was no relevant emission factor for the treatment available in the database.
- The emission of material usage and its transportation activities was taken as the emission for the treatment of emitted gas from manufacturing processes because the factory was not able to provide sufficient running data of emission absorbent towers.
- Data estimation and calculations were made by the engineers of the company.
- All distances in each material transportation route taken were obtained from Google Map.

13.2.1.8.3 Assumptions in "Transportation" Stage

- Distances of transportation activities were obtained from Google Map.
- Data estimation and calculation were made by the engineers of the company.

13.2.1.9 Treatment of Electricity

The emission factor of electricity used for manufacturing is from Chinese Government National Development and Reform Commission (NDRC). The emission factor for southern China in 2013 was 0.9223 kg CO_2 eq./kWh (Syndyne Industrial Co. Ltd., 2014).

13.3 CARBON FOOTPRINT ANALYSIS

13.3.1 Carbon Emission Calculation

According to the data provided by Shenzhen Sunshine Circuits Technology Co., Ltd. calculation of the "cradle-to-gate" assessment of PCF is based on the LCA approach, which includes the raw material stage, the manufacturing stage, and the distribution stage. The analysis is based on the concept of LCA and ISO/TS 14067:2013 (International Organization for Standardization (ISO), 2013) and allows computation of the total amount of carbon emission within the selected system boundary. This PCF study, according to ISO/TS 14067:2013, includes four phases of LCA, i.e., goal and scope definition, life cycle inventory (LCI), life cycle impact assessment (LCIA) and life cycle interpretation. Fig. 13.3 shows the framework of LCA from ISO 14040 (Wikipedia, 2013).

The unit processes comprising the product system are grouped into selected life cycle stages in this study, i.e., raw material acquisition, manufacturing, and transportation. GHG emissions and removals from the product's life cycle are assigned to the life cycle stage in which the GHG emissions and removals occur. Partial PCFs may be added together to quantify the total PCF, provided the same methodology is used, and there are no gaps and overlaps. The PCF is calculated and presented as a carbon dioxide equivalent (kg CO_2 eq.) using the relevant 100-year GWP. Findings in the impact assessment and the sensitivity and contribution analysis, performed as part of the interpretation, help identify the most relevant and contributing ("key") processes and basic flows of the system.

$$\text{Carbon Emission} = \text{Activity Data} \times \text{Emission Factor} \times \text{Global Warming Potential}$$

The "GWP" should be excluded, where the unit of the "emission factor" was "kg CO_2 eq./unit" (Su Zhou He-Chuan Chemical Technology Service Co. Ltd., n.d.). The activity data collected from each unit process were

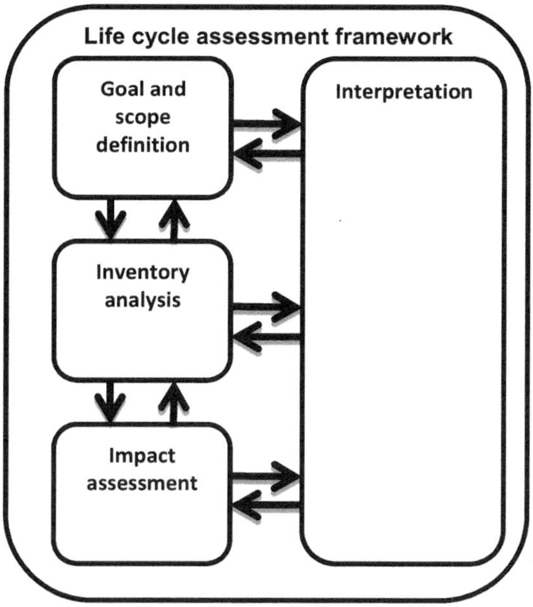

Figure 13.3 Framework of life cycle assessment (Wikipedia, 2013).

inputted into an online PCF analyzer (Green Bill of Materials (G-BOM) Analyzer provided by The Hong Kong Polytechnic University) and computed as in (Su Zhou He-Chuan Chemical Technology Service Co. Ltd., (n.d.)). The working principle, the methodology behind the software and access can be found in the following link: http://www.pctech.ise.polyu.edu.hk/ecodesign/gbom_analyzer.html

The G-BOM analyzer is online software based on the structure of the bill of materials. The G-BOM analyzer allows manufacturers to input their product information/data according to the product life cycle, which includes the raw materials, manufacturing, distribution, usage, and end-of-life stages, and simply calculates the PCF using the GHG emissions on the basis of the database provided. Users can define the material names, emission factors, and quantities according to their LCI. The GHG emission database is pregrouped into five life cycle stages. The emission factor can be selected manually in life cycle groups or by a keyword search engine. During the quantification of this study, values were rounded off to four decimal places in the LCI data input. The results of the PCF in the impact assessment were rounded off to two significant figures (PCBcity, (n.d.)).

13.4 LIFE CYCLE INVENTORY FOR PRODUCT CARBON FOOTPRINT

13.4.1 Life Cycle Inventory of "Raw Material" Stage

1. Sources of Data
   a. NDRC-Chinese Government National Development and Reform Commission
   b. ecoinvent—Swiss ecoinvent data version 2.2
   c. Journal of Tsinghua University (Science and Technology)
2. Database selection was based on credibility and geographical, technological, and time-related considerations. Emission factor for electricity consumption is obtained from NDRC 2013, the data used for components, consumables, packaging materials, and transportation are obtained from ecoinvent version 2.2, and the emission factor for sewage treatment of discharged effluent from the factory is obtained from the Journal of Tsinghua University (Science and Technology).
3. Materials used, weight and emission factor of raw materials, and product packaging materials are listed in Table 13.2.
4. Data of packaging of raw materials and product packaging (Table 13.3).
5. Transportation details of raw materials and product packaging are shown in Table 13.4.

13.4.2 Life Cycle Inventory of "Manufacturing" Stage

1. Process flow in manufacturing stage is shown in Fig. 13.4.
2. All activity data of resources and energy consumption in the "manufacturing" stage were provided by Shenzhen Sunshine Circuits Technology Co., Ltd.
3. The emission factor of electricity used is from Chinese Government NDRC. The emission factor for southern China in 2013 was 0.9223 kg CO_2 eq./kWh.
4. The electricity consumption provided by Shenzhen Sunshine Circuits Technology Co., Ltd. in manufacturing stage is shown in Table 13.5. It is based on allocation of the total electricity consumption and production of the selected product in the total production from Feb 2014 to Aug 2014. Electricity consumption for staff accommodation and canteen buildings is excluded. Table 13.5 shows the production volumes and electricity consumption from Feb 2014 to Aug 2014.

Table 13.2 Weight, materials used, and emission factors of raw materials and product packaging materials

No.	Name of materials	Materials	Weights per m^2 PCB production (kg)	Emission factor (kg CO_2 eq./kg)	Name in database (ecoinvent v2.2)
Raw materials					
1	Laminate	Epoxy resin 45% Woven glass 40% Copper foil 15%	3.0283	13.82	Epoxy resin, liquid, at plant [RER]; copper, primary, at refinery [GLO]; glass fiber, at plant [RER]; yarn production, cotton fibers [GLO]; weaving, cotton [GLO]; sheet rolling, copper [RER]
2	Copper ball	99.9% Cu 0.1% P	0.6256	5.03	Copper, primary, at refinery [GLO]; metal product manufacturing, average metal working [RER]
3	Electroless nickel EN-52M	Sodium hypophosphite ≤ 26%; in water	0.1809	0.48	Chemicals inorganic, at plant [GLO]; water, deionized, at plant [CH]
4	Electroless nickel EN-52A	Nickel (II) sulfate ≤ 46%; in water	0.1322	0.86	Chemicals inorganic, at plant [GLO]; water, deionized, at plant [CH]
5	Electroless nickel EN-52C	Sodium hydroxide ≤ 13%	0.0595	1.1	Sodium hydroxide, 50% in H_2O, production mix, at plant [RER]
6	Electroless nickel EN-52D	Lactic acid ≤ 1%; in water	0.1092	0.02	Chemicals inorganic, at plant [GLO]; water, deionized, at plant [CH]
7	Electroless nickel EN-52B	Sodium hypophosphite ≤ 58%; in water	0.1288	1.08	Chemicals inorganic, at plant [GLO]; water, deionized, at plant [CH]

Continued

Table 13.2 Weight, materials used, and emission factors of raw materials and product packaging materials—cont'd

No.	Name of materials	Materials	Weights per m^2 PCB production (kg)	Emission factor (kg CO_2 eq./kg)	Name in database (ecoinvent v2.2)
8	Potassium dicyanoaurate (ENIG process)	$KAu(CN)_2$ ($Au \geq 68.3\%$)	0.0004098	9013	Sodium cyanide, at plant [RER]; gold, at regional storage [RER]
9	Potassium dicyanoaurate (electroplating process)	$KAu(CN)_2$ ($Au \geq 68.3\%$)	0.007855	9013	Sodium cyanide, at plant [RER]; gold, at regional storage [RER]
Product packaging materials					
10	LDPE film	LDPE	0.04192	2.7	Packaging film, LDPE, at plant [RER]
11	Bubble wrap	LDPE	0.007765	3.79	Packaging film, LDPE, at plant [RER]; blow molding [RER]
12	Printed carton box	Carton board	0.07961	3.09	Kraft paper, unbleached, at plant [RER]; packaging, corrugated board, mixed fiber, single wall, at plant [RER]; printing color, offset, 47.5% solvent, at plant [RER]
13	Poly straps	PP	0.006392	2.36	Polypropylene, granulate, at plant [RER]; extrusion, plastic pipes [RER]
14	Inner packaging film	PE	0.008039	2.7	Packaging film, LDPE, at plant [RER]

15	Silica gel	Silica	0.001059	4.30	Sodium silicate production, spray powder, 80% [RER]; silicone product, at plant [RER]
16	Humidity card	Paper	0.001294	3.50	Kraft paper, unbleached, at plant [RER]; printing color, offset, 47.5% solvent, at plant [RER]
17	QA sticker	Paper	7.843E-05	4.06	Paper, wood containing, LWC, at plant [RER]; adhesive for metals, at plant [DE]; printing color, offset, 47.5% solvent, at plant [RER]
18	Product sticker	Paper	0.0001176	4.06	Paper, wood containing, LWC, at plant [RER]; adhesive for metals, at plant [DE]; printing color, offset, 47.5% solvent, at plant [RER]

The names of materials are provided by Shenzhen Sunshine Circuits Technology Co., Ltd. The materials used are provided by Shenzhen Sunshine Circuits Technology Co., Ltd. *CH*, Switzerland; *DE*, Germany; *ENIG*, electroless nickel/immersion gold; *GLO*, Global; *LDPE*, low-density polyethylene; *LWC*, light weight coated; *PCB*, printed circuit board; *PE*, polyethylene; *PP*, polypropylene; *QA*, quality assurance; *RER*, Europe.

Table 13.3 The weight and emission factor of packaging of raw materials and product packaging

No.	Packaging material	Weight (kg) of packaging material	Emission factor (kg CO_2 eq./kg)	Name in database (ecoinvent 2.2)
Packaging of raw materials				
1	Kraft paper	0.03357	0.85	Kraft paper, unbleached, at plant [RER]
2	LDPE plastic bag	0.0003450	2.7	Packaging film, LDPE, at plant [RER]
3	Poly strap	0.0005612	2.36	Polypropylene, granulate, at plant [RER]; extrusion, plastic pipes [RER]
4	Printed carton box	0.006251	3.09	Kraft paper, unbleached, at plant [RER]; packaging, corrugated board, mixed fiber, single wall, at plant [RER]; printing color, offset, 47.5% solvent, at plant [RER]
5	HDPE bucket	0.03251	3.28	Polyethylene, HDPE, granulate, at plant [RER]; injection molding [RER]
6	HDPE bottle	0.01499	3.28	Polyethylene, HDPE, granulate, at plant [RER]; injection molding [RER]
Packaging of product packaging materials				
7	EPE film	0.0004725	2.70	Packaging film, LDPE, at plant [RER]
8	PE bag	0.0001064	2.70	Packaging film, LDPE, at plant [RER]
9	Yellow back paper	0.0001373	2.05	Paper, wood containing, LWC, at plant [RER]; packaging film, LDPE, at plant [RER]

EPE, expanded polyethylene; *HDPE*, high-density polyethylene; *LDPE*, low-density polyethylene; *LWC*, light weight coated; *PE*, polyethylene; *RER*, Europe.

Table 13.4 Transportation of raw materials and product packaging materials

No.	Material	Weight (kg)	Location	Transportation tool	Distance (km)	Emission factor (kg CO_2 eq./tkm)	Name in database (ecoinvent 2.2)
Transportation of raw materials							
1	Laminate	3.0283	Wanjiang district, Dongguan city, Guangdong province	10 t lorry	56	0.22	Transport, lorry 7.5–16 t, EURO4 [RER]
2	Copper ball	0.6256	Taicang Economic Development Zone, Taicang, Suzhou, Jiangsu province	3 t lorry	1506	1.92	Transport, van < 3.5 t [RER]
3	Electroless nickel EN-52M	0.1809	Baomin road, Bao'an district, Shenzhen, Guangdong province	5 t lorry	21.1	0.47	Transport, lorry 3.5–7.5 t, EURO4 [RER]
4	Electroless nickel EN-52A	0.1322					
5	Electroless nickel EN-52C	0.0595					
6	Electroless nickel EN-52D	0.1092					
7	Electroless nickel EN-52B	0.1288					

Continued

Table 13.4 Transportation of raw materials and product packaging materials—cont'd

No.	Material	Weight (kg)	Location	Transportation tool	Distance (km)	Emission factor (kg CO_2 eq./tkm)	Name in database (ecoinvent 2.2)
8	Potassium dicyanoaurate (ENIG process)	0.0004098	Bao'an district, Shenzhen	Van	22.1	1.92	Transport, van < 3.5 t [RER]
9	Potassium dicyanoaurate (electroplating process)	0.007855					
Transportation of product packaging materials							
10	PE film	0.04192	Mingzhu road, Shijie town, Dongguan, Guangdong province	5 t lorry	65.2	0.47	Transport, lorry 3.5−7.5 t, EURO4 [RER]
11	PE bubble film	0.007765					
12	Printed carton box	0.07961	Songrui road, Songgang, Bao'an district, Shenzhen	10 t lorry	7.3	0.22	Transport, lorry 7.5−16 t, EURO4 [RER]
13	Poly strap	0.006392	Xitou, Songgang town, Bao'an district, Shenzhen	1.5 t lorry	10.4	1.92	Transport, van < 3.5 t [RER]

14	Inner packaging film	0.008039	Mingzhu road, Shijie town, Dongguan, Guangdong province	1.5 t lorry	65.2	1.92	Transport, van < 3.5 t [RER]
15	Silica gel	0.001059	98 Industrial district, Shajing town, Bao'an district, Shenzhen	1.5 t lorry	1.8	1.92	Transport, van < 3.5 t [RER]
16	Humidity card	0.001294	Licheng Technology Park, Shajing town, Bao'an district, Shenzhen	1.5 t lorry	8.3	1.92	Transport, van < 3.5 t [RER]
17	QA sticker	7.843E-05	Gonghe Tongchuang Technology Park, Shajing town, Bao'an district, Shenzhen	1.5 t lorry	7.5	1.92	transport, van < 3.5 t [RER]
18	Product sticker	0.0001176					

PE, polyethylene; *QA*, quality assurance; *RER*, Europe.

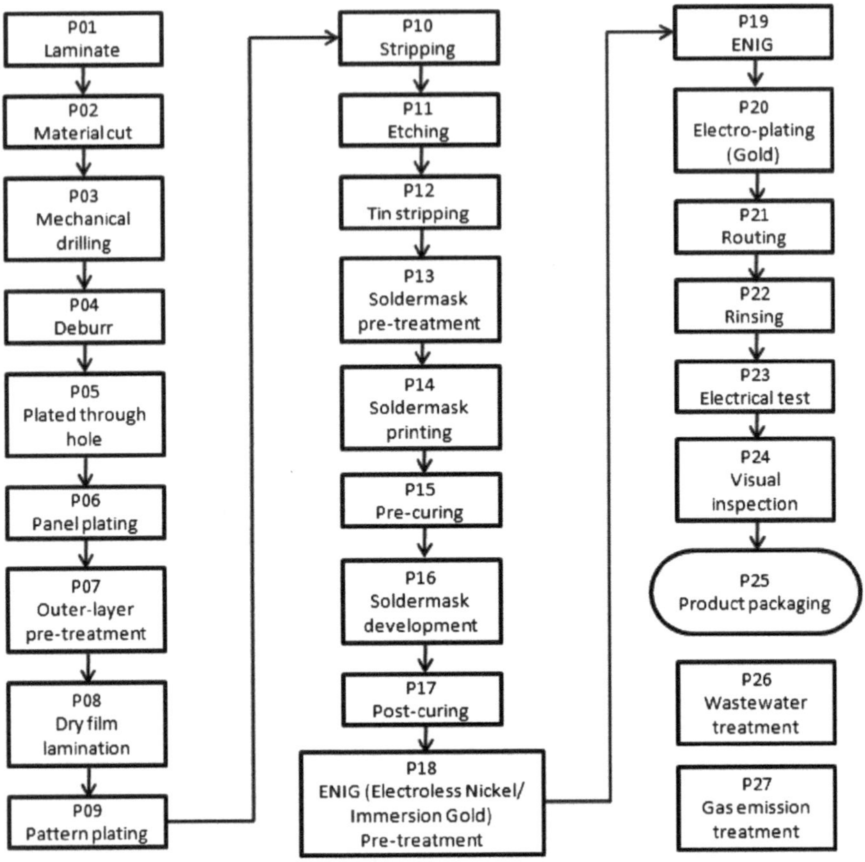

Figure 13.4 Process flow of printed circuit board manufacturing (information provided by the factory).

5. Average electricity consumption per m^2 of PCB was 83.86 kWh.
6. Water consumption (provided by Shenzhen Sunshine Circuits Technology Co., Ltd.) in manufacturing stage is shown in Table 13.6. It is based on the total water consumption and production volume of the selected product in the total production from Feb 2014 to Aug 2014. Water consumption for staff accommodation and canteen buildings is excluded. Table 13.6 shows the production volumes and water consumption from Feb 2014 to Aug 2014.
7. Water consumption per m^2 of PCB was 1.09 t.

Table 13.5 Production volumes and electricity consumption from Feb 2014 to Aug 2014

	Monthly electricity consumption (excluding accommodative area) kWh	Total monthly production for all models		Emission factor	Unit	Monthly energy consumption per m^2 of production (kWh)
Feb	1,874,671	Feb	22,076	0.9223	kg CO_2 eq./kWh (NDRC)	84.92
Mar	2,290,561	Mar	29,021			78.93
Apr	2,277,757	Apr	28,322			80.42
May	2,390,270	May	29,560			80.86
Jun	2,416,439	Jun	28,024			86.23
Jul	2,552,266	Jul	29,006			87.99
Aug	2,437,651	Aug	27,816			87.63
Average energy consumption per unit of production						83.86

NDRC, National Development and Reform Commission.

Table 13.6 Production volumes and water consumption from Feb 2014 to Aug 2014

	Monthly water consumption (excluding accommodative area) (t)	Total monthly production for all models (m^2)		Emission factor	Emission factor value	Monthly energy consumption per m^2 of production (t)
Feb	28,548	Feb	22,076	Tap water, at user [RER] (ecoinvent 2.2)	0.000319 kg CO_2 eq./kg	1.29
Mar	27,403	Mar	29,021			0.94
Apr	30,437	Apr	28,322			1.07
May	28,163	May	29,560			0.95
Jun	33,004	Jun	28,024			1.18
Jul	33,012	Jul	29,006			1.14
Aug	28,983	Aug	27,816			1.04
Average water consumption per unit of production						1.09

RER, Europe.

Table 13.7 Details of internal transportation tools used at manufacturing stage

Transportation tool	Energy source	Energy consumption	Emission factor	Name in database (ecoinvent 2.2)
Elevator	Electricity	Included in manufacturing electricity consumption		
Manual stacker	Electricity	Included in manufacturing electricity consumption		
Forklift truck	Diesel	0.0008585 (kg/m^2 PCB production)	0.51 (kg CO_2 eq./kg)	Diesel, at refinery [RER]
		0.03700 (MJ/m^2 PCB production)	0.09 (kg CO_2 eq./MJ)	Diesel, burned in building machine [GLO]

The density and the volumetric energy density of diesel used in the production is obtained from Diesel Fuel. Available from: http://en.wikipedia.org/wiki/Diesel_fuel. *GLO*, Global; *PCB*, printed control board; *RER*, Europe.

8. There are three types of internal transportation tools that cause carbon emissions, used for material or product delivery. Detailed descriptions and emission factors are shown in Table 13.7.

9. From Feb 2014 to Aug 2014, the total online production volume of selected PCB was 4856.2 m^2, the volume of final product was 4010.08 m^2; the rate of product waste was 16.8%, and the scrap rate was 0.624%.

10. The consumables used in the manufacturing stage were provided by Shenzhen Sunshine Circuits Technology Co., Ltd. Materials used, weight and emission factor of consumables are listed in Table 13.8.

11. Table 13.9 shows the packaging of consumables.

12. Table 13.10 shows the transportation of all consumables.

13. Weight and transportation details of solid wastes (packaging leftovers and cutoffs) generated from manufacturing processes and sludge produced in wastewater treatment facilities are based on the factory's internal calculation methods. All the solid wastes are transported to three different recycling plants, according to the category. The emission from transportation of waste was included, but the environmental burdens from the recycling and sludge treatment process were excluded from the calculation. The following Table 13.11 shows the details of solid waste management.

Table 13.8 Materials used, weight and emission factor of consumables

Process	No	Item (ISO, 2013)	Materials (Syndyne Industrial Co. Ltd., 2014)	Weight (kg) for each m^2 production	Emission factor (kg CO_2 eq./kg)	Name in database (ecoinvent 2.2)
P01 laminate P02 material cut	1	Material cutter	Steel alloy	0.0002	3.56	Steel, low alloyed, at plant [RER]; steel product manufacturing, average metal working [RER]
	2	Drilling needles	Plastic 12%; hard metal alloy (88%), WC 80%; Co 7%; V 0.35%; Zr 0.53%	0.0011	5.32	Polypropylene, granulate, at plant [RER]; injection molding [RER]; boron carbide, at plant [GLO]; cobalt, at plant [GLO]; chromium, at regional storage [RER]; zircon, 50% zirconium, at plant [AU]; metal product manufacturing, average metal working [RER]
P03 mechanical drilling	3	Aluminum plate	Aluminum alloy	0.093	2.23	Aluminum, primary, at plant [RER]; aluminum product manufacturing, average metal working [RER]
	4	Backup board	Phenolic resin 40%; unbleached kraft 60%	0.264	2.17	Phenolic resin, at plant [RER]; kraft paper, unbleached, at plant [RER]

Continued

Table 13.8 Materials used, weight and emission factor of consumables—cont'd

Process	No	Item (ISO, 2013)	Materials (Syndyne Industrial Co. Ltd., 2014)	Weight (kg) for each m^2 production	Emission factor (kg CO_2 eq./kg)	Name in database (ecoinvent 2.2)
	5	Crepe paper tape	Paper 35%; adhesive 20%; silicone 30%; polyester film 15%	0.00532	3.41	Paper, wood containing, LWC, at plant [RER]; adhesive for metals, at plant [DE]; silicone product, at plant [RER]; polyester resin, unsaturated, at plant [RER]; extrusion, plastic film [RER]
P04 deburr	6	Brush roller	Silicon carbide 20%; nylon 20%; steel 60%	0.0016	5.76	Silicon carbide, at plant [RER]; nylon 66, at plant [RER]; extrusion, plastic pipes [RER]; blow molding [RER]; steel, low alloyed, at plant [RER]; steel product manufacturing, average metal working [RER]
P05 plated through hole	7	Sulfuric acid	Sulfuric acid 98%	0.1328	0.12	Sulfuric acid, liquid, at plant [RER]
	8	Hydrogen peroxide	35.5% H_2O_2; 64.5% water	0.00013	1.12	Hydrogen peroxide, 50% in H_2O, at plant [RER]
	9	Leavening agent	Ethylene glycol ethers 45%; water 55%	0.03	1.46	Ethylene glycol diethyl ether, at plant [RER]; water, deionized, at plant [CH]

10	Desmear agent	Potassium permanganate 50–65 g/L; sodium hydroxide 35–55 g/L; water	0.02	1.14	Potassium permanganate, at plant [RER]; sodium hydroxide, 50% in H_2O,
11	Neutralizer	Neutralizer 8%–12%; sulfuric acid 6%–10%; water	0.03	0.47	Methylamine, at plant [RER]; sulfuric acid, liquid, at plant [RER]; water, deionized, at plant [CH]
12	Degreasing agent	Nitric acid 30%–38%; anionic surfactant 2%–4%; in water	0.013	1.17	Chemicals inorganic, at plant [GLO]; water, deionized, at plant [CH]
13	Active agent	Hydrochloric acid 10%–15%; tin (II) chloride 35%–45%; in water	0.002	0.91	Hydrochloric acid, from the reaction of hydrogen with chlorine, at plant [RER]; chemicals inorganic, at plant [GLO]; water, deionized, at plant [CH]
14	Predip	Organic acid 1%–10% in water	0.0416	0.096	Chemicals organic, at plant [GLO]; water, deionized, at plant [CH]
15	Accelerant	Sodium chlorite 30%–40% in water	0.00535	1.12	Sodium chlorate, powder, at plant [RER]; water, deionized, at plant [CH]

Continued

Table 13.8 Materials used, weight and emission factor of consumables—cont'd

Process	No	Item (ISO, 2013)	Materials (Syndyne Industrial Co. Ltd., 2014)	Weight (kg) for each m^2 production	Emission factor (kg CO_2 eq./kg)	Name in database (ecoinvent 2.2)
	16	Electroless copper plating liquid	Copper (II) sulfate; formic acid; sodium hypophosphite; dimethylamine borane; EDTA; tartaric acid; glycolic acid; sulfuric acid; potassium hydroxide; thiocarbamide	0.162	1.88	Chemicals inorganic, at plant [GLO]; chemicals organic, at plant [GLO]
	17	Nitric acid	Nitric acid	0.069	3.17	Nitric acid, 50% in H_2O, at plant [RER]
	18	Filter	Polypropylene filter 82%; polypropylene rack 18%	0.0019	10.90	Polypropylene, granulate, at plant [RER]; extrusion, plastic pipes [RER]; yarn production, cotton fibers [GLO]; injection molding [RER]
P06 panel plating	19	$CuSO_4$	98% $CuSO_4$	0.013	1.86	Chemicals inorganic, at plant [GLO];
	20	Sulfuric acid	98% sulfuric acid	0.16	0.12	Sulfuric acid, liquid, at plant [RER]

	21	Glazing agent	Sulfuric acid 0.2%–0.8%; copper (II) sulfate pentahydrate 0.5%–1%; formic acid 2%–5%; in water	0.0342	0.12	Sulfuric acid, liquid, at plant [RER]; chemicals inorganic, at plant [GLO]; formic acid, at plant [RER]; water, deionized, at plant [CH]
	22	Filter	Polypropylene filter 82%; polypropylene rack 18%	0.0005	10.90	Polypropylene, granulate, at plant [RER]; extrusion, plastic pipes [RER]; yarn production, cotton fibers [GLO]; injection molding [RER]
P07 outer layer pretreatment	23	Sulfuric acid	98% sulfuric acid	0.0249	0.12	Sulfuric acid, liquid, at plant [RER]
	24	Pumice	Silicon dioxide; aluminum oxide	0.0288	0.00076	Pumice, at mine [DE]; crushing, rock [RER]
	25	Nylon brush	Silicon carbide 20%; nylon 20%; steel 60%	0.00195	5.332	Silicon carbide, at plant [RER]; nylon 66, at plant [RER]; extrusion, plastic pipes [RER]; steel, low alloyed, at plant [RER]; steel product manufacturing, average metal working [RER]

Continued

Table 13.8 Materials used, weight and emission factor of consumables—cont'd

Process	No	Item (ISO, 2013)	Materials (Syndyne Industrial Co. Ltd., 2014)	Weight (kg) for each m^2 production	Emission factor (kg CO_2 eq./kg)	Name in database (ecoinvent 2.2)
P08 dry film lamination	26	Dry film	PET film 33.33%; photoresist 33.33%; LDPE film 33.33%	0.08	2.71	Polyethylene terephthalate, granulate, amorphous, at plant [RER]; extrusion, plastic film [RER]; acrylic acid, at plant [RER]; chemicals organic, at plant [GLO]; packaging film, LDPE, at plant [RER]
	27	Sticky roller	Silicone 50%; rubber 50%	0.00079	3.35	Silicone product, at plant [RER]; synthetic rubber, at plant [RER]; injection molding [RER]
	28	Diazo film	PET film; diazo	0.00035	2.56	Polyethylene terephthalate, granulate, amorphous, at plant [RER]; extrusion, plastic film [RER]; chemicals organic, at plant [GLO]
	29	Sodium carbonate	Na_2CO_3	0.0546	1.05	Sodium carbonate from ammonium chloride production, at plant [GLO]
	30	Filter	Polypropylene filter 82%; polypropylene rack 18%	0.0018	10.90	Polypropylene, granulate, at plant [RER]; extrusion, plastic pipes [RER]; yarn production, cotton fibers [GLO]; injection molding [RER]

P09 pattern plating	31	Degreaser	Surface activator 0.5%–1.0%; AES 1.0%–3.0%; water 85%–90%; aiding agent 7.0%–9.0%	0.04357	0.21	Chemicals inorganic, at plant [GLO]; chemicals organic, at plant [GLO]; water, deionized, at plant [CH]
	32	Sulfuric acid	98% sulfuric acid	0.3352	0.12	Sulfuric acid, liquid, at plant [RER]
	33	Sodium persulfate	Sodium persulfate: 99%	0.059	1.27	Sodium persulfate, at plant [GLO]
	34	Glazing agent	Polyethylene glycol; polypropylene glycol; water; aiding agent	0.02	2.25	Triethylene glycol, at plant [RER]; propylene glycol, liquid, at plant [RER]; water, deionized, at plant [CH]; chemicals inorganic, at plant [GLO]
	35	Copper (II) sulfate	98% $CuSO_4$	0.007	1.86	Chemicals inorganic, at plant [GLO]
	36	Tin (II) sulfate	98% $SnSO_4$	0.004	1.86	Chemicals inorganic, at plant [GLO]
	37	Tin ball	Tin 99%	0.07	19.02	Tin, at regional storage [RER]; metal product manufacturing, average metal working [RER]
	38	Nitric acid	Nitric acid: 35%	0.056	3.17	Nitric acid, 50% in H_2O, at plant [RER]
	39	Filter	Polypropylene filter 82%; polypropylene rack 18%	0.056	10.90	Polypropylene, granulate, at plant [RER]; extrusion, plastic pipes [RER]; yarn production, cotton fibers [GLO]; injection molding [RER]

Continued

Table 13.8 Materials used, weight and emission factor of consumables—cont'd

Process	No	Item (ISO, 2013)	Materials (Syndyne Industrial Co. Ltd., 2014)	Weight (kg) for each m^2 production	Emission factor (kg CO_2 eq./kg)	Name in database (ecoinvent 2.2)
P10 stripping	40	Sodium hydroxide	Sodium hydroxide 98%–99%	0.087	1.1	Sodium hydroxide, 50% in H_2O, production mix, at plant [RER]
P11 etching	41	Etching liquid	Ammonia 50%–60%; ammonium chloride 35%–38%; accelerant: 1.0%–2.0%	2.356	1.70	Ammonia, liquid, at regional storehouse [RER]; ammonium chloride, at plant [GLO]; chemicals inorganic, at plant [GLO]
P12 tin stripping	42	Tin stripping solution	Nitric acid 95%–98%; organic solderability preservatives (OSP) 1.0%–2.0%	0.557	3.14	Nitric acid, 50% in H_2O, at plant [RER]; chemicals organic, at plant [GLO]
P13 soldermask pretreatment	43	Sulfuric acid	98% sulfuric acid	0.014	0.12	Sulfuric acid, liquid, at plant [RER]
	44	Nylon brush	Silicon carbide 20%; nylon 20%; steel 60%	0.178	5.33	Silicon carbide, at plant [RER]; nylon 66, at plant [RER]; extrusion, plastic pipes [RER]; steel, low alloyed, at plant [RER]; steel product manufacturing, average metal working [RER]
	45	Pumice	Silicon dioxide; aluminum oxide	0.03716	0.00076	Pumice, at mine [DE]; crushing, rock [RER]

P14 soldermask printing	46	Component ink	Epoxy resin 20%–30%; resin 2.4%–4.4%; glycol ether solvent 6.0%–16.0%; naphtha 6.9%–11.9%; pigment 40.0%–50.0%; photoinitiator 2.7%–4.7%; additives 1.2%–3.2%	0.109	3.09	Epoxy resin, liquid, at plant [RER]; chemicals organic, at plant [GLO]; chemicals inorganic, at plant [GLO]
	47	Screen	PET	0.006	25.73	Polyethylene terephthalate, granulate, amorphous, at plant [RER]; yarn production, cotton fibers [GLO]; weaving, cotton [GLO]
	48	Stencil cleaner	Alcohols	0.0024	0.74	Methanol, at plant [GLO]
P15 precuring	49	Diazo film	PET film; diazo	0.000258	3.35	Silicone product, at plant [RER]; synthetic rubber, at plant [RER]; injection molding [RER]

Continued

Table 13.8 Materials used, weight and emission factor of consumables—cont'd

Process	No	Item (ISO, 2013)	Materials (Syndyne Industrial Co. Ltd., 2014)	Weight (kg) for each m^2 production	Emission factor (kg CO_2 eq./kg)	Name in database (ecoinvent 2.2)
P16 soldermask developing	50	Sodium carbonate	Sodium carbonate: 99.8%	0.048	1.05	Sodium carbonate from ammonium chloride production, at plant [GLO]
	51	Filter	Polypropylene filter 82%; polypropylene rack 18%	0.00174	10.90	Polypropylene, granulate, at plant [RER]; extrusion, plastic pipes [RER]; yarn production, cotton fibers [GLO]; injection molding [RER]
	52	Defoamer	Ethers 1.8%–2.2%; emulsifier 0.05% –0.08%; in water	0.00016	0.04	Chemicals organic, at plant [GLO]; water, deionized, at plant [CH]
P17 postcuring	–	–	–	–	–	–
P18 Electroless nickel/ immersion gold (ENIG) pretreatment	53	Sulfuric acid (AR class)	$H_2SO_4 \geq 98\%$	0.0806	0.12	Sulfuric acid, liquid, at plant [RER]
	54	Sodium persulfate	$Na_2S_2O_8 \geq 98\%$	0.0332	1.27	Sodium persulfate, at plant [GLO]
	55	Sulfuric acid (CP class)	$H_2SO_4 \geq 98\%$	0.0129	0.12	Sulfuric acid, liquid, at plant [RER]
	56	Brown corundum for PCB	Al_2O_3 100%	0.0223	1.24	Aluminum oxide, at plant [RER]

P19 ENIG	57	AC-10 Acidic cleaner	Sulfuric acid 2%; citric acid 2%; in water	0.0058	0.04	Sulfuric acid, liquid, at plant [RER]; chemicals organic, at plant [GLO]; water, deionized, at plant [CH]
	58	CA-40M activator	$CuSO_4 \geq 98\%$	0.002	0.19	Chemicals inorganic, at plant [GLO]; water, deionized, at plant [CH]
	59	CA-40R Activator	$H_2SO_4 \leq 15\%$	0.0381	0.02	Sulfuric acid, liquid, at plant [RER]; water, deionized, at plant [CH]
	60	CD-30 regulator	hydroxylamine sulfate $\leq 30\%$	0.0456	0.56	Chemicals inorganic, at plant [GLO]; water, deionized, at plant [CH]
	61	Electroless gold EG-70	Citric acid 25%; EDTA-4Na 5%; in water	0.043	0.57	Chemicals organic, at plant [GLO]; water, deionized, at plant [CH]
	62	Potassium dicyanoaurate	$KAu(CN)_2$ ($Au \geq 68.3\%$)	0.00019	9013	Sodium cyanide, at plant [RER]; gold, at regional storage [RER]
	63	Ammonia (AR class)	$NH_4OH \geq 27\%$ in water	0.0027	0.57	Ammonia, liquid, at regional storehouse [RER]; water, deionized, at plant [CH]
	64	Sodium hydroxide	NaOH 96%–99%	0.0132	1.1	Sodium hydroxide, 50% in H_2O, production mix, at plant [RER]

Continued

Table 13.8 Materials used, weight and emission factor of consumables—cont'd

Process	No	Item (ISO, 2013)	Materials (Syndyne Industrial Co. Ltd., 2014)	Weight (kg) for each m^2 production	Emission factor (kg CO_2 eq./kg)	Name in database (ecoinvent 2.2)
	65	Blue tape	Rubber type pressure sensitive adhesives 10.5%; PVC film 89.5%	0.0336	2.52	Adhesive for metals, at plant [DE]; polyvinyl chloride, suspension polymerized, at plant [RER]
	66	Nitric acid	Nitric acid ≥ 68%	0.0403	3.17	Nitric acid, 50% in H_2O, at plant [RER]
	67	Filter	Polypropylene filter 82%; polypropylene rack 18%	0.0156	10.90	Polypropylene, granulate, at plant [RER]; extrusion, plastic pipes [RER]; yarn production, cotton fibers [GLO]; injection molding [RER]
P20 electroplating (gold)	68	Potassium dicyanoaurate	$KAu(CN)_2$ (Au ≥ 68.3%)	0.0036	9013	Sodium cyanide, at plant [RER]; gold, at regional storage [RER]
	69	AURUNA 529 plating bath solution	Citric acid 20%; potassium citrate 25%; cobalt (II) citrate 5%–10%; in water	0.0206	0.99	Chemicals organic, at plant [GLO]; water, deionized, at plant [CH]
	70	Filter	Polypropylene filter 82%; polypropylene rack 18%	0.000642	10.90	Polypropylene, granulate, at plant [RER]; extrusion, plastic pipes [RER]; yarn production, cotton fibers [GLO]; injection molding [RER]

P21 routing	71	Milling cutter	Hard metal alloy, WC-93.11%; Co-6.0%; V-0.31%; Zr-0.33%; Mo-0.25%	0.00977	15.48	Boron carbide, at plant [GLO]; cobalt, at plant [GLO]; chromium, at regional storage [RER]; zircon, 50% zirconium, at plant [AU]; molybdenum, at regional storage [RER]; metal product manufacturing, average metal working [RER]
P22 rinsing	72	Sulfuric acid	98% sulfuric acid	0.00826	0.12	Sulfuric acid, liquid, at plant [RER]
P23 electrical test	—	—	—	—	—	—
P24 visual inspection	—	—	—	—	—	—
P25 product packaging	—	—	—	—	—	—
P26 wastewater treatment	73	Sulfuric acid	98% sulfuric acid	0.7077	0.12	Sulfuric acid, liquid, at plant [RER]
	74	Sodium hydroxide	Sodium hydroxide: 98%–99%	0.4074	1.1	Sodium hydroxide, 50% in H_2O, production mix, at plant [RER]
	75	Lime	Lime	1.4565	2.43	Lime from lithium carbonate hydration [GLO]
	76	NaClO	NaClO	2.9186	0.89	Sodium hypochlorite, 15% in H_2O, at plant [RER]

Continued

Table 13.8 Materials used, weight and emission factor of consumables—cont'd

Process	No	Item (ISO, 2013)	Materials (Syndyne Industrial Co. Ltd., 2014)	Weight (kg) for each m^2 production	Emission factor (kg CO_2 eq./kg)	Name in database (ecoinvent 2.2)
	77	Iron (II) sulfide	FeS	0.7604	1.86	Chemicals inorganic, at plant [GLO]
	78	Polyacrylamide (PAM)	Polyacrylamide	0.04232	1.9	Chemicals organic, at plant [GLO]
	79	Polyaluminum chloride (PAC)	Polyaluminum chloride	0.3347	1.86	Chemicals inorganic, at plant [GLO]
	80	Sodium sulfide	sodium sulfide	0.6488	1.86	Chemicals inorganic, at plant [GLO]
P27 gas emission treatment	81	Sodium hydroxide	Sodium hydroxide 98%–99%	0.02275	1.1	Sodium hydroxide, 50% in H_2O, production mix, at plant [RER]

Details of consumables are provided by Shenzhen Sunshine Circuits Technology Co., Ltd. The materials used are provided by Shenzhen Sunshine Circuits Technology Co., Ltd. *CH*, Switzerland; *DE*, Germany; *GLO*, Global; *LDPE*, low-density polyethylene; *LWC*, light weight coated; *PET*, polyethylene terephthalate; *PVC*, polyvinyl chloride; *RER*, Europe.

Table 13.9 The packaging of consumables

No.	Packaging material	Weight (g) of packaging material for each m^2 PCB production	Emission factor (kg CO_2 eq./kg)	Name in database (ecoinvent 2.2)
1	Kraft paper	0.002303	0.85	Kraft paper, unbleached, at plant [RER]
2	Wrapping paper/ packaging paper	0.0006477	1.19	Paper, wood containing, supercalendered (SC), at regional storage [RER]
3	LDPE plastic bag (no handle)/film	0.008337	2.7	Packaging film, LDPE, at plant [RER]
4	HDPE plastic bag (no handle)	8.4036E-05	2.56	Polyethylene, HDPE, granulate, at plant [RER]; extrusion, plastic film [RER]
5	Poly strap	0.0001952	2.36	Polypropylene, granulate, at plant [RER]; extrusion, plastic pipes [RER]
6	Printed carton box	0.4514	3.09	Kraft paper, unbleached, at plant [RER]; packaging, corrugated board, mixed fiber, single wall, at plant [RER]; printing color, offset, 47.5% solvent, at plant [RER]
7	Carton box (no print)	0.0001400	1.99	Kraft paper, unbleached, at plant [RER]; packaging, corrugated board, mixed fiber, single wall, at plant [RER]
8	PVC bucket	0.1227	3.28	Polyethylene, HDPE, granulate, at plant [RER]; injection molding [RER]
9	PVC core	0.5698	2.90	Polyvinylchloride, emulsion polymerized, at plant [RER]; extrusion, plastic pipes [RER]
10	Black plastic bag	0.1196	2.8	Packaging film, LDPE, at plant [RER]; pigments, paper production, unspecified, at plant [RER]

Continued

Table 13.9 The packaging of consumables—cont'd

No.	Packaging material	Weight (g) of packaging material for each m^2 PCB production	Emission factor (kg CO_2 eq./kg)	Name in database (ecoinvent 2.2)
11	PP woven bag	0.01469	20.22	Polypropylene, granulate, at plant [RER]; fleece production, polyethylene terephthalate [RER]; weaving, cotton [GLO]
12	Iron bucket	0.0005220	3.43	Pig iron, at plant [GLO]; metal product manufacturing, average metal working [RER]
13	Plastic bag (silver) (Wikipedia, 2014b)	1.2149E-05	3.22	Polyethylene terephthalate, granulate, amorphous, at plant [RER]; extrusion, plastic film [RER]
14	HDPE bottle	0.0005850	3.28	Polyethylene, HDPE, granulate, at plant [RER]; injection molding [RER]
15	Paper core	0.0007191	2.29	Kraft paper, unbleached, at plant [RER]; packaging, corrugated board, mixed fiber, single wall, at plant [RER]

GLO, Global; *HDPE*, high-density polyethylene; *LDPE*, low-density polyethylene; *PCB*, printed circuit board; *PP*, polypropylene; *PVC*, polyvinyl chloride; *RER*, Europe.

Table 13.10 Transportation details of consumables

No.	Name	Supplier address	Distance (km)	Transportation tool	Emission factor value (kg CO_2 eq./tkm)	Name in database (ecoinvent 2.2)
1	Material cutter	Nancheng, Dongguan city, Guangdong province	51	1.5 t lorry	1.92	Transport, van < 3.5 t [RER]
2	Drilling needle	Shuibei 2nd road, Shenzhen; Meixian, Meizhou city, Guangdong province	45.9 + 3 45	Minivan	1.92	Transport, van < 3.5 t [RER]
3	Aluminum plate	Fenghuang Industrial Zone, Fuyong own, Bao'an district, Shenzhen	5.5	Minivan	1.92	Transport, van < 3.5 t [RER]
4	Backup board					
5	Crepe paper tape	Hongxing West Industrial Zone, Songgang town, Bao'an district, Shenzhen	8.5	Van	1.92	Transport, van < 3.5 t [RER]
6	Brush roller	Baxiang Business Street, Batou community, Wanjiang district, Dongguan city, Guangdong	50.5	10 t lorry	0.22	Transport, lorry 7.5–16 t, EURO4 [RER]
7	Sulfuric acid	Xinhe East 2nd Xiang, Fuyong town, Bao'an district, Shenzhen	3.4	8 t lorry	0.22	Transport, lorry 7.5–16 t, EURO4 [RER]
8	Hydrogen peroxide	Xinhai Building, Nanshan Ave, Nan Shan Qu, Shenzhen, Guangdong province	24.6	15 t lorry	0.22	Transport, lorry 7.5–16 t, EURO4 [RER]

Continued

Table 13.10 Transportation details of consumables—cont'd

No.	Name	Supplier address	Distance (km)	Transportation tool	Emission factor value (kg CO_2 eq./tkm)	Name in database (ecoinvent 2.2)
9	Leavening agent	Luo Pu road, Panyu district, Guangzhou, Guangdong province	104	15 t lorry	0.22	Transport, lorry 7.5–16 t, EURO4 [RER]
10	Desmear agent					
11	Neutralizer					
12	Degreasing agent					
13	Active agent					
14	Predip					
15	Accelerant					
16	Electroless copper plating liquid					
17	Nitric acid (%)	Qingshuihe, Shenzhen	44.4	5 t lorry	0.47	Transport, lorry 3.5–7.5 t, EURO4 [RER]
18	Filter	Songshan Lake Technology Zone, Dongguan city, Guangdong province	30.3	5 t lorry	0.47	Transport, lorry 3.5–7.5 t, EURO4 [RER]
19	$CuSO_4$	Xinhai Building, Nanshan Ave, Nan Shan Qu, Shenzhen, Guangdong province	24.6	8 t lorry	0.22	Transport, lorry 7.5–16 t, EURO4 [RER]

20	Sulfuric acid	Xinhe East 2nd Xiang, Fuyong town, Bao'an district, Shenzhen	3.4	10 t lorry	0.22	Transport, lorry 7.5 −16 t, EURO4 [RER]
21	Glazing agent	Panyu district, Guangzhou, Guangdong province	70.5	15 t lorry	0.22	Transport, lorry 7.5−16 t, EURO4 [RER]
22	Filter	Songshan Lake Technology Zone, Dongguan city, Guangdong province	30.3	5 t lorry	0.47	Transport, lorry 3.5−7.5 t, EURO4 [RER]
23	Sulfuric acid	Xinhe East 2nd Xiang, Fuyong town, Bao'an district, Shenzhen	3.4	10 t lorry	0.22	Transport, lorry 7.5−16 t, EURO4 [RER]
24	Pumice	Henggang district, Shenzhen	59.7	Minivan	0.22	Transport, lorry 7.5−16 t, EURO4 [RER]
25	Nylon brush	Donglin town, Huzho city, Zhejiang province	1364	Van	1.92	Transport, van < 3.5 t [RER]
26	Dry film	Guangyi road, Nansha district, Guangzhou	46.3	5 t lorry	0.47	Transport, lorry 3.5−7.5 t, EURO4 [RER]
27 28	Sticky roller Diazo film	Songshan Lake Technology Zone, Dongguan city, Guangdong province	30.3	1.5 t lorry	0.22	Transport, lorry 7.5−16 t, EURO4 [RER]
29	Sodium carbonate	Binhai Economic Development Zone, Weifang city, Shandong province	2062	8 t lorry	0.22	Transport, lorry 7.5−16 t, EURO4 [RER]
30	Filter	Songshan Lake Technology Zone, Dongguan city, Guangdong province	30.3	8 t lorry	0.22	Transport, lorry 7.5−16 t, EURO4 [RER]

Continued

Table 13.10 Transportation details of consumables—cont'd

No.	Name	Supplier address	Distance (km)	Transportation tool	Emission factor value (kg CO_2 eq./tkm)	Name in database (ecoinvent 2.2)
31	Degreaser	Luo Pu road, Panyu district, Guangzhou, Guangdong province	104	3 t lorry	0.22	Transport, lorry 7.5–16 t, EURO4 [RER]
32	Sulfuric acid	Xinhe East 2nd Xiang, Fuyong town, Bao'an district, Shenzhen	3.4	10 t lorry	0.22	Transport, lorry 7.5–16 t, EURO4 [RER]
33	Sodium persulfate	Xinhai Building, Nanshan Ave, Nan Shan Qu, Shenzhen, Guangdong province	24.6	8 t lorry	0.22	Transport, lorry 7.5–16 t, EURO4 [RER]
34	Glazing agent	Times Center, Zhongxin road, Shajing, Bao'an district, Shenzhen	7.9	3 t lorry	1.92	Transport, van < 3.5 t [RER]
35	$CuSO_4$	Xinhai Building, Nanshan Ave, Nan Shan Qu, Shenzhen, Guangdong province	24.6	8 t lorry	0.22	Transport, lorry 7.5–16 t, EURO4 [RER]
36	Stannous sulfate	Huogu Du, Gejiu city, Yunnan province	1380	10 t lorry	0.22	Transport, lorry 7.5–16 t, EURO4 [RER]
37	Tin ball					
38	Nitric acid	Qingshuihe, Shenzhen	44.4	5 t lorry	0.47	Transport, lorry 3.5–7.5 t, EURO4 [RER]
39	Filter	Songshan Lake Technology Zone, Dongguan city, Guangdong province	30.3	8 t lorry	0.22	Transport, lorry 7.5–16 t, EURO4 [RER]
40	Sodium hydroxide	Qingshuihe, Shenzhen	44.4	5 t lorry	0.47	Transport, lorry 3.5–7.5 t, EURO4 [RER]

41	Etching liquid	Kaiping city, Guangdong province	195	Mobile tank	1.92	Transport, van < 3.5 t [RER]
42	Tin stripping solution					
43	Sulfuric acid	Xinhe East 2nd Xiang, Fuyong town, Bao'an district, Shenzhen	3.4	10 t lorry	0.22	Transport, lorry 7.5–16 t, EURO4 [RER]
44	Nylon brush	Donglin town, Huzho city, Zhejiang province	1364	Van	1.92	Transport, van < 3.5 t [RER]
45	Pumice	Henggang district, Shenzhen	59.7	Minivan	1.92	Transport, van < 3.5 t [RER]
46	Component ink	Kut Shing street, Chai Wan, Hong Kong	93.8	10 t lorry	0.22	Transport, lorry 7.5–16 t, EURO4 [RER]
47	Screen	Tian'an digital city, Tairan 4th road, Futian district, Shenzhen	35.4	5 t lorry	0.47	Transport, lorry 3.5–7.5 t, EURO4 [RER]
48	Stencil cleaner	Ban Zhang Hu Yi Heng Jie, Huicheng district, Huizhou city, Guangdong province	92.4	1 t lorry	1.92	Transport, van < 3.5 t [RER]
49	Diazo film	Songshan Lake Technology Zone, Dongguan city, Guangdong province	30.3	1.5 t lorry	1.92	Transport, van < 3.5 t [RER]
50	Sodium carbonate	Binhai Economic Development Zone, Weifang city, Shandong province	2062	8 t lorry	0.47	Transport, lorry 3.5–7.5 t, EURO4 [RER]
51	Filter	Songshan Lake Technology Zone, Dongguan city, Guangdong province	30.3	8 t lorry	0.47	Transport, lorry 3.5–7.5 t, EURO4 [RER]

Continued

Table 13.10 Transportation details of consumables—cont'd

No.	Name	Supplier address	Distance (km)	Transportation tool	Emission factor value (kg CO_2 eq./tkm)	Name in database (ecoinvent 2.2)
52	Defoamer	Yintian Industrial Zone, Xixiang town, Shenzhen	19.5	5 t lorry	0.47	Transport, lorry 3.5–7.5 t, EURO4 [RER]
53	Sulfuric acid (AR class)	Xinhe East 2nd Xiang, Fuyong town, Bao'an district, Shenzhen	3.4	10 t lorry	0.22	Transport, lorry 7.5–16 t, EURO4 [RER]
54	Sodium persulfate	Houting Cun Industrial Zone, Shajing town, Bao'an district, Shenzhen	36.8	8 t lorry	0.22	Transport, lorry 7.5–16 t, EURO4 [RER]
55	Sulfuric acid (CP class)	Xinhe East 2nd Xiang, Fuyong town, Bao'an district, Shenzhen	3.4	10 t lorry	0.22	Transport, lorry 7.5–16 t, EURO4 [RER]
56	Brown corundum for PCB	Longhe 9th Road, Nanlian, Longgang district, Shenzhen	61.6	5 t lorry	0.47	Transport, lorry 3.5–7.5 t, EURO4 [RER]
57	AC-10 acidic cleaner	Baomin road, Bao'an district, Shenzhen, Guangdong province	21.1	5 t lorry	0.47	Transport, lorry 3.5–7.5 t, EURO4 [RER]
58	CA-40M activator					
59	CA-40R activator					
60	CD-30 regulator					
61	Electroless gold EG-70					

62	Potassium dicyanoaurate (31.7%)	Bao'an district, Shenzhen	22.1	Van	1.92	Transport, van < 3.5 t [RER]
63	Ammonia (AR class)	Xinhai Building, Nanshan Ave, Nan Shan Qu, Shenzhen, Guangdong province	24.6	8 t lorry	0.22	Transport, lorry 7.5–16 t, EURO4 [RER]
64	Sodium hydroxide	Xinhe East 2nd Xiang, Fuyong town, Bao'an district, Shenzhen	3.4	10 t lorry	0.22	Transport, lorry 7.5–16 t, EURO4 [RER]
65	Blue tape	Hongxing West Industrial Zone, Songgang town, Bao'an district, Shenzhen	8.5	Van	1.92	Transport, van < 3.5 t [RER]
66	Nitric acid	Qingshuihe, Shenzhen	44.4	5 t lorry	0.47	Transport, lorry 3.5–7.5 t, EURO4 [RER]
67	Filter	Songshan Lake Technology Zone, Dongguan city, Guangdong province	30.3	8 t lorry	0.22	Transport, lorry 7.5–16 t, EURO4 [RER]
68	Potassium dicyanoaurate	Bao'an district, Shenzhen	22.1	Van	1.92	Transport, van < 3.5 t [RER]
69	AURUNA 529 Plating bath solution	Zhongshan West Rd, Xuhui district, Shanghai	1480	5 t lorry	0.47	Transport, lorry 3.5–7.5 t, EURO4 [RER]
70	Filter	Songshan Lake Technology Zone, Dongguan city, Guangdong province	30.3	8 t lorry	0.22	Transport, lorry 7.5–16 t, EURO4 [RER]

Continued

Table 13.10 Transportation details of consumables—cont'd

No.	Name	Supplier address	Distance (km)	Transportation tool	Emission factor value (kg CO_2 eq./tkm)	Name in database (ecoinvent 2.2)
71	Milling cutter	Honghua Tao Development Zone, Yidu city, Hubei province	1182	Van	1.92	Transport, van < 3.5 t [RER]
72	Sulfuric acid	Xinhe East 2nd Xiang, Fuyong town, Bao'an district, Shenzhen	3.4	10 t lorry	0.22	Transport, lorry 7.5–16 t, EURO4 [RER]
73	Sulfuric acid	New Green Island Building, Nanshan Ave, Nan Shan district, Shenzhen	27.8	5 t lorry	0.47	Transport, lorry 3.5–7.5 t, EURO4 [RER]
74	Sodium hydroxide	Xinkai road South, Tanggu district, Tianjin city	2265	5 t lorry	0.47	Transport, lorry 3.5–7.5 t, EURO4 [RER]
75	Lime	Longxin community, Longgang district, Shenzhen	60.1	5 t lorry	0.47	Transport, lorry 3.5–7.5 t, EURO4 [RER]
76	NaClO	Fuqiao Industrial Zone 2, Fuyong town, Bao'an district, Shenzhen	7.9	5 t lorry	0.47	Transport, lorry 3.5–7.5 t, EURO4 [RER]
77	FeS	Longxin community, Longgang district, Shenzhen	60.1	5 t lorry	0.47	Transport, lorry 3.5 –7.5 t, EURO4 [RER]

78	PAM	Zhenxing Road, Zhongtang town, Dongguan city, Guangdong province	66	5 t lorry	0.47	Transport, lorry 3.5 –7.5 t, EURO4 [RER]
79	PAC	Gongyi city, Henan province	1625	5 t lorry	0.47	Transport, lorry 3.5 –7.5 t, EURO4 [RER]
80	Sodium sulfide	Mingshan district, Ya'an city, Sichuan province	1943	5 t lorry	0.47	Transport, lorry 3.5 –7.5 t, EURO4 [RER]
81	Sodium hydroxide	Xinkai road South, Tanggu district, Tianjin city	2265	5 t lorry	0.47	Transport, lorry 3.5 –7.5 t, EURO4 [RER]

RER, Europe.

Table 13.11 Weight and transportation details of solid wastes

Solid waste type	Amount (kg/m^2 PCB production)	Distance to recycling plant (km)	Transportation tool	Emission factor value (kg CO_2 eq./tkm)	Name in database (ecoinvent 2.2)
Packaging waste, material cutoffs	0.7312	3.2	Van or 3 t lorry	1.92	Transport, van < 3.5 t [RER]
Filters	0.02802	37			
Pumice, dry film waste	0.1066	5.5			
Sludge	3.124	37	6 t lorry	0.47	Transport, lorry 3.5–7.5 t, EURO4 [RER]

PCB, printed circuit board; *RER*, Europe.

14. Weight, transportation details, and emission factors of discharged effluent after wastewater treatment and hazardous liquids are shown in Table 13.12. The emissions of sewage water treatment of effluent and transportation activities for hazardous liquid are included, but the treatment of hazardous liquid is excluded in the calculation.
15. All the gas generated from the manufacturing process is treated by absorbent towers. The emission of material usage and its transportation activities was taken as the emission for the treatment of emitted gas because the factory was not able to provide sufficient running data of emission absorbent towers.

13.4.3 Life Cycle Inventory of "Transportation" Stage

1. Transportation routes and tools used for distributing finished products (provided by Shenzhen Sunshine Circuits Technology Co., Ltd.) are as shown in Table 13.13: Mass of the final product is 4.2725 kg/m^2 and packaging is 0.1463 kg/m^2, therefore the total mass of 4.4187 kg/m^2 is taken for calculating the carbon footprint in the transportation stage.
2. Distances of routes are obtained from Google Map.

13.5 PRODUCT CARBON FOOTPRINT ANALYSIS

13.5.1 Carbon Footprint Results in Component/Activity

Calculation of the carbon footprint of the PCB was based on a comprehensive coverage of raw material stage, manufacturing stage, and transportation stage. By assigning quantities of the input and output in each process, the carbon footprint of each input, output, and process were obtained (Table 13.14).

13.6 LIFE CYCLE INTERPRETATION

13.6.1 Results of Life Cycle Interpretation

1. The carbon footprint per functional unit (each m^2 of PCB with packaging) is 282.9331 kg CO_2 eq. The summary of carbon footprint in each life cycle stage is presented in Table 13.15, Figs. 13.5 and 13.6.
2. "Manufacturing" stage accounts for more than one half of the total carbon emission (56.74%) in the whole life cycle of the product. In the "manufacturing" stage, electricity consumption has a significant contribution, as shown in Table 13.15 (27.34% of total PCF).
3. The "raw material" stage contributes 43.26% of total product carbon emissions in the whole life cycle. In the "raw material" stage, usage of potassium dicyanoaurate in electroplating process has a significant contribution (25.02% of total PCF).

Table 13.12 Descriptions, weights, and transportation details of liquid wastes

Item	Description	Amount (kg/m^2 PCB production)	Emission factor (kg CO_2 eq./m^3)	Name in database		Source
Water effluent	Water discharged to sewage system after treatment	740	0.947 (Appendix 2)	Sewage treatment anaerobic-anoxic-oxic (Northern China, 600 Km^3/d)		Xie and Wang (2012)
Item	**Description**	**Amount (kg/m^2 PCB production)**	**Distance to recycling plant (km)**	**Transportation tool**	**Name in database (ecoinvent 2.2)**	**Emission factor value (kg CO_2 eq./tkm)**
Hazardous liquid	ENIG process liquid waste	1.2110 0.05702	37 5.5	3 t lorry	Transport, van < 3.5 t [RER]	1.92

ENIG, electroless nickel/immersion gold; *PCB*, printed circuit board; *RER*, Europe.

Table 13.13 Traveling distance and transportation tools of finished product to the United States

	Routes	Transportation tool	Distance (km)	Emission factor (kg CO_2 eq./tkm)	Sources of emission factor (ecoinvent 2.2)
1	Shenzhen Sunshine Circuits Technology Co., Ltd. → Huanggang Port, Shenzhen	1.5 t lorry	41.6	1.92	Transport, van < 3.5 t [RER]
	Huanggang Port → Hong Kong International Airport	1.5 t lorry	40.7	1.92	Transport, van < 3.5 t [RER]
	Hong Kong International Airport → Indianapolis International Airport	Aircraft	16,000	1.1	Transport, aircraft, freight [RER]
	Indianapolis International Airport → Harlingen-valley International Airport	Aircraft	2143	1.1	Transport, aircraft, freight [RER]
	Harlingen-valley International Airport → Customer end, Texas 78567	1.5 t lorry	22.69	1.92	Transport, van < 3.5 t [RER]

RER, Europe.

Table 13.14 Carbon footprint result of all materials from "cradle-to-gate" life cycle stages

No.	Item	Carbon footprint (kg CO_2 eq./f.u.)	Percentage of total PCF
Raw material stage			
Raw materials			
1	Laminate	41.8470	14.79
2	Copper ball	3.1468	1.11
3	Electroless nickel (EN-52M)	0.08683	0.03
4	Electroless nickel (EN-52A)	0.1137	0.04
5	Electroless nickel (EN-52C)	0.06545	0.02
6	Electroless nickel (EN-52D)	0.002184	0.00
7	Electroless nickel (EN-52B)	0.1391	0.05
8	Potassium dicyanoaurate (ENIG process)	3.6935	1.31
9	Potassium dicyanoaurate (electroplating process)	70.7926	25.02
	Subtotal	119.8871	42.37
Product packaging			
10	LDPE film	0.1132	0.04
11	Bubble wrap	0.02943	0.01
12	Printed carton box	0.2460	0.09
13	Poly straps	0.01509	0.01
14	Inner packaging film	0.02171	0.01
15	Silica gel	0.004553	0.00
16	Humidity card	0.004529	0.00
17	QA sticker	0.0003184	0.00
18	Product sticker	0.0004776	0.00
	Subtotal	0.4353	0.15
Packaging of raw materials			
1	Kraft paper	0.02853	0.01
2	LDPE plastic bag	0.0009314	0.00
3	Poly strap	0.001325	0.00
4	Printed carton box	0.01932	0.01
5	HDPE bucket	0.1066	0.04
6	HDPE bottle	0.04915	0.02
	Subtotal	0.2059	0.07
Packaging of packaging materials			
7	EPE film	0.001276	0.00
8	PE bag	0.0002873	0.00
9	Yellow back paper	0.0002814	0.00
	Subtotal	0.001844	0.00

Table 13.14 Carbon footprint result of all materials from "cradle-to-gate" life cycle stages—cont'd

No.	Item	Carbon footprint ($kg\ CO_2$ eq./f.u.)	Percentage of total PCF
Transportation of raw materials and packaging materials			
1	Laminate	0.0373	0.01
2	Copper ball	1.8089	0.64
3	Electroless nickel (EN-52M)	0.001794	0.00
4	Electroless nickel (EN-52A)	0.001311	0.00
5	Electroless nickel (EN-52C)	0.0005901	0.00
6	Electroless nickel (EN-52D)	0.001083	0.00
7	Electroless nickel (EN-52B)	0.001277	0.00
8	Potassium dicyanoaurate (ENIG process)	1.7389E-05	0.00
9	Potassium dicyanoaurate (electroplating process)	0.0003333	0.00
10	LDPE film	0.001285	0.00
11	Bubble wrap	0.0002379	0.00
12	Printed carton box	0.0001279	0.00
13	Poly straps	0.0001276	0.00
14	Inner packaging film	0.001006	0.00
15	Silica gel	3.6593E-06	0.00
16	Humidity card	2.0623E-05	0.00
17	QA sticker	1.1294E-06	0.00
18	Product sticker	1.6941E-06	0.00
	Subtotal	1.8556	0.66
Subtotal of raw material stage		122.3857	43.26
Manufacturing stage			
Energy consumption			
1	Electricity consumption for each m^2 PCB production	77.3441	27.34
2	Water consumption for each m^2 PCB production	0.3477	0.12
3	Diesel consumption for each m^2 PCB production	0.003751	0.00
	Subtotal	77.6922	27.46
Consumables			
1	Material cutter	0.000712	0.00
2	Drilling needles	0.005852	0.00
3	Aluminum plate	0.2074	0.07
4	Backup board	0.5729	0.20

Continued

Table 13.14 Carbon footprint result of all materials from "cradle-to-gate" life cycle stages—cont'd

No.	Item	Carbon footprint (kg CO_2 eq./f.u.)	Percentage of total PCF
5	Crepe paper tape	0.01814	0.01
6	Brush roller	0.009222	0.00
7	Sulfuric acid	0.01594	0.01
8	Hydrogen peroxide	0.0001456	0.00
9	Leavening agent	0.04380	0.02
10	Desmear agent	0.02277	0.01
11	Neutralizer	0.0141	0.00
12	Degreasing agent	0.01521	0.01
13	Active agent	0.00182	0.00
14	Predip	0.003994	0.00
15	Accelerant	0.005992	0.00
16	Electroless copper plating liquid	0.3046	0.11
17	Nitric acid	0.2187	0.08
18	Filter	0.02131	0.01
19	$CuSO_4$	0.02418	0.01
20	Sulfuric acid	0.0192	0.01
21	Glazing agent	0.004104	0.00
22	Filter	0.005608	0.00
23	Sulfuric acid	0.002988	0.00
24	Pumice	2.18304E-05	0.00
25	Nylon brush	0.01038	0.00
26	Dry film	0.2168	0.08
27	Sticky roller	0.002642	0.00
28	Diazo film	0.000896	0.00
29	Sodium carbonate	0.05733	0.02
30	Filter	0.02019	0.01
31	Degreaser	0.009150	0.00
32	Sulfuric acid	0.04022	0.01
33	Sodium persulfate	0.07493	0.03
34	Glazing agent	0.045	0.02
35	Copper (II) sulfate	0.01302	0.00
36	Tin (II) sulfate	0.00744	0.00
37	Tin ball	1.3314	0.47
38	Nitric acid	0.1775	0.06
39	Filter	0.6280	0.22
40	Sodium hydroxide	0.0957	0.03
41	Etching liquid	4.0052	1.42

Table 13.14 Carbon footprint result of all materials from "cradle-to-gate" life cycle stages—cont'd

No.	Item	Carbon footprint (kg CO_2 eq./f.u.)	Percentage of total PCF
42	Tin stripping solution	1.7490	0.62
43	Sulfuric acid	0.00168	0.00
44	Nylon brush	0.9491	0.34
45	Pumice	2.81673E-05	0.00
46	Component ink	0.3368	0.12
47	Screen	0.1544	0.05
48	Stencil cleaner	0.001776	0.00
49	Diazo film	0.0008643	0.00
50	Sodium carbonate	0.0504	0.02
51	Filter	0.01958	0.01
52	Defoamer	0.0000064	0.00
53	Sulfuric acid (AR class)	0.009672	0.00
54	Sodium persulfate	0.04216	0.01
55	Sulfuric acid (CP class)	0.001548	0.00
56	Brown corundum for PCB	0.02765	0.01
57	AC-10 acidic cleaner	0.000232	0.00
58	CA-40M activator	0.00038	0.00
59	CA-40R activator	0.000762	0.00
60	CD-30 regulator	0.02554	0.01
61	Electroless gold EG-70	0.02451	0.01
62	Potassium dicyanoaurate	1.7143	0.61
63	Ammonia (AR class)	0.001539	0.00
64	Sodium hydroxide	0.01452	0.01
65	Blue tape	0.08904	0.03
66	Nitric acid	0.1278	0.05
67	Filter	0.1750	0.06
68	Potassium dicyanoaurate	32.8569	11.61
69	AURUNA 529 plating bath solution	0.02039	0.01
70	Filter	0.007200	0.00
71	Milling cutter	0.1512	0.05
72	Sulfuric acid	0.0009912	0.00
73	Sulfuric acid	0.08493	0.03
74	Sodium hydroxide	0.4481	0.16
75	Lime	3.5393	1.25
76	NaClO	2.5976	0.92
77	Iron (II) sulfide	1.4144	0.50

Continued

Table 13.14 Carbon footprint result of all materials from "cradle-to-gate" life cycle stages—cont'd

No.	Item	Carbon footprint (kg CO_2 eq./f.u.)	Percentage of total PCF
78	Polyacrylamide (PAM)	0.08040	0.03
79	Polyaluminum chloride (PAC)	0.6226	0.22
80	Sodium sulfide	1.2067	0.43
81	Sodium hydroxide	0.02503	0.01
	Subtotal	56.8485	20.09
Packaging of consumables			
1	Kraft paper	0.001958	0.00
2	Wrapping paper/packaging paper	0.00077078	0.00
3	LDPE plastic bag (no handle)/film	0.022518	0.01
4	HDPE plastic bag (no handle)	0.00012151	0.00
5	Poly strap	0.0004606	0.00
6	Printed carton box	1.3950	0.49
7	Carton box (no print)	0.0002787	0.00
8	HDPE bucket	0.4026	0.14
9	PVC core	1.6524	0.58
10	Black plastic bag	0.3349	0.12
11	PP woven bag	0.2970	0.10
12	Iron bucket	0.001791	0.00
13	Plastic bag (silver)	2.9121E-05	0.00
14	HDPE bottle	0.001919	0.00
15	Paper core	0.001647	0.00
	Subtotal	4.1134	1.45
Transportation of consumables			
1	Material cutter	0.00001958	0.00
2	Drilling needles	0.0008256	0.00
3	Aluminum plate	0.0009821	0.00
4	Backup board	0.002788	0.00
5	Crepe paper tape	8.6822E-05	0.00
6	Brush roller	0.00001778	0.00
7	Sulfuric acid	9.9334E-05	0.00
8	Hydrogen peroxide	7.0356E-07	0.00
9	Leavening agent	0.0006864	0.00
10	Desmear agent	0.0004576	0.00
11	Neutralizer	0.0006864	0.00
12	Degreasing agent	0.0002974	0.00
13	Active agent	0.00004576	0.00
14	Predip	0.0009518	0.00
15	Accelerant	0.0001224	0.00
16	Electroless copper plating liquid	0.003707	0.00
17	Nitric acid	1.4399	0.51

Table 13.14 Carbon footprint result of all materials from "cradle-to-gate" life cycle stages—cont'd

No.	Item	Carbon footprint (kg CO_2 eq./f.u.)	Percentage of total PCF
18	Filter	0.02706	0.01
19	$CuSO_4$	0.07036	0.02
20	Sulfuric acid	0.1197	0.04
21	Glazing agent	0.5304	0.19
22	Filter	0.007121	0.00
23	Sulfuric acid	0.01863	0.01
24	Pumice	3.3012	1.17
25	Nylon brush	5.1068	1.81
26	Dry film	7.1117	2.52
27	Sticky roller	0.04589	0.02
28	Diazo film	2.0362E-05	0.00
29	Sodium carbonate	0.02477	0.01
30	Filter	1.1999E-05	0.00
31	Degreaser	0.008700	0.00
32	Sulfuric acid	0.0002507	0.00
33	Sodium persulfate	0.0003193	0.00
34	Glazing agent	0.0003034	0.00
35	Copper (II) sulfate	0.00003788	0.00
36	Tin (II) sulfate	0.01060	0.00
37	Tin ball	0.1855	0.07
38	Nitric acid	0.001169	0.00
39	Filter	0.0003733	0.00
40	Sodium hydroxide	0.001816	0.00
41	Etching liquid	0.8821	0.31
42	Tin stripping solution	0.2085	0.07
43	Sulfuric acid	0.00001047	0.00
44	Nylon brush	0.4662	0.16
45	Pumice	0.004259	0.00
46	Component ink	0.002249	0.00
47	Screen	0.00009983	0.00
48	Stencil cleaner	0.00042578	0.00
49	Diazo film	1.5009E-05	0.00
50	Sodium carbonate	0.02177	0.01
51	Filter	1.1599E-05	0.00
52	Defoamer	1.4664E-06	0.00
53	Sulfuric acid (AR class)	6.0289E-05	0.00
54	Sodium persulfate	0.0002688	0.00
55	Sulfuric acid (CP class)	9.6492E-06	0.00
56	Brown corundum for PCB	0.0006456	0.00
57	AC-10 acidic cleaner	5.7519E-05	0.00
58	CA-40M activator	0.00001983	0.00

Continued

Table 13.14 Carbon footprint result of all materials from "cradle-to-gate" life cycle stages—cont'd

No.	Item	Carbon footprint (kg CO_2 eq./f.u.)	Percentage of total PCF
59	CA-40R activator	0.0003778	0.00
60	CD-30 regulator	0.0004522	0.00
61	Electroless gold EG-70	0.0004264	0.00
62	Potassium dicyanoaurate	8.0706E-06	0.00
63	Ammonia (AR class)	1.46124E-05	0.00
64	Sodium hydroxide	9.8736E-06	0.00
65	Blue tape	0.0005484	0.00
66	Nitric acid	0.0008410	0.00
67	Filter	0.0001040	0.00
68	Potassium dicyanoaurate	0.0001547	0.00
69	AURUNA 529 plating bath solution	0.01433	0.01
70	Filter	4.2796E-06	0.00
71	Milling cutter	0.02217	0.01
72	Sulfuric acid	6.1785E-06	0.00
73	Sulfuric acid	0.009247	0.00
74	Sodium hydroxide	0.43368	0.15
75	Lime	0.04114	0.01
76	NaClO	0.01084	0.00
77	Iron (II) sulfide	0.02148	0.01
78	Polyacrylamide (PAM)	0.001313	0.00
79	Polyaluminum chloride (PAC)	0.2557	0.09
80	Sodium sulfide	0.5925	0.21
81	Sodium hydroxide	0.02422	0.01
	Subtotal	21.0405	7.44
Waste management			
1	Solid waste from manufacturing processes	0.007609	0.00
2	Sludge from wastewater treatment	0.05433	0.02
3	Discharged water effluent	0.7008	0.25
4	Hazardous liquid	0.08663	0.03
	Subtotal	0.8494	0.30
Subtotal of manufacturing stage		160.5474	56.74
Transportation stage (for reference only)			
1	Transportation of final product	89.0760	NA
Total product carbon footprint (excluding distribution) (kg CO_2 eq./m^2)		282.9331	100

ENIG, electroless nickel/immersion gold; *EPE*, expanded polyethylene; *HDPE*, high-density polyethylene; *LDPE*, low-density polyethylene; *PCF*, product carbon footprint; *PP*, polypropylene; *PVC*, polyvinyl chloride; *QA*, quality assurance.

Table 13.15 Contribution to the total carbon footprint at each stage

Life cycle stages		Carbon footprint (kg CO_2 eq./f.u.)		Contribution (%)	
Raw material stage	Raw materials	119.8871	122.3857	42.37	43.26
	Product packaging	0.4353		0.15	
	Packaging of raw materials	0.2059		0.07	
	Packaging of product packaging	0.0018		0.00	
	Transportation of raw materials and product packaging	1.8556		0.66	
Manufacturing stage	Electricity consumption of production	77.3441	160.5474	27.34	56.74
	Water consumption of production	0.3477		0.12	
	Diesel consumption of production	0.004		0.00	
	Consumables	56.8485		20.09	
	Packaging of consumables	4.1134		1.45	
	Transportation of consumables	21.0405		7.44	
	Waste management	0.8494		0.30	
Transportation stage (for reference)		89.0760		NA	
Total product carbon footprint (kg CO_2 eq./m^2)		282.9331		100	

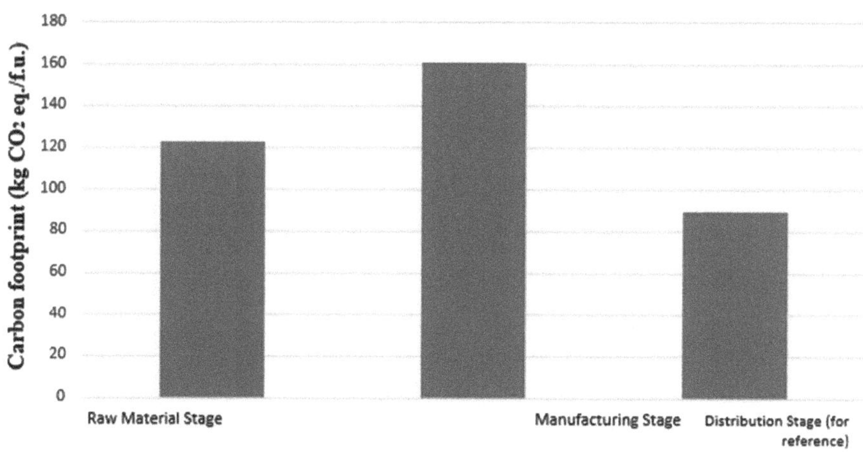

Figure 13.5 Contribution to carbon footprint from each life cycle stage.

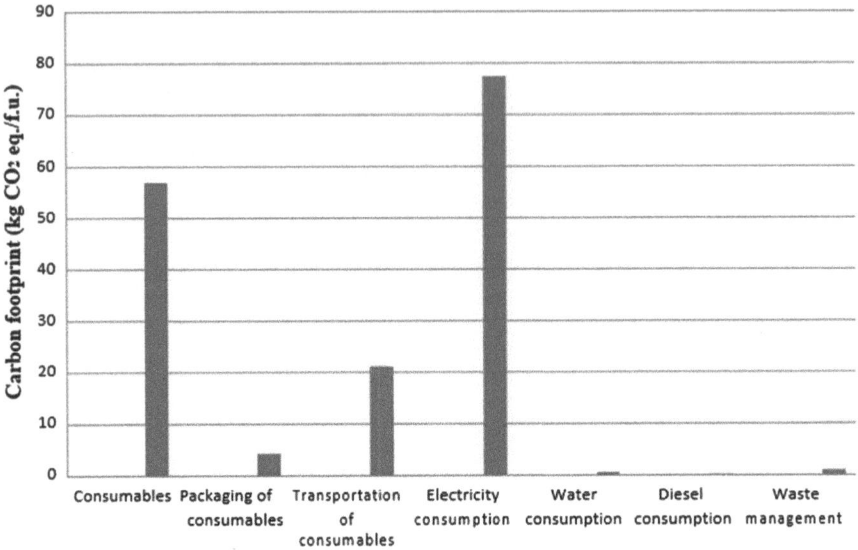

Figure 13.6 Contribution to carbon footprint from each process in the manufacturing stage.

13.6.2 Specific Greenhouse Gas Emissions and Removals

Table 13.16 separately shows the specific GHG emissions and removals that arise in this study.

13.6.3 Limitations

13.6.3.1 Focus on a Single Environmental Issue

The PCF only reflects the sum of GHG emissions and removals of a product system, expressed as CO_2 eq, associated with raw material acquisition, manufacturing, and transportation of a product. Although the PCF can be an important environmental aspect of the life cycle of a product affecting

Table 13.16 Specific greenhouse gas emissions and removals that arise in this study

Source	Emission and removals
Fossil carbon sources and sinks	N.A.
Biogenic carbon sources and sinks	N.A.
Direct land use change (dLUC)	N.A.
Aircraft transportation	N.A.

N.A., Not applicable.

"climate," a product's life cycle may have other environmental impacts that are also of concern (e.g., resource depletion, air, water, soil, and ecosystems).

13.6.3.2 Limitations Related to the Assumptions

Assumptions were made at each stage of this study and that may affect the absolute accuracy of the final result.

13.6.3.3 Limitations Related to the Methodology

The PCF was calculated based on LCA methodology. ISO 14040 and ISO 14044 address the inherent limitations and trade-offs of LCA methodology. These include the establishment of a functional unit and the system boundary, the availability and selection of appropriate data sources, allocation rules, and assumptions regarding transportation. Some of the chosen data may be limited to a specific geographical area (e.g., national electricity grid) and/or may vary in time (e.g., seasonal variations). The choice of the functional unit and allocation rules may also be important in more accurately modeling the life cycle.

13.6.4 Disclosure and Justification of Value Choices

One of the important choices is the database selection. The priority for the database selection was based on credibility, geographical, technological, and time-related considerations. In this report, emission factor for electricity consumption is obtained from NDRC 2013, emission factor for sewage treatment is obtained from XIE Tao and WANG Chengwen of School of Environment of Tsinghua University (Xie and Wang, 2012), and the data used for components, consumables, packaging materials, and transportation are obtained from ecoinvent version 2.2.

13.6.5 Sensitivity Analysis

This section describes the sensitivity analysis regarding the influence of choice of emission factor database on the elements with most significant contributions to the GHG emissions, and electricity, on the final result.

13.6.5.1 Sensitivity Analysis of Significant Input Data in Material Stage

As shown in Table 13.15 in Section 13.6.1, potassium dicyanoaurate used in the electroplating process in raw material stage contributes the largest amounts of carbon emissions at this stage, and its emission factor is a secondary data, which is not localized. Sensitivity analysis was conducted for this inventory. Different emission factors were used to analyze the sensitivity. Table 13.17 and Fig. 13.7 show the difference in carbon footprint results on changing the emission factor of potassium dicyanoaurate from ecoinvent to Korea LCI and JEMAI.

Table 13.17 Result of changing emission factor of the indium tin oxide glass used from different databases

Material	Weight (kg/m^2 PCB production	Database	Name in database	Emission factor (kg CO_2 eq./kg)	Carbon emission (kg CO_2 eq.)/f.u.	Final carbon footprint result (kg CO_2 eq./f.u.)
Potassium dicyanoaurate	0.007855	ecoinvent 2.2	Sodium cyanide, at plant [RER]; gold, at regional storage [RER]	9013	70.7926	282.9331
		Korean LCI 2012, JEMAI	Sodium cyanide; gold mine	7121	55.9355	268.4766

JEMAI, Japan Environmental Management Association for Industry; *LCI*, life cycle inventory; *PCB*, printed circuit board; *RER*, Europe.

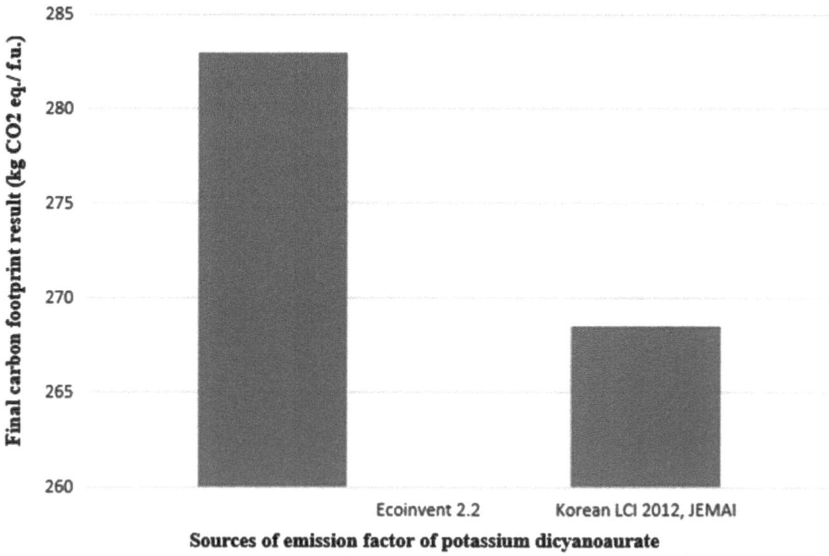

Figure 13.7 Result of changing emission factor of potassium dicyanoaurate used from different databases.

13.6.5.2 Sensitivity Analysis of Significant Input Data in Manufacturing Stage

As shown in Table 13.15 in Section 13.6.1, electricity used in the manufacturing stage contributes large amounts of carbon emissions, and its emission factor is a secondary data, not localized. Sensitivity analysis was conducted for this inventory. Different emission factors were used to analyze the sensitivity. Table 13.18 and Fig. 13.8 show the difference in carbon footprint results on changing the emission factor of electricity

Table 13.18 Result of changing emission factor of electricity used from different databases

Electricity consumption (kWh/m^2 PCB production)	Database	Emission factor (kg CO_2 eq./ kWh)	Carbon emission (kg CO_2 eq./f.u.)	Final carbon footprint result (kg CO_2 eq./f.u.)
83.86	NDRC	0.9223	77.3441	282.9331
	ecoinvent 2.2	1.51	126.6286	332.2176
	Hou Ping et al. (2012)	0.853	71.5326	277.1216

NDRC, National Development and Reform Commission; *PCB*, printed circuit board.

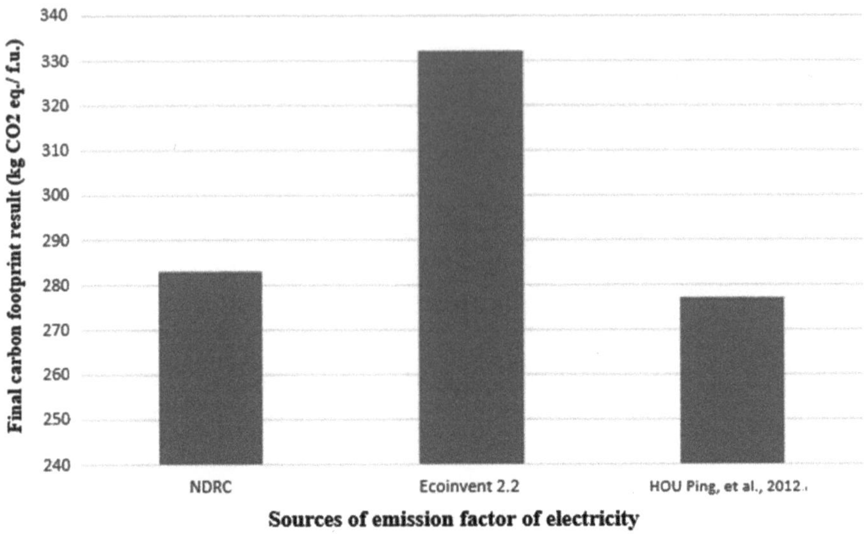

Figure 13.8 Result of changing emission factor of electricity used from different databases. *NDRC,* National Development and Reform Commission.

Table 13.19 Summary of data types and the possible errors that may contribute to the result uncertainty

Type of data	Type of error contributing uncertainty to results
Material data	Variation of the packaging materials from different suppliers
Process data	Allocation of electricity for manufacturing one unit of PCB
Process data	Seasonal factor of production
Emission factor	Localized emission factor is the first choice and adopts other emission factors only if localized emission factor is not applicable or is unavailable

PCB, printed circuit board.

provided by NDRC, ecoinvent 2.2 and Hou Ping et al. (2012) (from a Chinese research journal), for use in the manufacturing stage.

13.6.6 Uncertainty Analysis

In this study, the data include primary and secondary data. Some uncertainties may exist. Table 13.19 summarizes the type of data and the possible errors that may contribute to the uncertainty of the results.

13.6.7 Conclusion

According to the results, electricity consumption (27.34%) accounts for a large part of the total GHG emissions, from "cradle-to-gate" stages.

Alternative electricity production methods may be employed as substitutes for reducing emissions on account of the consumption of electricity to reduce the GHG emissions. In raw material stage, potassium dicyanoaurate usage (25.02%) also accounts for a significant part of the total GHG emissions, alternative materials might be used to reduce the GHG emissions. The carbon footprint result reflects that improvement in the manufacturing stage (56.74%), especially in electricity consumption may effectively reduce the carbon footprint of the product. Thus, a reduction of emission from electricity usage and a better design and development of material is critical for reducing the PCF. Hence a better design at manufacturing level, choice of appropriate materials at raw material stage and use of alternative electricity production can play a major role in reducing carbon emissions. These case studies can guide manufacturers to produce low-carbon products and get an edge over similar products in the market. These studies are also a source of large data inventory (primary data source), which can be in future used for a comprehensive LCA analysis covering all environmental indicators.

APPENDIX 1

Table A1.1 Usage of drilling needles and milling cutters

钻咀数量统计 Drilling needle usage (unit)		含包装重量 kg/盒 Weight including packaging kg/box	包装盒重量 (kg/盒) Weight of packaging box (kg/box)	消耗钻咀总重(kg) Total weight of drilling needle usage (kg)
0.5	930	0.159	0.062	1.804
0.95	307	0.164	0.062	0.627
1.6	231	0.162	0.062	0.462
2	223	0.171	0.062	0.487
0.85	728	0.164	0.062	1.486
3.25	207	0.217	0.062	0.641
1	36	0.165	0.062	0.074
钻咀合计 Subtotal	2662	–	–	5.580

备注: 每个钻咀所带胶粒重量为 0.0026 kg, 其包装盒为0.13 kg. Note: The weight of plastic ring on drilling needle is 0.0026 kg, its packaging box weighs 0.13 kg.

Table A1.2 Usage of drilling needles and milling cutters

铲刀数量统计 Milling cutter usage (unit)		含包装重量 kg/盒 Weight including packaging kg/box	包装盒重量 (kg/盒) Weight of packaging box (kg/box)	消耗铲刀总重量(kg) Total weight of milling cutter usage (kg)
2.4	3,406	0.256	0.062	13.215
1.5	5,109	0.233	0.062	17.473
1.8	1,277	0.248	0.062	4.751
0.8	3,406	0.238	0.062	11.989
铲刀合计 Subtotal	13,198	–	–	47.428

APPENDIX 2

Calculation of sewage treatment of factory end effluent emission factor. Source: Xie, T., Wang, C.W., 2012. Wastewater treatment plant GHG emission assessment. Journal of Tsinghua University (Science and Technology) 52 (4), 473–477.

Sewage treatment plant in northern China

Treatment method: A2O

Daily treatment volume: 600,000 m^3

GHG emission in northern China wastewater treatment plant: 568,000 kg CO_2 eq./kg per day

EF value = 568,000/600,000 = 0.947 kg CO_2 eq./m^3

APPENDIX 3

Pictures 1–5 from factory site.

Picture 1 Factory manufacturing line.

Picture 2 Factory manufacturing line.

Picture 3 Factory manufacturing line.

Picture 4 Factory wastewater treatment facility.

Picture 5 Factory manufacturing end effluent after wastewater treatment.

ACKNOWLEDGMENTS

This work is supported by the Trade and Industry Department of HKSAR Government under **SME Development Fund**, "Development of an embedded Greenhouse Gas (GHG) emissions database with a G-BOM analyzer and a SME advisory kit for electrical and electronic industries to respond to the implementation and compliance of ISO 14067 (carbon footprint of products)."

We would like to express our thanks to SGS Hong Kong Limited for the support provided to the Project "*Development of an embedded Greenhouse Gas (GHG) emissions database with a G-BOM analyzer and a SME advisory kit for electrical and electronic industries to respond to the implementation and compliance of ISO 14067 (carbon footprint of products).*"

This case study report of the product carbon footprint of an LCD module (Brand: Clover Display Limited; Product model: C51486A) was made possible by the technical and professional contribution of our collaborating organization: SGS Hong Kong Limited.

We would like to thank the above organization for giving valuable technical support and professional advice and for participating in site assessment.

REFERENCES

EDP Korea, 2005. Printed Circuit Board [EDP 2005-108(0)]. Available from: http://pcr-library.edf.org.tw/data/korea/printed_circuit_board.pdf.

Hou, P., Wang, H.T., Zhang, H., Fan, C.D., Huang, N., 2012. Greenhouse gas emission factors of Chinese power grids for organization and product carbon footprint. China Environmental Science 32, 961–967.

International Organization for Standardization (ISO), 2013. ISO/TS 14067:2013 Greenhouse Gases – Carbon Footprint of Products – Requirements and Guidelines for Quantification and Communication.

PCBcity, (n.d.). About the Analysis of PCB High Concentrated Organic Liquid Waste. Available from: http://news.pcbcity.com.cn/meeting/ShuiChuLi_2006/pcb_04.asp (in Chinese).
Su Zhou He-Chuan Chemical Technology Service Co., Ltd., (n.d.). Water-Based Degreasing Agent Composition Analysis. Available from: http://www.plahcc.com/ProductView.asp?ID=59 (in Chinese).
Syndyne Industrial Co., Ltd., 2014. Phenolic Resin Back-up Laminate SD-BU26: Commonly Known as Back-up Board, Entry Board, PCB Drilling Board. Available from: http://www.syndyne.com.tw/index.php?page=SD_BU26&lang=en-gb.
Wikipedia, 2013. Diazo — Wikipedia, the Free Encyclopedia. Available from: http://en.wikipedia.org/wiki/Diazo.
Wikipedia, 2014a. Diesel Fuel. Available from: http://en.wikipedia.org/wiki/Diesel_fuel.
Wikipedia, 2014b. Antistatic Bag. Available from: http://en.wikipedia.org/wiki/Antistatic_bag.
Xie, T., Wang, C.W., 2012. Wastewater treatment plant greenhouse gas emission assessment. Journal of Tsinghua University (Science and Technology) 52 (4), 473—477.

FURTHER READING

British Standards Institution (BSI), 2011. PAS 2050:2011 Specification for the Assessment of the Life Cycle Greenhouse Gas Emission of Goods and Services.
European Commission, 2012. Directive 2012/19/EU of the European Parliament and of the Council on Waste Electrical and Electronic Equipment (WEEE) Recast.
International Organization for Standardization (ISO), 2006. ISO 14040:2006 Environmental Management — Life Cycle Assessment — Principles and Framework.
National Development and Reform Commission, 2012. 2012 Baseline Emission Factors for Regional Power Grids in China. Available from: cdm.ccchina.gov.cn/WebSite/CDM/UpFile/.

INDEX

"*Note:* Page numbers followed by "f" indicate figures and "t" indicate tables."

A

C

D

E

R

V

FSC
www.fsc.org
MIX
Papier | Fördert gute Waldnutzung
FSC® C083411

Kolibri 360 GmbH
Ferdinand-Jühlke-Straße 7
99095 Erfurt, Deutschland
produktsicherheit@kolibri360.de